GREAT APES AND HUMANS

ZOO AND AQUARIUM BIOLOGY AND CONSERVATION SERIES

Published in cooperation with the American Zoo and Aquarium Association

This series publishes innovative works in the field of zoo and aquarium biology and conservation, with priority given to books that focus on the interface between captive and field conservation and seek to merge theory with practice. Topics appropriate for this series include but are not limited to zoo- and aquarium-based captive breeding and reintroduction programs, animal management science, philosophy and ethics, public education, professional training and technology transfer, and field conservation efforts.

EDITED BY BENJAMIN B. BECK, TARA S. STOINSKI,
MICHAEL HUTCHINS, TERRY L. MAPLE, BRYAN NORTON,
ANDREW ROWAN, ELIZABETH F. STEVENS,
AND ARNOLD ARLUKE

SMITHSONIAN INSTITUTION PRESS • Washington and London

Copy editor: Debbie K. Hardin
Production editor: Ruth Spiegel
Designer: Janice Wheeler

Library of Congress Cataloging-in-Publication Data

Great apes and humans : the ethics of coexistence / edited by Benjamin B. Beck ...[et al].
p. cm.
Includes bibliographical references (p.).
ISBN 1-56098-969-6 (alk. paper)
1. Apes—Congresses. 2. Animal welfare—Moral and ethical aspects—Congresses. 3. Wildlife conservation—Congresses. I. Beck, Benjamin B.

QL737.P96 G733 2001
599.88—dc21 2001020943

British Library Cataloguing-in-Publication Data available

Manufactured in the United States of America
08 07 06 05 04 03 02 01 5 4 3 2 1

♾ The paper used in this publication meets the minimum requirements of the American National Standard for Information Sciences—Permanence of Paper for Printed Library Materials ANSI Z39.48-1984.

CONTENTS

Contributors **vii**

Introduction and Acknowledgments **ix**

Foreword

PROBLEMS FACED BY WILD AND CAPTIVE CHIMPANZEES: FINDING SOLUTIONS **xiii**

JANE GOODALL

SECTION 1: GREAT APES IN THE WILD

1. Africa's Great Apes **3**

THOMAS M. BUTYNSKI

2. The Orangutan and the Conservation Battle in Indonesia **57**

HERMAN D. RIJKSEN

3. Bushmeat Hunting and the Great Apes **71**

KARL AMMANN

4. Bushmeat Trade in the Congo Basin **86**

DAVID S. WILKIE

SECTION 2: GREAT APES IN CAPTIVITY

5. Captive Apes and Zoo Education **113**

TARA S. STOINSKI, JACQUELINE J. OGDEN, KENNETH C. GOLD, AND TERRY L. MAPLE

6. Sanctuaries for Ape Refugees **133**

GEZA TELEKI

7. The Retirement of Research Apes 150
THOMAS L. WOLFLE

SECTION 3: HISTORY AND EVOLUTION

8. Negotiating the Ape–Human Boundary 163
RAYMOND CORBEY

9. Phylogenies, Fossils, and Feelings 178
RUSSELL H. TUTTLE

10. Darwinian Reflections on Our Fellow Apes 191
ROGER FOUTS

SECTION 4: ETHICS, MORALITY, AND LAW

11. Conceptual Capacities of Chimpanzees 215
SARAH T. BOYSEN AND VALERIE KUHLMEIER

12. Moral Decisions about Wild Chimpanzees 230
RICHARD W. WRANGHAM

13. The Grand Apes 245
DUANE M. RUMBAUGH, E. SUE SAVAGE-RUMBAUGH, AND MICHAEL J. BERAN

14. Cognitive Relatives and Moral Relations 261
COLIN ALLEN

15. A Great Shout: Legal Rights for Great Apes 274
STEVEN M. WISE

16. Inclusivist Ethics 295
PAUL WALDAU

17. The Moral Status of Great Apes 313
MARY ANNE WARREN

18. Rights or Welfare: A Response to the Great Ape Project 329
MICHAEL HUTCHINS, BRANDIE SMITH, RANDY FULK, LORI PERKINS, GAY REINARTZ, AND DAN WHARTON

19. Perspectives on the Ethical Status of Great Apes 367
ARNOLD ARLUKE

Index 379

CONTRIBUTORS

Colin Allen
Department of Philosophy
Texas A&M University

Karl Ammann
World Society for the Protection of Animals

Arnold Arluke
Department of Sociology
Northeastern University

Benjamin B. Beck
Smithsonian National Zoological Park

Michael J. Beran
Language Research Center
Georgia State University

Sarah T. Boysen
Department of Psychology
Ohio State University

Thomas M. Butynski
Africa Biodiversity Conservation Program
Zoo Atlanta

Raymond Corbey
Department of Philosophy
Tilburg University
and
Department of Archaeology
Leiden University, The Netherlands

Roger Fouts
The Chimpanzee and Human Communication Institute
Central Washington University

Randy Fulk
North Carolina Zoological Park

Kenneth C. Gold
Chicago, Illinois

Jane Goodall
The Jane Goodall Institute

Michael Hutchins
Conservation and Science Department
American Zoo and Aquarium Association

Valerie Kuhlmeier
Department of Psychology
Ohio State University

Terry L. Maple
Zoo Atlanta
and
School of Psychology, Georgia Institute of Technology

Bryan Norton
School of Public Policy
Georgia Institute of Technology

Jacqueline J. Ogden
Walt Disney World Animal Programs

Lori Perkins
Department of Conservation Technology
Zoo Atlanta

Gay Reinartz
Zoological Society of Milwaukee County

Herman D. Rijksen
Rijksen Conservation Consult

Andrew Rowan
Research, Education, and International Issues
Humane Society of the United States

Duane M. Rumbaugh
Language Research Center
Georgia State University

E. Sue Savage-Rumbaugh
Language Research Center
Georgia State University

Brandie Smith
Conservation and Science Department
American Zoo and Aquarium Association

Elizabeth F. Stevens
Walt Disney World Animal Programs

Tara S. Stoinski
TECHlab
Zoo Atlanta
and
School of Psychology, Georgia Institute of Technology

Geza Teleki
Department of Anthropology
George Washington University

Russell H. Tuttle
Department of Anthropology
University of Chicago

Paul Waldau
Department of Environmental and Population Health
Tufts University School of Veterinary Medicine
and
The Great Ape Project—International

Mary Anne Warren
Philosophy Department
San Francisco State University

Dan Wharton
Wildlife Conservation Society

David S. Wilkie
Psychology Department
Boston College
and
Living Landscape Program
Wildlife Conservation Society

Steven M. Wise
Center for the Expansion of Fundamental Rights, Inc.

Thomas L. Wolfle
Retired, Institute for Animal Laboratory Research
and
National Research Council

Richard W. Wrangham
Department of Anthropology
Peabody Museum
and
Harvard University
Kibale Chimpanzee Project

INTRODUCTION AND ACKNOWLEDGMENTS

This book grew from a workshop held at the Disney Institute in Lake Buena Vista, Florida, on June 21 to 24, 1998. The editors and most of the authors attended, as well as Birute Galdikas, John Grandy, and Oliver Ryder, who were unable to submit chapters for this volume.

The workshop was stimulated by *The Great Ape Project* (Cavalieri and Singer 1994). The Great Ape Project declares that the great apes (gorillas, bonobos, chimpanzees, and orangutans) should be classified with living humans in the genus *Homo,* and thus are entitled to the same rights—specifically that they should not be harmed, killed, or held captive without cause. The American Zoo and Aquarium Association's (AZA) Ape Taxon Advisory Group (Ape TAG), recognizing the challenges posed by this position, recommended that the underlying assumptions and conclusions of the project be honorably considered by experts in ape biology, conservation, protection, research and exhibition, as well as experts in ethical, legal, and moral issues. We recognized that a workshop needed to be small to foster direct communication and personal acquaintance and private to foster frankness and issue-based discussion. These requirements implied exclusion of many qualified experts and observers; we apologize to those who did not have the opportunity to attend and hope this book compensates to some degree.

Another stimulus to the workshop was an earlier workshop held in Atlanta, Georgia, in 1992 to

> explore concerns regarding the welfare of animals maintained in captivity for breeding programs, [recognizing] that a comprehensive understanding of [the] ethical dilemmas will require a broader exploration of the role of captive breeding programs in conserva-

tion, and also the entire role of zoos and aquariums in modern society. (Norton 1995, xxii)

The AZA board of directors sponsored the Atlanta workshop, a risky, pioneering effort to establish a scholarly dialogue among the zoo, animal protectionist, and wildlife conservation communities. The workshop and resultant volume, *Ethics on the Ark* (Norton et al. 1995), influenced and advanced the ways in which zoos and aquariums are managed and the ways in which they are perceived. It established links of professional regard among previously unacquainted antagonists, and founded an atmosphere of thoughtful dialogue. It inspired our "apes and humans" workshop.

The editors acknowledge that the title of the book implies the conventional bias that apes and humans are different. We did not achieve a consensus on this or the other declarations of the Great Ape Project, but we are more aware of the many perspectives and the complicated data that can be applied to the issues and serve as precedents for their resolution. This volume is an attempt to share that awareness. We (workshop participants) discovered that we were unanimously opposed to the commercial exploitation of apes as food (bushmeat) and the destruction of natural ape habitats. The AZA subsequently took the initiative to found a Bushmeat Crisis Task Force, a coalition of many conservation and protection organizations and involved individuals, including the Humane Society of the United States, that has already made progress in addressing this grave and complex threat. We discovered our unanimous opposition to cruelty, pain, and humiliation of apes in any context. The Ape TAG is working to publicize the unsuitability of apes (and other primates) as household pets or entertainers. We agreed that although apes are no more unique than slugs or dandelions, they are good "flagship" species to grab public attention to issues of animal protection and conservation. We agreed that zoos and the research community need not import more apes from the wild, and should control breeding to prevent a "surplus." And we discovered that nearly all of us talked to our dogs.

The workshop and costs of manuscript preparation were completely underwritten by Disney's Animal Kingdom. We are most grateful to Judson Green, then-president of Walt Disney World Attractions, and Bob Lamb, vice president, Disney's Animal Kingdom, for their support. We wish to stress that the sponsors gave us complete organizational and intellectual freedom. The staff of the Disney Institute and Disney's Animal Kingdom, especially Kim Sams and Michelle Caroli, provided magnificent and totally unobtrusive hospitality. Peter Cannell, editor of the Smithsonian Institution Press at the outset of the project, and his successor, Vince Burke, have been patient, prodding, and ever-helpful.

REFERENCES

Cavalieri, P., and Singer, P. (eds.). 1994. *The Great Ape Project: Equality beyond humanity.* New York: St. Martin's Press.

Norton, B. 1995. Preface. In B. G. Norton, M. Hutchins, E. F. Stevens, and T. L. Maple (eds.), *Ethics on the ark: Zoos, animal welfare, and wildlife conservation* (pp. xxi–xxiii). Washington, D.C.: Smithsonian Institution Press.

Norton, B. G., Hutchins, M., Stevens, E. F., and Maple, T. L. (eds.). 1995. *Ethics on the ark: Zoos, animal welfare, and wildlife conservation.* Washington, D.C.: Smithsonian Institution Press.

JANE GOODALL

FOREWORD

Problems Faced by Wild and Captive Chimpanzees: Finding Solutions

In 1960 I began a study of the chimpanzees living in the Gombe National Park in Tanzania. Today I am seldom able to visit more than three or four times a year, for two weeks at a time, but the work continues. Data are collected daily by a team of researchers, making the Gombe project the longest unbroken study of any group of wild animals. Information from this research and from other chimpanzee study sites has provided a wealth of data about these apes. Rich data have also accumulated from studies of gorillas and bonobos in Africa and orangutans in Asia. This information, together with behavioral, psychological, and physiological data from a variety of studies of captive great apes around the world, has served to emphasize their close evolutionary relationship to ourselves. How shocking, then, to learn that these amazing beings are vanishing in the wild and being subjected to abuse in many captive situations.

All long-term field studies of chimpanzees have revealed enduring and supportive bonds between family members—that is, between mother and growing offspring and between siblings. These may persist throughout life. Chimpanzees in the wild use many different objects as tools; they hunt cooperatively and share the kill; many postures and gestures of their communication repertoire are uncannily similar to ours, such as kissing, embracing, holding hands, patting on the back, swaggering, punching, and tickling.

From the age of 2 years onward youngsters spend a great deal of time playing. When possible they play with other youngsters, but often mothers travel and feed away from other adults. Then, although some mothers readily respond to the play invitations of their youngsters, there are many occasions when the children must occupy themselves, particularly first-born infants who have no older sibling to serve as a playmate. But there is plenty to do in the wild. They practice acrobatics in the trees and use many different objects as toys, such as rocks, bunches of fruit, sticks, and so on. Infants are very inventive in their play.

Chimpanzees, like other big-brained animals, are very curious, interested in what is going on around them. This is especially true of infants. We now know that chimpanzees, like humans, can learn certain behaviors simply by watching the performance of another, then imitating that action. And so, because a new behavior "invented" by one individual is so fascinating to others in the group, particularly youngsters, it may, especially if it is adaptive, be passed on to the next generation through observation, imitation, and practice. In this way the identity of foods traditionally eaten by members of a community will be acquired by the youngsters. The same food item, present in two well-separated chimpanzee ranges, may be eaten in one area and not the other, and vice versa. It is the same with tool using. In all areas where chimpanzees have been studied, different kinds of tool-using behaviors have been observed. It is the inherent curiosity of other group members, especially during childhood when behavior is most flexible, that enable new patterns to be passed on. Their curiosity, their ability to manipulate objects, and their persistence when trying to accomplish a desired goal has obviously led to these different kinds of tool using. In other words, chimpanzees have developed primitive cultures.

As they move into maturity chimpanzees continue to be fascinated by the world around them, as exemplified by adult males at Gombe who may sit on a rock in the stream watching the water pouring over the cliff 80 feet above, following with their eyes as it rushes past them, gazing up again.

Chimpanzees show intellectual abilities once thought unique to our own species. They have excellent memories, and they can plan for the immediate future. They are capable of cross-modal transfer of information, generalization and abstraction, and simple problem solving. They are aware of themselves as individuals, and they can interpret the moods and identify the wants and needs of others. They have demonstrated a sense of humor. Moreover, although harder to prove, they undoubtedly feel and express emotions similar to those that we label happiness, sadness, rage, irritation, fear, despair, and mental as well as physical suffering. None of this should surprise us in view of the remarkable similarity between the anatomy of the brain and central nervous system of chimpanzee and human. All of this helps to blur the line, once perceived as so sharp, between humans and the rest of the animal kingdom. Once science admits that it is not, after all, only humans who have personalities, are capable of rational thought, and know emotions similar to happiness, sadness, anger, despair, this should lead to a new respect for other animals with whom we share the planet, especially for the great apes, our closest living relatives. In fact that respect is seldom apparent.

In this context I want to look at the various situations in which chimpanzees live, both in the wild and in captivity. Only when we understand their day to day problems can we decide whether we should try to intervene to improve their situation, and if so, how this should best be accomplished.

PROBLEMS FACED BY CHIMPANZEES IN THE WILD

There are still some chimpanzees living in utterly remote wilderness areas who seldom if ever encounter humans—for example, those in the Ndoke National Park area in the People's Republic of Congo (Brazzaville). There are a number of areas, spread across the range of the chimpanzee, that have been given protected status to preserve wildlife. In some countries (e.g., Tanzania and Uganda) efforts are made by wildlife authorities to patrol such areas. Protection is also afforded by wildlife research teams working within these forests. Too often, though, poachers with guns, snares, or spears have easy access, and there are many illegal logging operations with pit saws and illegal mining.

Situations where chimpanzees live in well-protected areas are rare, and even then their lives are not without pain and danger (chapters 1, 12, this volume). Just as the lives of human hunter–gatherers are seldom as idyllic as some would like to believe, so it is with all animals in nature. At Gombe the chimpanzees sometimes suffer painful, even horrifying, wounds. In 1966, there was an epidemic of a paralytic disease, probably poliomyelitis, that affected many members of the community. Some died, others survived with severe and debilitating paralysis. Most recently in 1998 there was an epidemic of sarcoptic mange that affected about 50 percent of all known chimpanzees. It caused much discomfort and claimed the lives of three small infants. The top-ranking male became so sick that he lost his position. Infants lose their mothers and may die, with grief a possible cause of death. The growth of those that survive is severely retarded, and they may suffer psychosocial dwarfism.

There were four years when two females, Passion and Pom, hunted, killed, and ate the newborn infants of other community females. Ten babies were thus killed, or vanished, during this period. Only one survived. Females may be brutally attacked and their infants killed by males of neighboring communities. During another four-year period the males of one community penetrated the range of a smaller neighboring group and perpetrated a series of brutal gang attacks that effectively wiped out an entire community.

The most severe threat to the Gombe chimpanzees is human population growth in the areas around the tiny 30-sq. mi national park. The 120 or so chimpanzees, in three different communities, are isolated, cut off from other conspecifics by cultivated hillsides on three sides, the lake on the fourth. Even 15 years ago chimpanzee habitats stretched far along the eastern shore of Lake Tanganyika. Today the trees have gone as more and more desperate people, including large numbers of refugees from Burundi and Congo, try to grow food on the very steep slopes. In the rainy season the precious thin layer of topsoil is washed down into the lake. In some places the shoreline looks like rocky desert, and the fish breeding grounds have become silted up.

How can we hope to save the forest jewel that is the Gombe National Park, and its famous chimpanzees, when the local people are facing starvation? There are now more people living there than the land can support, there is almost nowhere for them to move, and they mostly cannot afford to buy food from other areas. In many places the women have to dig up the roots of previously cut trees to get wood to cook their food.

The Jane Goodall Institute has initiated a project in the Kigoma region to try to address this problem. Tree nurseries have been established in 33 villages around Gombe and along the lakeshore. Fruit trees and fast growing trees for building poles, firewood, and charcoal are nurtured as seedlings, then planted in the villages. George Strunden, project manager, has picked a team of qualified Tanzanians who introduce the program into the villages. He has also trained women who demonstrate appropriate tree growing methods. Farming methods that help control and prevent soil erosion are introduced. And there is a strong conservation education element that includes taking small groups of secondary school students to Gombe. There is a big push to increase the self-esteem of women, teaching them skills that will enable them to earn money for themselves. A number of scholarships are offered annually to enable girls from primary schools to benefit from further education. A small microcredit program has been introduced based on the Grameen Bank system. By working with the local medical authorities, TACARE (Lake Tanganika Catchment Reforestation and Education Project) is able to bring primary health care to the village women, along with family planning and AIDS education. Most recently we have formed a partnership with UNICEF that will enable us to bring hygenic latrines and freshwater wells to 33 villages in the area. It should be stressed that the villagers are consulted about their needs, and only projects that have their absolute support are introduced.

Only if we work with the villagers, helping to improve the standard of living of some of the poorest people in Tanzania, do we have a chance to protect the Gombe chimpanzees. Without the goodwill of the local people, the last forests within the park itself, and the tiny remnant forests outside, would surely disappear. A significant factor in our battle to save the Gombe chimpanzees is our employing field staff from communities around the park since 1988. These men follow the chimpanzees, make detailed reports, use 8-mm video cameras, and are proud of their work. They talk about it to family and friends. They care about the chimpanzees as individuals. I believe this is why, until the recent influx of refugees from eastern Congo (people who traditionally eat the meat of monkeys and apes), we had only one case of poaching at Gombe.

Across Africa the great apes face problems caused by the relentless growth of human populations, habitat destruction, and fragmentation of populations. Peasants clear-cut forests to create fields for crops and grazing. They cut down hundreds of trees for the charcoal industry. The forest soils are fragile and soon become infertile and barren when the tree cover is destroyed. So the desert spreads.

In some parts of Africa apes are hunted for the live animal trade and for food. In addition they may be caught in the snares set by village hunters for antelopes and bush pigs. They can usually break the wire, but the tightened noose causes great pain and typically results in gangrene and the loss of the affected hand or foot and sometimes ends in death. Between 40 and 50 percent of all adult chimpanzees in the study communities at Budongo and in the Tai Forest have lost a hand or foot in this way.

Wildlife is sometimes endangered as a result of the ethnic violence so tragically prevalent in many parts of the chimpanzees' range. These conflicts may displace hundreds of refugees who flee their homes, as in Liberia, the Democratic Republic of Congo, Sierra Leone, and Rwanda. Typically they are starving and forced to hunt wild animals for food. Those chimpanzees remaining in Cabinda are endangered by the land mines that have been placed throughout the forests in northwest Angola.

The great apes are also threatened by the live animal trade, when dealers pay hunters to shoot females simply to steal their infants for export. This trade is by no means as extensive as it was in the days before the Convention on International Trade in Endangered Species of Wild Fauna and Flora (CITES), but there is still brisk business in some parts of the world, such as the United Arab Emirates, various countries in South America, and parts of eastern Europe. For every infant that arrives at its final destination alive, about ten chimpanzees are estimated to have died in Africa: mothers who escaped only to die later of their wounds, along with their infants; infants killed during capture; other individuals who tried to protect the victims; and captured infants who die of wounds, dehydration, malnutrition, or shock and depression.

Chimpanzees and other wildlife in remote unprotected forests are seriously and increasingly threatened by commercial activities, particularly logging. Even companies that practice sustainable logging have a highly adverse effect on much animal life. Roads made for transportation of logs open up the forests for settlements. People then cut down trees to grow crops, for firewood, for building poles. They set snares to catch antelopes and other animals for food. And they carry human diseases into areas where they have never been before, and the great apes are susceptible to almost all of our infectious diseases. Most serious of all, the roads provide easy access to previously inaccessible areas for commercial hunters who ride the logging trucks. The roads and trucks provide, for the first time, the means for meat, dried or even fresh, to be transported from the heart of the forest to towns far away. Subsistence hunting permitted indigenous people to live in harmony with the forests for hundreds of years. It is the new commercial hunting that threatens the animals of many of the remaining forests. This is the infamous bushmeat trade, exposed by Karl Ammann (chapter 3, this volume).

Thus it is clear that wild chimpanzees, gorillas, and bonobos are, only too often, persecuted by their closest relatives, the human apes. The chimpanzee population, that must have numbered more than one million at the beginning of the twentieth century, has been reduced to 200,000 at the very most, spread through 21 countries. It is the rate of decrease of all the great apes that is so alarming. If nothing is done to halt the bushmeat trade, it is estimated that almost no great apes will remain in the Congo Basin in 10 to 15 years. Many organizations have joined The Bushmeat Crisis Task Force in the United States and the Ape Alliance in Europe, which are working on developing methods to slow down and ultimately eliminate this trade.

SANCTUARIES

There is not much meat on an infant chimpanzee. Orphans, whose mothers have been shot and sold for meat, are sometimes offered for sale in native or tourist markets. In some areas mothers are shot only so that their infants can be stolen for sale. These pathetic orphans are sometimes bought as pets, to attract customers to a hotel or other place of business, or simply because people feel sorry for them. Paying money for any wild animal for sale serves only to perpetuate a cruel trade. Yet it is hard to turn away from a small infant who looks at you with eyes filled with pain and hopelessness. A solution to this moral dilemma is to persuade government officials to confiscate these victims, because in most African countries there is a law prohibiting the hunting and sale of endangered species, such as the great apes, without a license.

After confiscation, the orphan must be cared for. The Jane Goodall Institute has established sanctuaries in a number of locations. The biggest is in Congo (Brazzaville), where Graziella Cotman cares for 80 at the time of writing (October 2000). The Tchimpounga sanctuary, north of Pointe-Noire on the coast, was built by the petroleum company Conoco, in 1991. It was designed for 25 chimpanzees at most. It has become urgent to add additional enclosures, but we have been delayed by civil war.

In this area, as at Gombe, we employ individuals from the surrounding villages to care for the chimpanzees (and other animals) and as support staff. We also buy fruit and vegetables locally, and this boosts the economy. In addition, we use these orphans as the focus of an environmental education program. The local people are amazed and fascinated when they see the chimpanzees close up. We are trying to establish a wildlife reserve to protect the remaining forest–savanna mosaic in the area. Most of the savanna has been destroyed by eucalyptus plantations, a project of Shell Oil working with a Congolese company, but there is a beautiful unspoiled area around our sanctuary. With permission from the central government we are working with local government officials and also employing ecoguards from each of the seven nearby villages. There are more wild chimpanzees in the area than we believed. When the fighting stops it may be possible to attract tourists and thus bring foreign exchange into the country. Although the building and maintenance of chimpanzee sanctuaries is very expensive, we are not only caring for abandoned orphans but raising awareness through conservation education, and trying to protect the wild chimpanzees.

We do not believe it is possible to return our orphans to the wild unless we find an area where there are no wild chimpanzees. Our chimpanzees trust people and would almost certainly wander into a village and be hurt, or hurt someone. Thus we have taken on a major commitment. We are caring for three orphan chimpanzees in the Kitwe-Point sanctuary in the Kigoma region, confiscated when being smuggled from eastern Democratic Republic of Congo (DRC) to, it seemed, the Middle East. We are also involved in three other sanctuaries, one in Uganda, one in Kenya, and one that is planned in South Africa. Most of the chimpanzees in the Sweetwaters Sanctuary in Kenya also originated in the DRC, were confiscated in Burundi, then flown to Kenya because of the ethnic violence in Burundi.

ZOOS

There are approximately 250 chimpanzees living in zoos accredited by the American Zoo and Aquarium Association (AZA) and participating in the chimpanzee Species Survival Plan (chapter 18, this volume). There are about 1,700 in all zoos worldwide. There is much controversy regarding zoos, with many animal rights activists believing that they should be closed. Of course, chimpanzees belong in the wild, and if they are lucky enough to live in a protected area, or one remote from people, that is the best life. That life cannot be replicated in captive situations. In the forest they have a great deal of freedom of choice. They can choose whether to travel on their own, in a small group, or to join large excitable gatherings. They can usually choose which individuals to associate with. Females can wander off, with their dependent young, and stay feeding peacefully and grooming together for hours, or even days. Close companions meet often, others may avoid each other. They know the excitement of participating in hunts or boundary patrols, and even aggressive, almost war-like encounters with individuals of neighboring social groups. To survive they must spend much time searching for and sometimes preparing their food—they are occupying their brains, using their skills. They are free. Nevertheless, when compared with the life of chimpanzees living in danger zones in Africa, it sometimes seems to me that those in the really good zoos—those in which there are large enclosures, rich social groups, and an enriched environment—may in fact be better off.

On the other hand, there are still many zoos that should be closed—zoos where chimpanzees are forced to live alone or in pairs in tiny cement-floored, iron-barred, old-fashioned cages. There they suffer terribly from boredom. In African zoos, where sometimes even the keepers can only eat one meal a day, if that, conditions are often appalling for all the animals, and there is much suffering. Lack of water is often a major problem, because there may be no running water and water is delivered very sporadically by keepers who are in the business simply for a job.

PETS

In Africa people sometimes buy chimpanzees to rescue them from the roadside or market. Or, as in America, they may buy them as surrogate children or as attractions to a hotel. However, by the time these "pets" are 5 or 6 years old they begin to resent discipline. They are by then as strong as a man, and can deliver a nasty bite. They are, after all, not human children, and usually it becomes impossible to keep them in the house. Good zoos seldom want to take ex-pets, who have not been able to learn proper chimpanzee behavior from their peers. What becomes of these unwanted orphans?

THE CIRCUS CHIMPANZEE AND OTHERS IN ENTERTAINMENT

It is now well-accepted that the pretraining of exotic animals used in entertainment—the circus, movies, and advertising—is almost always harsh, if not downright cruel. There are accounts of infant chimpanzees taken out of earshot and beaten into submission with iron bars. The bars are wrapped in newspaper rolls so that when the trainer is on the set, when an officer of the humane society is present, the rolled papers are sufficient to instill instant obedience. Of course chimpanzees can be taught to do almost anything with kindness, patience, and reward. But instant obedience is required if he or she is to perform stupid tricks, again and again, on demand. So most trainers establish a relationship based on dominance and fear. When chimpanzee actors pass the age when they can be usefully exploited in the entertainment field they too become surplus animals. Where do they go?

THE MEDICAL RESEARCH LAB

Many ex-pet and ex-entertainment chimpanzees end their days in medical research. It is hard for me to visit the laboratories to see chimpanzees, who have committed no crime, locked into 5 ft × 5 ft × 7 ft high prison cages. They are there because their biology and physiology is so like ours that they can be infected with almost all human diseases. Hundreds have been used in hepatitis and AIDS research. Admittedly some laboratories are improving, developing programs to enrich their prisoners' lives, giving them more space. But there are still hundreds in the United States and other parts of the world in such cells.

SURPLUS CHIMPANZEES

A major problem today is the so-called "surplus" chimpanzee population. This was highlighted when the U.S. Air Force announced that it had to divest itself of the chimpanzee colony descended from the original group formed for space research in the early 1960s. At about the same time the National Institutes of Health announced that it had many chimpanzees (originally the figure was around 500) that were no longer required for experimentation (the result of an aggressive captive breeding program initiated in the 1980s when it was thought that chimpanzees were essential to the development of a vaccine against AIDS). Finding suitable placement for so many chimpanzees is a daunting task (chapter 7, this volume). Most of the more than 300 that were housed at Laboratory for Experimental Medicine and Surgery in Primates (LEMSIP) were placed, by Dr. Jim Mahoney, in various zoos and sanctuaries around North America, including 15 at a new sanctuary built for them by the Fauna Foundation near Montreal. Because euthanasia has been ruled out, it seems that many more sanctuaries must be created for chimpanzees that have been stockpiled by the research community. Moreover, as scientists increasingly have turned away from chimpanzee models in AIDS research, the laboratories will no longer be a repository for ex-entertainment chimpanzees or ex-pets, compounding that already big and costly problem. Nor is North America alone in this regard. The European Union's chimpanzee breeding and research facility in the Netherlands and the laboratory in Austria are also trying to get rid of surplus chimpanzees. In zoos throughout the world many young males also become surplus as they move toward maturity and threaten the stability of small captive groups. Other chimpanzees are confiscated from dealers who attempt to smuggle them across borders or are seized from circus owners accused of cruelty. Only too often there is nowhere for these confiscated individuals to go.

The following stories of three chimpanzees, two born in the wild and one in captivity, serve to remind us that, when we talk of the "surplus" problem we are actually talking of the fate of individuals, each with his or her own personality, each having been exploited by humans.

Gregoire was born in the wild, in the northern forests of Congo (Brazzaville). When I met him he was alone in a dark cage, one in a row of similarly caged, solitary primates at the Brazzaville Zoo. Gregoire had been given to the zoo when his owners left the country, and he had been there since about 1949, some 40 years. He was almost hairless, and I could see nearly every bone in his body. Most of the animals at that zoo were starving; it was cheaper to replace animals who died of malnutrition than to buy an adequate diet. I knew I had to help Gregoire even though he had, somehow, survived without help for so long. A small group of people got together and agreed to save up food and deliver it to the zoo. The Jane Goodall Institute employed its own keeper to care for Gregoire and the other primates. Gregoire put on weight and his hair began to grow. Then the Brigitte Bardot Foundation gave us a small grant (after she saw video of Gregoire), and we were able to build a small "patio" for him. By this time Graziella Cotman was living in Brazzaville, and she was able to introduce three small orphans to the old male. One was a 2-year-old female, whom I named Cherie. A wonderful relationship, a bit like a grandfather and granddaughter, developed between this little girl and old Gregoire. Things were going well, until civil war broke out again. The zoo, near the airport, was in the middle of the war zone. Fortunately Gregoire, his young companions, and the two adult chimpanzees could be airlifted to the Tchimpounga Sanctuary (along with a group of young gorillas and bonobos). When Gregoire arrived his back was raw, apparently because he had rushed under his low bed shelf whenever the shelling got too close. But once again this old man adapted, and his hair grew back. Today he is in a group with two adult females and three youngsters.

Sebastian was brought to Kenya (where there are no wild chimpanzees) from West Africa. He ended up in the orphanage run by the Kenya Wildlife Service. There he lived for more than 20 years, becoming the star attraction. When I met him his quarters consisted of a small indoor and private cage that led into a circular mesh enclosure. He lived alone because he had seriously hurt females who had been introduced to him. He was very gentle with humans whom he liked, and loved to manicure my nails with a piece of twig. But when crowds of visitors arrived, especially when these were children who would make faces and tease him, he would display wildly, back and forth in the enclosure, throwing anything he could find. Yet when he was put in a newly built enclosure that prevented the public from approaching closely, he became seriously depressed and refused to eat. Eventually he was returned to his original home where he quickly recovered. Several years later he again became depressed when the orphanage was temporarily closed for reconstruction. Not until it was again opened to the public, and the daily teasing and displaying sessions resumed, did he recover. Clearly, the crowds provided stimulation and entertainment.

Lucy was born in captivity. As a tiny baby she was adopted by the Jane and Maurice Temerlins, a psychoanalyst and his zoologist wife. Lucy was brought up like a human child, clothes and all. The original plan was to find out whether a chimpanzee brought up with love and affection would be able to nurture her first baby despite having no experience of other chimpanzees. But as she reached adolescence the Termerlins decided that their lives had been ruled by Lucy for too long. After considering all options, they decided, with the best of intentions, to give her her freedom, to send her to Africa. Although she went with a trusted human, all that she had learned in her "human" days had to be forgotten. She had learned sign language, but her signs were ignored by the only person she knew, the person with whom, until the nightmare began, she had communicated in sign language. Lucy was introduced to two rambunctious young wild-born chimpanzees. She wanted nothing to do with them. She fell into deep depression. Although she eventually began to behave more like a chimpanzee, I personally believe the exercise was very cruel. Lucy died, years after arriving in Africa and ultimately being released on an island. Her body was found on the island with hands and feet removed. The whole exercise can be compared with taking a middle-class American girl of about 14 years old to live with a group of indigenous people in some far off part of the world. She would leave behind all her clothes, all her comforts, and all her culture. And her American companion would pretend not to understand a word she said.

These three chimpanzees had unnatural life styles to which they adapted. Humans created those situations. We have no right to try to effect change from our arrogant human perspective of "we know what is best." Rather we should try to get inside the mind of the individual chimpanzees and move slowly, a step at a time, toward a solution that is *best for them*.

CONCLUSIONS

Clearly chimpanzees today face many problems, both in the wild and in captivity. These problems are all different and need their own unique solutions that take into account all the variables; the country, the different people involved, the resources available, especially financial, and the personalities of the chimpanzees themselves. We cannot draw sweeping conclusions about the correct procedure in all zoos, in all sanctuaries, and in all situations.

Those trying to help the great apes have their own perspectives. There are many differences of opinion. But so long as we all have the same goals—the improvement of conditions for the great apes, in the world and in captivity—we should be able to work together. Each of us will bring to the table a unique set of experiences, different skills, and different ideas. Provided the apes are thriving, we must not be quick to condemn the way others manage things just because they are different from our own perception of what is right. Our solutions may not be perfect, but we each do our best.

I have encountered criticism for starting sanctuaries (they have been called a waste of money to help a few individuals when precious funds are needed to save the species), but for me there was no option. I simply could not look into the eyes of a pathetic orphan and leave it to its fate, because, for so many years, I have been able to look into the eyes of chimpanzees who are wild and free and in control of their own lives.

Let us move forward, united toward our goal of conserving the great apes in the wild, striving for the best treatment for all captive apes, and eliminating them from invasive medical research. Whether we care about the apes as species or as individuals, we all want solutions that will give them a chance to survive and to enjoy the best possible quality of life.

Section 1

GREAT APES IN THE WILD

THOMAS M. BUTYNSKI

1
AFRICA'S GREAT APES

Four of the six currently recognized species of (nonhuman) great ape live in tropical Africa. These are the robust chimpanzee (or common chimpanzee), *Pan troglodytes;* gracile chimpanzee (or bonobo or pygmy chimpanzee), *Pan paniscus;* western gorilla, *Gorilla gorilla;* and eastern gorilla, *Gorilla beringei.*

Africa's apes are of particular interest and importance because (1) they are essential components of the continent's tropical forest ecosystems, affecting the vegetation and accounting for a significant portion of the mammalian biomass (Redmond 1998); (2) they are vital to our understanding of human evolution (Fleagle 1982) and human diseases (Gao et al. 1999); and (3) they are a source of protein for many people in West and Central Africa (Bowen-Jones 1998).

The geographic ranges of the gracile chimpanzee, eastern gorilla, western gorilla, and robust chimpanzee are limited to within 4, 4, 7, and 13 degrees of the equator, respectively. Because they occupy the tropical forests, they are part of Africa's richest assemblage of plant and animal species. For example, these four apes are often sympatric with ten or more other species of nonhuman primates (Butynski 1997; Oates 1996a; Struhsaker 1997). As such, the chimpanzees and gorillas are among the most important "flagship species" for the conservation of African tropical forests. Unfortunately, despite their uniqueness and importance, all four species of African ape are in danger of extinction in the wild.

This chapter presents an overview of what is known about the current taxonomy, distribution, abundance, conservation status, and threats to chimpanzees and gorillas. This chapter also poses some ethical questions concerning the exploitation of the great apes and their habitats.

It is hoped that this overview will bring additional awareness to the plight of

Africa's apes, and thereby promote efforts on behalf of their welfare and conservation. Much of what is said concerning the conservation and exploitation of Africa's four great apes also applies to the world's other two great apes, the Bornean orangutan, *Pongo pygmaeus,* and the Sumatran orangutan, *Pongo abelii.* Both orangutans are endangered, numbering 15,000 to 20,000 and 5,000 to 7,000 individuals, respectively (Leiman and Ghaffar 1996).

PROBLEMS WITH THE DATABASE AND ESTIMATES OF NUMBERS AND GEOGRAPHIC RANGES OF AFRICA'S APES

The accuracy of the estimates presented in this chapter for numbers of apes and for the sizes of ape geographic ranges varies greatly. Some estimates, particularly for those taxa that have small populations and tiny distributions, are close to the actual numbers. In contrast, the estimates for those taxa with relatively large, widespread populations are often mere best guesses.

There are two reasons for this lack of accuracy for many of these estimates. First, the numbers and distributions of only a small portion of great ape populations have been adequately assessed. Second, many of the surveys on which the estimates are based were conducted more than a decade ago. With the widespread loss of ape habitat and the rapid decline in ape numbers, data more than a few years old often may be of limited value. For example, the estimate of robust chimpanzee numbers for the Democratic Republic of Congo (DRC, formerly Zaire), the country thought to hold the largest number of robust chimpanzees (Teleki 1991), is especially important but particularly speculative. The distribution of the robust chimpanzee in the DRC is poorly known, little of their potential habitat has ever been surveyed, and those surveys that have been conducted are now dated.

Given the limitations of the estimates, these values should be used with the utmost care and always with a cautionary note concerning accuracy. In addition, to provide as great a margin of safety as possible, we should always apply the "precautionary principle"—in other words, we should always assume the lowest estimates for the sizes of ape populations and geographic ranges when making conservation decisions.

In this chapter the term *geographic range* is equivalent to the term *extent of occurrence* as defined by the International Union for the Conservation of Nature and Natural Resources (IUCN; 2000). *Area of occupancy* (or *area occupied*) is used in this chapter as defined by the IUCN (2000). For those taxa of great ape that are widespread, the area of occupancy is certainly much smaller than the geographic range. For example, for the Grauer's gorilla, *G. beringei graueri,* the area of occupancy is estimated to be only 13 percent of the geographic range. I suspect that once we have more accurate information, the area of occupancy of the more widespread taxa will be found to be only 5 to 25 percent of the geographic range.

TAXONOMY AND DISTRIBUTION OF AFRICA'S APES

To set priorities for conservation action, it is important to determine the number of species and subspecies of African ape and their distributions.

Robust Chimpanzee

The robust chimpanzees, *P. troglodytes,* live in savanna–woodlands, mosaic grassland–forests, and tropical moist forests, and are found from sea level to about 9,200 ft (2,800 m) elevation (Groves 1971; Kortlandt 1983; Teleki 1989). The robust chimpanzee probably once spanned most of equatorial Africa, from south Senegal to southwest Tanzania, ranging over all or part of at least 25 countries (Hill 1969; Teleki 1989). Today the robust chimpanzee is the most widely distributed of Africa's apes, occurring in 22 countries from 13 degrees north to 7 degrees south latitude (Hill 1969; Kortlandt 1983; Lee, Thornback, and Bennett 1988; Teleki 1989; Table 1-1, Figure 1-1). With few exceptions, however, the past and present distributions of the robust chimpanzee within these countries are poorly known. The geographic range of the robust chimpanzee as shown in Figure 1-1 is approximately 936,000 sq. mi (2,340,000 sq. km, as measured by Map Info).

Three subspecies of robust chimpanzee have usually been recognized in recent decades: western chimpanzee, *P. troglodytes verus;* central chimpanzee, *P. troglodytes troglodytes;* and eastern chimpanzee, *P. troglodytes schweinfurthii* (Groves 2001; Napier and Napier 1967). Mitochondrial DNA studies, however, lend support to the recognition of the Nigeria chimpanzee as a distinct subspecies, *P. troglodytes vellerosus* (Gonder et al. 1997; IUCN/SSC PSG 2000).

A mitochondrial DNA study by Morin et al. (1994) found that *P. troglodytes verus* might be sufficiently different from *P. troglodytes troglodytes* and *P. troglodytes schweinfurthii* to warrant elevation to a full species (*Pan verus*). Recognition of *P. verus* is pending, primarily because intervening populations have not been adequately sampled and because morphological, ecological, or behavioral differences sufficient to merit species designation have not been demonstrated (Groves 2001; IUCN/SSC PSG 2000; Jolly, Oates, and Disotell 1995). This is obviously an area for research and consideration.

Hill (1967, 1969) recognized a fifth subspecies of robust chimpanzee, the koolokamba or gorilla-like chimpanzee, *P. troglodytes koolokamba,* for the montane forests of Cameroon and Gabon. This classification lacks current support among primate taxonomists; the specimens ascribed to *P. troglodytes koolokamba* all fall

Table 1-1

Estimated Number of Robust (Common) Chimpanzee, *Pan troglodytes,* in 2000 by Subspecies and Country

The taxonomy used is that of the IUCN/SSC Primate Specialist Group (2000) and the *2000 IUCN Red List of Threatened Animals* (Hilton-Taylor 2000).

	Number of Chimpanzees	
Subspecies and Country	Low	High
Western Chimpanzee, *P. troglodytes verus*	25,500	52,900
Benin	0	0
Gambia	0	0
Togo	0	0
Nigeria	0	?[a]
Burkina Faso	0	Few?
Guinea-Bissau	100	200
Senegal	200	400
Ghana	300	500
Sierra Leone	1,500	2,500
Mali	1,800	3,500
Liberia	3,000	4,000
Côte d'Ivoire	10,500	12,800
Guinea	8,100	29,000
Nigeria Chimpanzee, *P. troglodytes vellerosus*	4,000	6,000
Cameroon	1,500	3,500[b]
Nigeria	>2,500	>2,500[a]
Central Chimpanzee, *P. troglodytes troglodytes*	47,500	78,000
Democratic Republic of Congo (DRC)	?	?
Angola (Cabinda)	200	500
Central African Republic (CAR)	800	1,000
Equatorial Guinea (Rio Muni/Mbini)	1,000	2,000
Cameroon	8,500	11,500[b]
People's Republic of Congo (PRC)	10,000	10,000
Gabon	27,000	53,000[c]
Eastern Chimpanzee, *P. troglodytes schweinfurthii*	75,200	117,700
Central African Republic (CAR)	?	?
Sudan	200	400
Burundi	200	500
Rwanda	500	500
Tanzania	1,500	2,500

continued

Table 1-1
continued

Subspecies and Country	Number of Chimpanzees: Low	High
Uganda	2,800	3,800
Democratic Republic of Congo (DRC)	70,000	110,000
Total	152,200	254,600

Note: All data from Teleki (1991), except as follows: Burundi (Nishida 1994; Teleki 1991); Cameroon (L. Usongo, personal communication); Côte d'Ivoire (Marchesi et al. 1995); Equatorial Guinea (Teleki 1991; J. Sabater-Pi, personal communication, quoted in Nishida 1994); Gabon (Tutin and Fernandez 1984; L. White, personal communication, quoted in Stevens 1997); Guinea (Ham and Carter 2000); Guinea-Bissau (Gippoliti and Dell'Omo 1995; Teleki 1991; E. Feron, personal communication); Mali (Pavy 1993); Nigeria (J. Oates, personal communication); PRC (S. Kuroda, personal communication, quoted in Nishida 1994); Rwanda (Nishida 1994); Senegal (Galet-Luong et al. in press); Uganda (Edroma et al. 1997). Teleki (1991) does not provide references for the sources of his estimates, but most of these can be found in Lee et al. (1988).

[a]The chimpanzee in Nigeria west of the Niger River may belong to the subspecies *P. troglodytes verus.*

[b]An unknown number of between 10,000 and 15,000 chimpanzees in Cameroon in 1988 (L. Usongo, personal communication) belonged to the subspecies *P. troglodytes vellerosus* (Gonder et al. 1997). This table assumes that 1,500–3,500 of these are *P. troglodytes vellerosus.*

[c]L. White (personal communication, quoted in Stevens 1997) estimates that the number of chimpanzees in Gabon declined rapidly to 30,000 by 1997 as a result of activities associated with the expansion of logging in this country.

within the range of variation of *P. troglodytes troglodytes* (Cousins 1980; Groves 2001; Shea 1984).

The western chimpanzee, *P. troglodytes verus,* once occurred in 12 or 13 countries, but is currently patchily distributed in 9 or 10 countries from southeast Senegal (possibly once reaching the Gambia River) east probably to either the Dahomey Gap or the Niger River (Lee et al. 1988; Teleki 1989; E. Sarmiento and J. Oates, personal communication). Although now highly fragmented, the range of the western chimpanzee may have been almost continuous from Senegal to Togo until the mid-1900s (Jolly et al. 1995). Teleki (1989) suggested that the original population of western chimpanzee had a geographic range of nearly 800,000 sq. mi (2,000,000 sq. km), but it seems unlikely that it was ever this large. The geographic range presented in Figure 1-1 is 246,800 sq. mi (617,000 sq. km, as measured by Map Info computer mapping software).

The northern limit of the Nigeria chimpanzee, *P. troglodytes vellerosus,* is suspected to be either the Niger River or the Dahomey Gap, and the southern limit

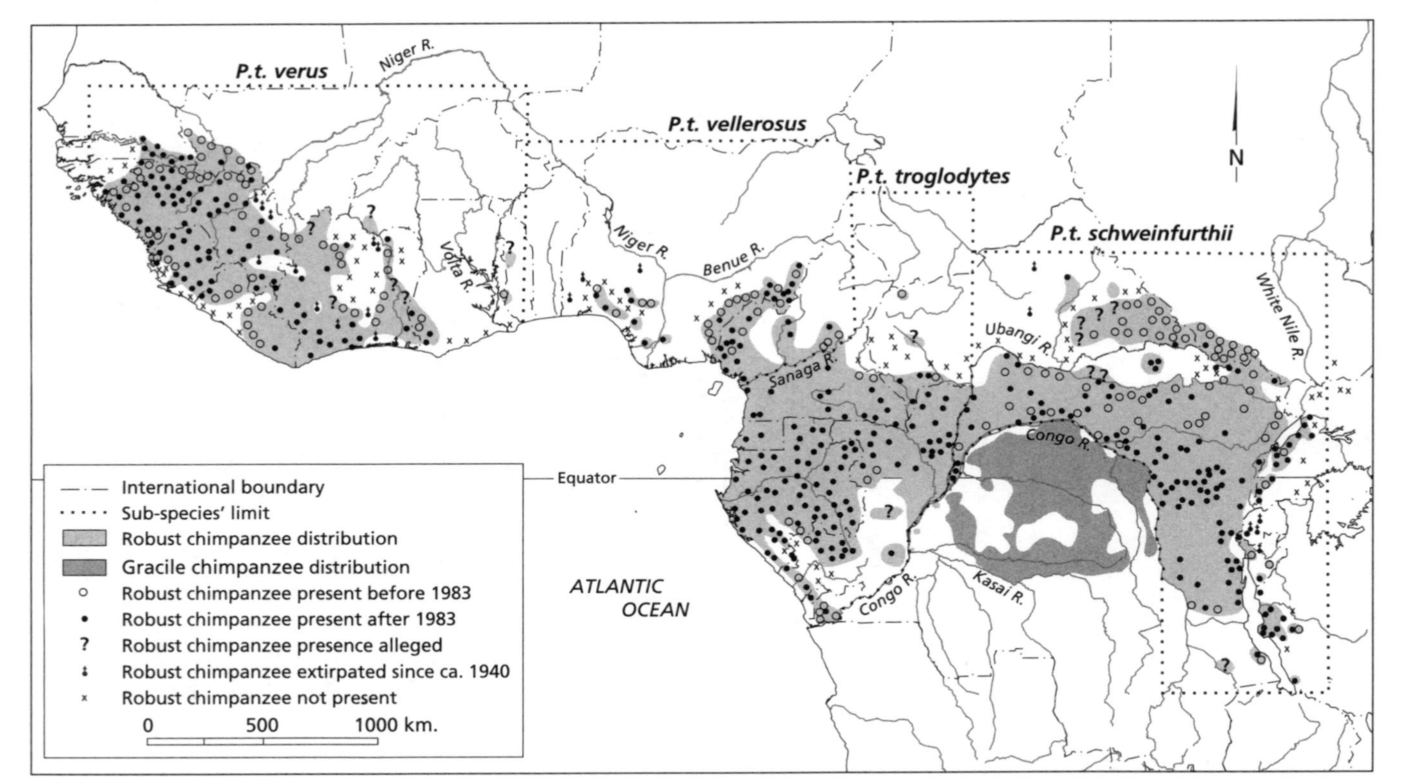
P.t. verus
Niger R.
P.t. vellerosus
P.t. troglodytes
N
P.t. schweinfurthii
Volta R.
Niger R.
Benue R.
Sanaga R.
Ubangi R.
White Nile R.
Congo R.
Congo R.
Kasai R.
Equator
ATLANTIC
OCEAN
International boundary
Sub-species' limit
Robust chimpanzee distribution
Gracile chimpanzee distribution
Robust chimpanzee present before 1983
Robust chimpanzee present after 1983
Robust chimpanzee presence alleged
Robust chimpanzee extirpated since ca. 1940
Robust chimpanzee not present
0
500
1000 km.

is probably the Sanaga River (M. K. Gonder and J. Oates, personal communication). Thus the geographic range of this subspecies lays in what was considered the southern range of the western chimpanzee (i.e., Benin and west Nigeria) and the northern range of the central chimpanzee, *P. troglodytes troglodytes* (i.e., east Nigeria and west Cameroon). The geographic range shown in Figure 1-1 is 60,800 sq. mi (152,000 sq. km, as measured by Map Info).

The geographic range of the central chimpanzee extends across seven countries, from the west bank of the Ubangi River southwest to near the mouth of the Congo River and north probably to the Sanaga River (Gonder et al. 1997), not to the Niger River as reported previously (Hill 1969; Lee et al. 1988; Teleki 1989; Tuttle 1986). In 1987, the central chimpanzee was known to range over an area of about 6,800 sq. mi (17,000 sq. km), with an additional 101,600 sq. mi (254,000 sq. km) of potentially suitable habitat in need of survey (Lee et al. 1988; Teleki 1989). The geographic range shown in Figure 1-1 is 280,000 sq. mi (700,000 sq. km, as measured by Map Info).

The eastern chimpanzee, *P. troglodytes schweinfurthii,* occurs in seven countries. The geographic range presumably extends from the east bank of the Ubangi River

---

Figure 1-1. Distribution of the robust (common) chimpanzee, *Pan troglodytes,* and gracile chimpanzee (bonobo), *Pan paniscus.* The pre-1983 localities for the robust chimpanzee are taken from Vandebroek (1958), Hillman (1982), Kortlandt (1983), Thys Van den Audenaerde (1984), and Tutin and Fernandez (1984). Much of the data on sites where robust chimpanzees were confirmed to be present post-1983 were compiled by E. van Adrichem (unpublished data) in 1998, but come also from Fay et al. (1989), Mwanza and Yamagiwa (1989), Massawe (1992, 1995), Hart and Sikubwabo (1994), Nicholas (1995), Anderson (1997), Gonder et al. (1997), Ogawa, Kanamori, and Mukeni (1997), Abedi-Lartey (1998), Hall, White, et al. (1998), Omari et al. (1999), Ham and Carter (2000), Allan (in press), and Galat-Luong et al. (in press). In addition, unpublished data were provided by A. Blom, M. Languy, R. Fotso, and S. Gartlan (Cameroon); A. Blom, M. Colyn (CAR); D. Messinger, J. Hart, K. Smith, F. Smith, and D. Wilkie (DRC); S. Lahm (Gabon); M. K. Gonder, J. Oates, and S. Gartlan (Nigeria); S. Blake, J. Moore, M. Colyn, and A. Blom (PRC); J. Kingdon (Sudan); and J. Moore (Tanzania).

Given the high levels of habitat loss and hunting since 1930, the robust chimpanzee is not now found over the entire range shown. There are undoubtedly sites where the robust chimpanzee occurs but that have yet to be documented. Note that there are large parts of the geographic range of the robust chimpanzee that have never been surveyed (e.g., southeast CAR and north DRC) or that have not been surveyed since before 1985 (e.g., the vast forests of Gabon). Although the robust chimpanzee almost certainly still occurs over much of Gabon, there are no current data to confirm this. The distribution of the gracile chimpanzee is taken from Figure 1-2.

across much of the DRC north of the Congo River and east of the Lualaba River, to southeast Central African Republic (CAR) and extreme southwest Sudan, to west Uganda, Rwanda, and Burundi, to southwest Tanzania (Kortlandt 1983; Lee et al. 1988). There were 11,600 sq. mi (29,000 sq. km) of habitat known to be occupied by the eastern chimpanzee in 1987 to 1989, with an additional 189,000 sq. mi (473,000 sq. km) of potential habitat (Lee et al. 1988; Teleki 1989, 1991). The geographic range shown in Figure 1-1 is 348,400 sq. mi (871,000 sq. km, as measured by Map Info).

Gracile Chimpanzee

Considered a subspecies of the robust chimpanzee until 1933 (Coolidge 1933), the gracile chimpanzee, *Pan paniscus,* is now widely acknowledged to be a distinct species with many morphological, ecological, behavioral, and genetic characteristics not shared with the robust chimpanzee (Groves 2001; Horn 1979; Ruvolo et al. 1994; Thompson 1997; Thompson-Handler, Malenky, and Reinartz 1995; Tuttle 1986; Uchida 1996).

The gracile chimpanzee is endemic to the mosaic grassland–forest, lowland forest, and swamp forest of the central Congo Basin in the DRC. The range lies south and east of the Congo River, north of the Kwa Kasai and Sankuru Rivers, and west of the Lualaba River at elevations ranging from about 980 to 1,640 ft (300 to 500 m; see Figure 1-2; Kano 1984; Kortlandt 1995; Lee et al. 1988; Thompson 1997; Thompson-Handler et al. 1995; Thys Van den Audenaerde 1984; Vandebroek 1958). In no place are the gracile chimpanzee and the robust chimpanzee known to be sympatric. The geographic range of the gracile chimpanzee appears to be separated from that of the robust chimpanzee by the Congo, Lualaba, and perhaps Lukuga Rivers. It remains uncertain whether the chimpanzee south of the Lukuga River in the Marungu Mountains (off the west side of Lake Tanganyika) is *P. paniscus* or *P. troglodytes* (Thompson 1997). If they are *P. paniscus,* this would considerably extend the known geographic range of this ape.

Kano (1984, 1992) estimated the "potential" geographical range of the gracile chimpanzee to be between 54,000 sq. mi (135,000 sq. km) and 80,000 sq. mi (200,000 sq. km), and Kortlandt (1976, 1995), Thompson-Handler et al. (1995), and Thompson (1997) estimated the potential geographic range to be 140,000 sq. mi (350,000 sq. km), 336,160 sq. mi (840,400 sq. km), and 188,800 sq. mi (472,000 sq. km), respectively. Given the rapid decline and disappearance of populations of the gracile chimpanzee over at least the past 30 years, the actual geographic range today is certainly much less than the potential range.

Note that in Kano (1984) there is a typographical error. The geographic range should be 54,000 sq. mi (135,000 sq. km), not 5,400 sq. mi (13,500 sq. km;

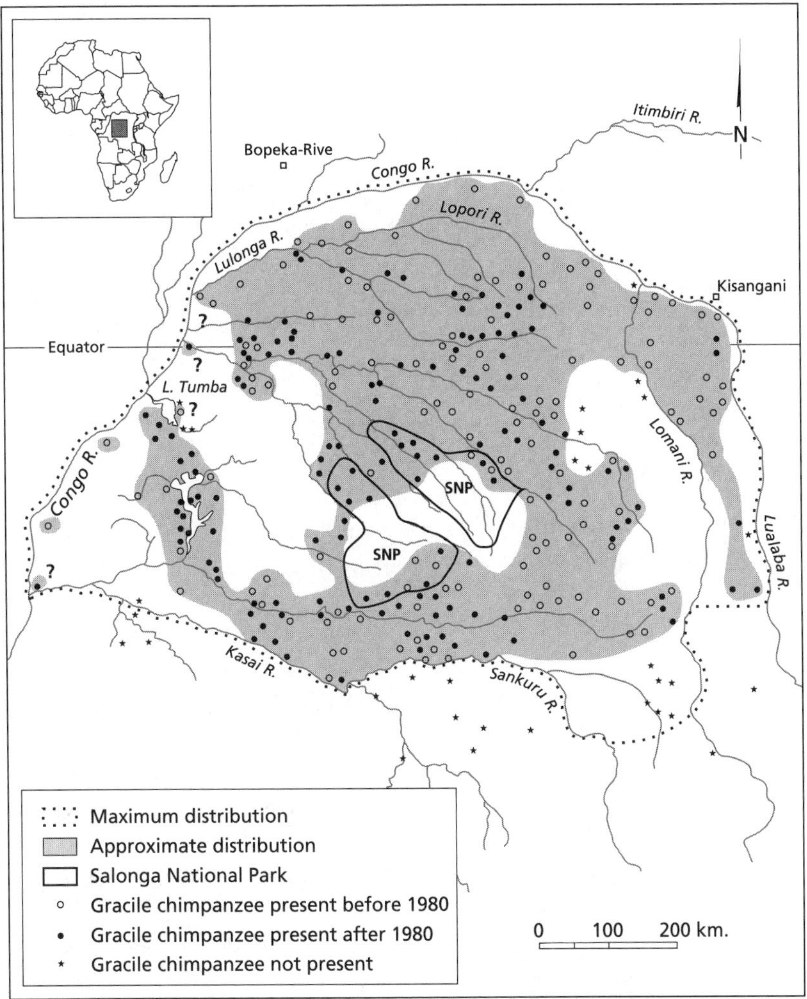

Figure 1-2. Distribution of the gracile chimpanzee (bonobo), *Pan paniscus,* based on Kano (1984), Thys Van den Audenaerde (1984), Kortlandt (1995), Thompson (1997), Van Krunkelsven, Bila-Isia, and Draulans (2000), D. Messinger (personal communication), J. Hart (personal communication), M. Colyn (personal communication), J. Thompson (personal communication), and S. Gartlan (personal communication). About half of all of the dots represent exact localities where the gracile chimpanzee was collected or observed. Villages and towns were sometimes given by Kortlandt (1995) as the locality of collected or observed gracile chimpanzee. These data are not plotted on this map because the distance and direction of the actual collection or observation site from the village or town are not known. The limits of "maximum distribution" and of "approximate distribution" are the author's interpretation of the available data.

Note that many of the plotted localities are based on data obtained pre-1980, whereas a good number of these are pre-1960. Habitat loss and hunting have since extirpated the gracile chimpanzee from some, perhaps many, of these sites. As such, this map may depict more closely the known historical distribution of this species rather than the current distribution. Also note that there are large parts of the geographic range of the gracile chimpanzee that have not been surveyed since before 1980, and even larger areas that have never been surveyed.

Kortlandt 1996b; Thompson 1997). The geographic range presented in Figure 1-2 is 137,200 sq. mi (343,000 sq. km, as measured by Map Info). The area that the gracile chimpanzee actually occupies is substantially less as the distribution is highly fragmented and discontinuous (Kano 1984, 1992; Thompson 1997; Thompson-Handler et al. 1995). Thompson (1997) estimated the area occupied to be no more than 47,200 sq. mi (118,000 sq. km).

Western Gorilla

Gorillas occur in two widely separated regions: one in west Central Africa and the other in east Central Africa (see Table 1-2, Figure 1-3). The two regions are separated by about 560 mi (900 km) of lowland forest and extensive swamp forest in the Congo Basin (Coolidge 1929; Schaller 1963). Mitochondrial DNA sequence data (but not the nuclear DNA sequence data; Jensen-Seaman 2000) indicate that genetic differences between gorillas living in these two regions are slightly greater than the differences between the robust chimpanzee and gracile chimpanzee and approach the level of genetic divergence exhibited between *Pan* and *Homo* (Garner and Ryder 1996; Morell 1994; Ruvolo et al. 1994; Ryder, Garner, and Burrows 1999; Uchida 1996). The genetic distance, together with a number of morphological, ecological, and behavioral differences, provide support for recognizing two species, the western gorilla (with two subspecies) and the eastern gorilla (with two or three subspecies; Groves 1996, 2001; IUCN/SSC PSG 2000; Ruvolo et al. 1994; Ryder et al. 1999).

The western gorilla, *G. gorilla,* inhabits lowland forest, swamp forest, and montane forest from sea level to about 5,250 ft (1,600 m). The two recognized subspecies are the western lowland gorilla, *G. gorilla gorilla,* and the Cross River gorilla, *G. gorilla diehli* (Groves 2001; IUCN/SSC PSG 2000; Sarmiento and Oates 2000).

The western lowland gorilla occupies lowland forest and swamp forest from sea level to 5,250 ft (1,600 m). This subspecies is distributed over six or seven countries from south Cameroon and southwest CAR, through Equatorial Guinea and Gabon, into parts of the People's Republic of Congo (PRC) and extreme north Angola (Cabinda enclave; see Figure 1-3). This subspecies is now probably absent from its former range in extreme western DRC, north of the Congo River. Harcourt (1996) estimated that the western lowland gorilla has a geographic range of approximately 178,000 sq. mi (445,000 sq. km). The geographic range shown in Figure 1-3 is about 283,600 sq. mi (709,000 sq. km, as measured by Map Info).

The Cross River gorilla is a recently resurrected subspecies inhabiting the upper Cross River region on the Nigeria–Cameroon border, about 160 mi (260 km) north of the range of the western lowland gorilla (Harcourt, Stewart, and Inahoro

Table 1-2

Estimated Numbers of Western Gorilla, *Gorilla gorilla,* and Eastern Gorilla, *Gorilla beringei,* in 2000 by Subspecies and Country

The taxonomy used here is that of the IUCN/SSC Primate Specialist Group (2000) and the *2000 IUCN Red List of Threatened Animals* (Hilton-Taylor 2000).

Species, Subspecies, and Country	Number of Gorillas
Western Gorilla, *G. gorilla*	94,700
Western Lowland Gorilla, *G. gorilla gorilla*[a]	94,500
Democratic Republic of Congo (DRC)	0
Angola (Cabinda)	present
Equatorial Guinea (Rio Muni/Mbini)	1,500
Central African Republic (CAR)	9,000
Cameroon	15,000
People's Republic of Congo (PRC)	34,000
Gabon	35,000
Cross River Gorilla, *G. gorilla diehli*	200
Nigeria	100
Cameroon	100
Eastern Gorilla, *G. beringei*	17,500
Mountain Gorilla, *G. beringei beringei*	324
Uganda	12
Rwanda	129
Democratic Republic of Congo (DRC)	183
Grauer's Gorilla, *G. beringei graueri*	16,900
Democratic Republic of Congo (DRC)	16,900
Bwindi Gorilla, *G. beringei* (ssp.?)	300
Uganda[b]	300

Note: All data for the western gorilla are from Harcourt (1996), except as follows: Cameroon (L. Usongo, personal communication); Equatorial Guinea (Gonzalez-Kirchner 1997); Gabon (Tutin and Fernandez 1984); PRC (Fay and Agnagna 1992); Cross River gorilla (Oates et al. 1999); Grauer's gorilla (Hall, Saltonstall, et al. 1998); mountain gorilla (Butynski and Kalina 1998a; Sholley 1991); Bwindi gorilla (Butynski and Kalina 1998a).

[a]Harcourt (1996) estimated that there are 44,000 western lowland gorillas in the PRC, 43,000 in Gabon, 12,500 in Cameroon, 9,000 in the CAR, and 3,000 in Equatorial Guinea.

[b]A few groups occasionally cross the border from Uganda to enter Sarambwe Forest of the DRC (personal observation).

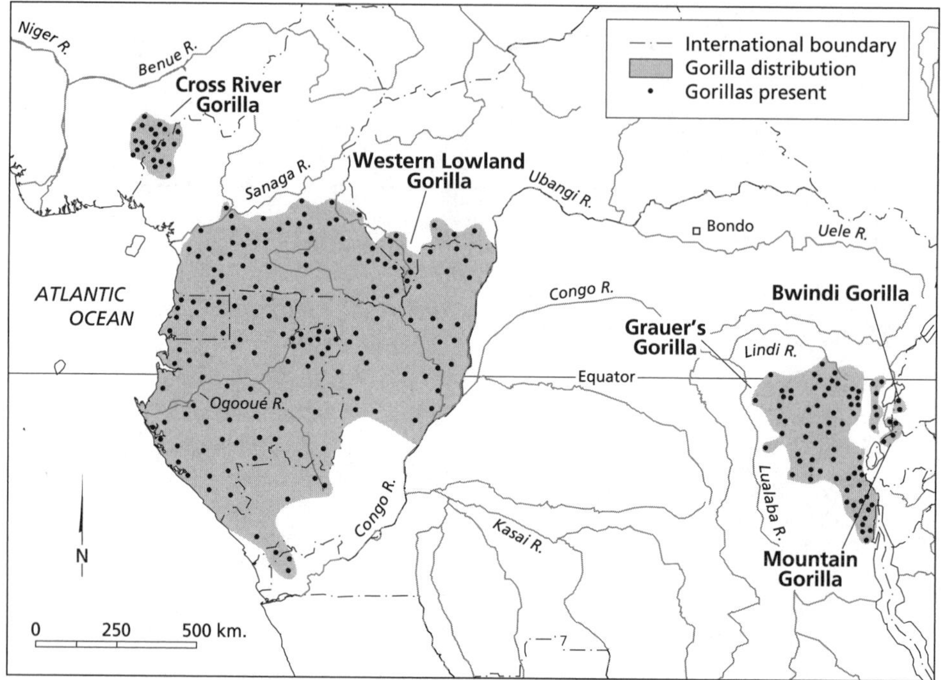

Figure 1-3. Distribution of the western gorilla, *Gorilla gorilla,* and eastern gorilla, *Gorilla beringei,* based on confirmed localities where gorillas were collected or observed as indicated in Emlen and Schaller (1960), Groves (1971), Goodall and Groves (1977), Tutin and Fernandez (1984), Mwanza and Yamagiwa (1989), Fay et al. (1989), Fay and Agnagna (1992), Hall and Wathaut (1992), Blake (1994), Hart and Sikubwabo (1994), Blake et al. (1995), Sarmiento and Butynski (1996), Gonzalez-Kirchner (1997), Hall, Saltonstall, et al. (1998a), Hall, White, et al. (1998), Omari et al. (1999), and Sarmiento and Oates (2000), and on unpublished data provided by E. Sarmiento, R. Dowsett, and S. Gartlan (Cameroon); S. Blake, M. Colyn, and A. Blom (CAR); S. Blake, M. Colyn, M. Agnagna, A. Blom, J. Moore, and D. Wilkie (PRC); J. Hart (DRC); and S. Lahm (Gabon).

Given the high levels of habitat loss and hunting over the past several decades, gorillas are not now found over the entire range as shown. For example, the present distribution of the Nigeria gorilla is very much more restricted than during the nineteenth century, as determined from the sites at which museum specimens were collected (Sarmiento and Oates 2000). On the other hand, there are undoubtedly some sites where the gorilla occurs but that have yet to be documented.

Bondo (Djabbir) on the Uele River represents the location where four gorilla skulls were obtained from villagers in 1908 (Coolidge 1929, 1933; Schouteden 1947). These four gorillas were given the name *G. gorilla uellensis,* but Groves (1971) could not distinguish them from *G. gorilla gorilla.* Bondo is about 370 mi (600 km) east of the present geographic range of *G. gorilla gorilla.*

1989; Lee et al. 1988; Oates 1998; Sarmiento and Oates 2000). Formerly referred to as western lowland gorillas, recent morphological studies show that the Cross River gorilla is as different from the western lowland gorilla as the mountain gorilla is from some populations of Grauer's gorilla (Oates 1998; Sarmiento and Oates 2000; Stumpf et al. 1998). The Cross River gorilla occurs in lowland and submontane forest over an elevation range of 660 to 5,250 ft (200 to 1,600 m; Oates et al. 1999). Their geographic range is at least 600 sq. mi (1,500 sq. km), and the area occupied is roughly 80 sq. mi (200 sq. km; J. Oates, personal communication). The geographic range shown in Figure 1-3 plots the sites where Cross River gorillas were collected (E. Sarmiento, personal communication). It appears that the geographic range of the Cross River gorilla during the late 1800s was roughly 8,800 sq. mi (22,000 sq. km).

Eastern Gorilla

Eastern gorillas, *G. beringei,* live in submontane and montane forest from about 2,300 to more than 13,100 ft (700 to 4,000 m; Harcourt 1996; Sarmiento, Butynski, and Kalina 1996; Schaller 1963). Two subspecies of eastern gorilla are recognized; mountain gorilla, *G. beringei beringei,* and Grauer's gorilla, *G. beringei graueri* (Groves 1970, 1996, 2001; Sarmiento and Butynski 1996; see Figure 1-3).

The mountain gorilla is restricted to one population in the Virunga Volcanoes, where the borders of Uganda, Rwanda, and the DRC meet (Sholley 1991). This population occupies an area of about 150 sq. mi (375 sq. km) at elevations from 7,200 to more than 13,100 ft (2,200 to 4,000 m; Harcourt and Fossey 1981; Vedder and Weber 1990).

The gorillas of the Bwindi–Impenetrable National Park in southwest Uganda are sometimes referred to as "mountain gorillas," but their taxonomy remains uncertain. The gorillas of this population may represent an undescribed subspecies (Hilton-Taylor 2000; IUCN/SSC PSG 2000; Sarmiento and Butynski 1996; Sarmiento et al. 1996). Until more data are obtained, these gorillas should be referred to simply as the "Bwindi gorillas" (i.e., *Gorilla beringei* ssp.?). The Bwindi gorillas occupy an area of about 86 sq. mi (215 sq. km) and range in elevation from 4,900 to 7,550 ft (1,500 to 2,300 m; Sarmiento et al. 1996).

Grauer's gorilla has a discontinuous distribution east of the Lualaba River and west of the Albertine (Western) Rift from the northwest corner of Lake Edward in the north to the northwest corner of Lake Tanganyika in the south (Hall, Saltonstall, et al. 1998; Figure 1-3). Grauer's gorilla is endemic to the DRC (unless the Bwindi gorillas of Uganda are of this subspecies). The subspecific status of some other gorilla populations generally referred to as Grauer's gorilla remains equivocal, such as the population on Mt. Tshiaberimu to the northeast of Lake

Table 1-3
Category of Threat Allocated to Each Species and Subspecies of African Great Ape in the *1988 IUCN Red Data Book* (Lee et al. 1988), in the *1996 IUCN Red List of Threatened Animals* (IUCN 1996), and in the *2000 IUCN Red List of Threatened Animals* (Hilton-Taylor 2000), and the estimated number of individuals for each species and subspecies in 2000.

The taxonomy used here is that of the IUCN/SSC Primate Specialist Group (2000) and the *2000 IUCN Red List of Threatened Animals* (Hilton-Taylor 2000).

Species and subspecies	1988[a]	1996	2000	Total Number[b]
Robust Chimpanzee, *P. troglodytes*	V	E	E	203,000
Western Chimpanzee, *P. troglodytes verus*	E	E	E	39,000
Nigeria Chimpanzee, *P. troglodytes vellerosus*	c	c	E	5,000
Central Chimpanzee, *P. troglodytes troglodytes*	V	E	E	63,000
Eastern Chimpanzee, *P. troglodytes schweinfurthii*	V	E	E	96,000
Gracile Chimpanzee (Bonobo), *P. paniscus*	V	E	E	35,000
Western Gorilla, *G. gorilla*	c	c	E	94,700
Western Lowland Gorilla, *G. gorilla gorilla*	V	E	E	94,500
Cross River Gorilla, *G. gorilla diehli*	c	C	C	200
Eastern Gorilla, *G. beringei*	c	c	E	17,500
Grauer's Gorilla, *G. beringei graueri*	E	E	E	16,900
Mountain Gorilla, *G. beringei beringei*	E	C	C	300
Bwindi Gorilla, *G. beringei* (ssp?)	c	c	C	300

[a]V = vulnerable; E = endangered; C = critically endangered.

[b]See Tables 1-1 and 1-2 for country-by-country estimates for 2000 and for references for the data on which these estimates are based.

[c]This taxon not recognized during this year. Thus no degree of threat assessment was made. Note that "the" gorilla, *G. gorilla,* was considered "vulnerable" in 1988 and "endangered" in 1996, before being recognized in 2000 as comprising two species, *G. gorilla* and *G. beringei.*

Edward (Sarmiento and Butynski 1996). The range of elevation used by Grauer's gorilla is from about 2,300 to 9,500 ft (700 to 2,900 m; Butynski and Sarmiento 1996). Harcourt (1996) estimated the geographic range at 16,800 sq. mi (42,000 sq. km). The geographic range shown in Figure 1-3 is 44,800 sq. mi (112,000 sq. km; as measured by Map Info). The area occupied by Grauer's gorilla in 1963 was estimated to be 8,400 sq. mi (21,000 sq. km; Schaller 1963), but it has declined considerably over the past 35 years to roughly 6,000 sq. mi (15,000 sq. km) in the early 1990s (Hall, Saltonstall, et al. 1998). Thus for this subspecies the area occupied is roughly 13 percent of the geographic range, reflecting its current highly fragmented, localized distribution.

Table 1-3 provides a summary of changes in the taxonomy of Africa's apes from 1988 to 2000.

ABUNDANCE OF AFRICA'S APES

The question of how many individuals there are in each taxon of African ape is a thorny one, as indicated by the considerable ranges in the estimates provided for some taxa over the past decade. Nonetheless, even rough estimates provide some indication of the level of endangerment for each taxon and, therefore, some basis on which to set priorities for conservation action.

Robust Chimpanzee

The western chimpanzee, *P. troglodytes verus,* is extinct in the wild in three countries (Benin, Gambia, Togo) and almost extinct in four other countries (Burkina Faso, Ghana, Guinea-Bissau, Senegal; IUCN 1996; Lee et al. 1988). Although the western chimpanzee is reported to be extirpated from Burkina Faso, J. Moore (personal communication) has good second-hand information that a few chimpanzees are still present in that country in riverine forest along the Volta River near "the bend" at the village of Douroula.

Teleki (1989) suggested that the original population of western chimpanzee numbered more than 600,000 individuals. In 1987 there were an estimated 2,000 in known habitats (3,600 sq. mi; 9,000 sq. km) and another 12,000 to 19,000 in potential habitats (15,600 sq. mi; 39,000 sq. km; Lee et al. 1988; Teleki 1989). Teleki (1989) argued that in Sierra Leone alone the population dropped from 20,000 in the late nineteenth century to 2,000 in 1987.

Teleki (1991) estimated that Côte d'Ivoire held 500 to 1,000 western chimpanzees in 1989. This estimate was contradicted by field surveys in 1989 and 1990 by Marchesi et al. (1995) that indicated that there were 10,500 to 12,800 chimpanzees in Côte d'Ivoire. These researchers found the highest densities of chimpanzees in the Marahoue National Park. Work in the Marahoue National Park in 1997 to 1998 strongly suggests, however, that the density of chimpanzees there has declined dramatically since 1990 (Barnes 1997; Struhsaker 1998). It is not known whether there has been a similar decline throughout Côte d'Ivoire.

Teleki (1991) suggested that there were 600 to 800 chimpanzees in Mali. More recent field work indicates that there are 1,800 to 3,500 in Mali (Pavy 1993).

Sugiyama and Soumah (1988) and Teleki (1991) placed the number of chimpanzees in Guinea at 1,400 to 6,600 and 2,000 to 4000, respectively. More recently, however, the results of a 15-month nationwide survey indicate that the number of chimpanzees in Guinea is roughly 17,600 (range 8,100 to 29,000; Ham and Carter 2000). Guinea and Côte d'Ivoire support the largest numbers of western chimpanzees.

The total number of western chimpanzees in 1989 was estimated at 8,000 to 13,000 (Teleki 1991). In 1997, the World Wide Fund for Nature (WWF) estimated that there were 12,000 western chimpanzees (Kemf and Wilson 1997). More recent surveys (see Table 1-1) indicate that this was a considerable underestimate and that today there are probably between 25,000 and 53,000 western chimpanzees.

Estimates of past and current numbers of the Nigeria chimpanzee, *P. troglodytes vellerosus,* are particularly difficult because this subspecies was subsumed within *P. troglodytes troglodytes* until recently and because the limits of its geographic range remain uncertain. Teleki (1991) estimated that there were 100 to 300 chimpanzees in Nigeria in 1989, but Hogarth (1997) found roughly 1,500 chimpanzees in the Gashaka Gumti National Park alone. J. Oates (personal communication), who along with his students has conducted the most extensive surveys of the forests of Nigeria, estimated that there were at least 2,500 chimpanzees in Nigeria in 1998. It is probable that at least a few thousand of the 10,000 to 15,000 chimpanzees estimated to be present in Cameroon (L. Usongo, personal communication) are Nigeria chimpanzees. A reasonable guess at this time is that there are 4,000 to 6,000 Nigeria chimpanzees.

The largest population of the central chimpanzee, *P. troglodytes troglodytes,* is found in Gabon. There are also substantial populations in the PRC and Cameroon. Smaller populations are present in Equatorial Guinea, the CAR, north Angola (Cabinda enclave), and extreme west DRC, north of the Congo River. The estimate of 51,000 to 77,000 chimpanzees in Gabon in 1980 to 1983 (Tutin and Fernandez 1984) is now outdated, given the substantial increases in logging and hunting in the country since the surveys were conducted. Indeed, L. White (quoted in Stevens 1997) suspects that the number of chimpanzees in Gabon has declined over the past 15 years by more than 50 percent, from about 64,000 to 30,000 individuals.

In 1991 Teleki estimated that there were 3,000 to 5,000 chimpanzees in the PRC. Since then, two to three times as many chimpanzees were found to be present (S. Kuroda, personal communication, quoted in Nisida 1994). L. Usongo (personal communication) reported that in 1998 there were probably more chimpanzees in Cameroon (10,000 to 15,000 animals) than estimated by Teleki (1991; 6,000 to 10,000 animals). The majority of these are central chimpanzees, whereas those in the north are probably Nigeria chimpanzees.

In 1987, there were about 5,000 central chimpanzees in known localities (6,800 sq. mi; 17,000 sq. km) and an estimated 57,000 to 91,000 in unsurveyed areas with potentially suitable habitat (101,600 sq. mi; 254,000 sq. km; Lee et al. 1988; Teleki 1989). Teleki (1991) estimated that there were 62,000 to 91,000 central chimpanzees in 1989. The data in Table 1-1 indicate that the present number of central chimpanzees is likely between 48,000 and 78,000.

The number of eastern chimpanzees, *P. troglodytes schweinfurthii,* from known habitats (11,600 sq. mi; 29,000 sq. km) was put at 10,000 in 1987 to 1989, with an additional 65,000 to 108,000 animals in potential habitats (189,200 sq. mi; 473,000 sq. km; Lee et al. 1988; Teleki 1989, 1991). This gives a total of 75,000 to 118,000 individuals for this subspecies in 1987 to 1989. The majority (an estimated 70,000 to 110,000) were in the DRC (Teleki 1991), and small populations totaling 5,100 to 8,600 animals were in Uganda, Tanzania, Rwanda, Burundi, and Sudan. There are no data on the number of eastern chimpanzees in southeast CAR.

The major problem with estimating the number of eastern chimpanzees is the paucity of information on densities and distributions of this species in the DRC. Based on extensive surveys, Hart and Hall (1996) estimated that there are 12,800 to 21,900 chimpanzees within the 12,210-sq. mi (30,530-sq. km) area in east DRC covered by the Maiko National Park, Kahuzi Biega National Park, and Okapi Wildlife Reserve. On this basis it seems reasonable to retain Teleki's (1991) estimate of 70,000 to 110,000 eastern chimpanzees for the DRC. Information from Burundi (Nishida 1994), Rwanda (Nishida 1994), and Uganda (Edroma, Rosen, and Miller 1997) suggests that the present number of eastern chimpanzees remains between 75,000 and 118,000 (Table 1-1).

Tables 1-1 and 1-3 indicate that roughly 3 percent of the robust chimpanzees are Nigeria chimpanzees, 20 percent are western chimpanzees, 31 percent are central chimpanzees, and 47 percent are eastern chimpanzees. Teleki (1991) has claimed that there were once millions of robust chimpanzees in Africa. Goodall (2000) has argued that there were about 2,000,000 robust chimpanzees in Africa at the beginning of the twentieth century, more than a million in 1960, and no more than about 150,000 today.

The total number of robust chimpanzees in known localities (22,000 sq. mi; 55,000 sq. km) in 1987 was about 17,000, and estimates based on area of potentially suitable habitat (306,400 sq. mi; 766,000 sq. km) suggested an additional 134,000 to 218,000 animals (Teleki 1989). This yields a total of 151,000 to 235,000 robust chimpanzees in 1987. Teleki (1991) provided a population range total of 145,000 to 228,000 for this species in 1989. The country-by-country estimates presented in Table 1-1 indicate that, despite a considerable decline in robust chimpanzee numbers since 1989 because of habitat loss and hunting, the current number of robust chimpanzees is probably between 152,000 and 255,000. This suggests that the estimates provided for 1987 and 1989 were somewhat low. The estimates presented in this chapter are not very different from the estimate of 200,000 robust chimpanzees given by the WWF in 1997 (Kemf and Wilson 1997).

Gracile Chimpanzee

The highest estimate of numbers of gracile chimpanzees, *P. paniscus,* provided almost three decades ago by Kortlandt (1976), is 100,000 to 200,000. At about the same time, Teleki and Baldwin (1979) provided the much lower estimate of 13,000. In 1973, Kano (1984, 1992) conducted the most extensive survey of the geographic range of the gracile chimpanzee. He estimated that there were 54,000 gracile chimpanzees in the northern half of their geographic range and a total population of fewer than 100,000. Based on a potential geographic range of 140,000 sq. mi (350,000 sq. km) and a density of 1 gracile chimpanzee per sq. mi (0.4 gracile chimpanzees/sq. km), Kortlandt (1995, 1996b), suggested that there were perhaps 100,000 in 1995. Using 47,200 sq. mi (118,000 sq. km) as the area occupied, and "the lowest potential density of 0.25 individuals per sq. km," Thompson (1997) provided an estimate of 29,500 gracile chimpanzees. More recently, Thompson (personal communication) suggested that there are only 10,000 to 15,000 remaining gracile chimpanzees, whereas Dupain, Van Elsacker, and Verheyen (in press) estimated that there may still be more than 50,000. Estimates over the past decade of the number of gracile chimpanzees, therefore, range from 10,000 to 100,000. Based on the limited information available, and given that this species is now widely believed to be suffering from an escalating rate of decline over a large portion of its range, the best guess is that there are currently 20,000 to 50,000 gracile chimpanzees (Thompson 1997; Thompson-Handler et al. 1995; Van Krunkelsven et al. 2000).

Western Gorilla

The majority of western lowland gorillas, *G. gorilla gorilla,* are in Gabon and the PRC. The number of gorillas in Gabon was estimated to be 28,000 to 42,000 in the early 1980s (Tutin and Fernandez 1984). The PRC had a similar number in 1989 to 1990 (Fay and Agnagna 1992). Cameroon held an estimated 15,000 gorillas in 1998 (L. Usongo, personal communication). There were probably more than 5,000 gorillas on mainland Equatorial Guinea (Rio Muni/Mbini) during the early 1960s (Sabater Pi 1966), but the population dropped to 1,000 to 2,000 by 1989 to 1990 (Gonzalez-Kirchner 1997).

The data presented in Table 1-2 suggest that the current number of western lowland gorillas is roughly 95,000. This is close to the recent estimates by Harcourt (1996), the WWF (Kemf and Wilson 1997), and Plumptre, McNeilage, and Hall (1999) of 111,500, 111,000, and 110,000 western lowland gorillas, respectively. These more recent figures are more than twice the estimate of 40,000 western lowland gorillas provided by the Gorilla Advisory Committee of the IUCN/SSC (Species Survival Commission) Primate Specialist Group (Vedder 1987) and the estimate of 35,000 to 45,000 given in the 1988 *IUCN Red Data Book* (Lee et al. 1988). These differences are the result of an improved database, not an increase in the number of gorillas.

There are approximately 200 to 250 Cross River gorillas, *G. gorilla diehli,* in four subpopulations in Nigeria and Cameroon (Oates et al. 1999). Only about 0.2 percent of the western gorillas are Cross River gorillas. This is the rarest of the gorilla subspecies.

Eastern Gorilla

At the time of the last census in 1989 there were about 324 mountain gorillas, *G. beringei beringei,* in the Virunga Volcanoes (Sholley 1991; Table 1-2, Figure 1-3). It is suspected that this population has not changed substantially since 1989.

There are about 300 Bwindi gorillas, *G. beringei* ssp.?, in the Bwindi–Impenetrable National Park (Butynski and Kalina 1998a).

Schaller (1963) estimated that there were 2,500 to 4,500 Grauer's gorillas, *G. beringei graueri,* in the late 1950s. More intensive surveys conducted in 1989 to 1995 indicate, however, that there are today approximately 16,900 Grauer's gorillas (range 8,660 to 25,500 animals) in at least 11 populations. An estimated 86 percent of these inhabit the Kahuzi-Biega National Park–Kasese region. Approximately two thirds are found within the Kahuzi-Biega and Maiko National Parks (Hall, Saltonstall, et al. 1998; Hall, White, et al. 1998).

The total number of eastern gorillas is probably in the vicinity of 17,500. This is considerably higher than the estimate of 10,500 provided by Harcourt (1996) or the estimate of 10,600 given by the WWF (Kemf and Wilson 1997). Of the total 17,500 eastern gorillas, roughly 96 percent are Grauer's gorillas, 2 percent are mountain gorillas, and 2 percent are Bwindi gorillas.

CONSERVATION STATUS OF AFRICA'S APES

All of Africa's apes are listed as "endangered species" under Section 4 of the U.S. Endangered Species Act of 1973. All four species are also listed under Appendix I of the Convention on International Trade in Endangered Species of Wild Fauna and Flora (CITES). That is, they are considered "species threatened with extinction which are or may be affected by trade" (CITES 1973). All 150 member countries of CITES are required to "take appropriate measures to enforce the provisions of the present Convention and to prohibit trade in specimens in violation thereof."

The most recent action plan for African primates, produced by the IUCN/SSC Primate Specialist Group (Oates 1996a), summarizes the primate conservation needs for Africa and sets priorities for conservation action. The robust chimpanzee, gracile chimpanzee, and gorilla are listed among the seven African primates with the highest "conservation priority ratings."

The 1988 *IUCN Red Data Book* (Lee et al. 1988) ranked the western chimpanzee, Grauer's gorilla, and mountain gorilla as "endangered taxa." All other taxa of African ape were ranked as "vulnerable" (Table 1-3). These two degrees of threat are defined in Lee et al. (1988, xvii) as follows:

> *Endangered:* Taxa in danger of extinction and whose survival is unlikely if the causal factors continue to operate. This category includes taxa whose numbers have been reduced to a critical level or whose habitats have been so drastically reduced that they are deemed to be in immediate danger of extinction. *Vulnerable:* Taxa believed likely to move into the "endangered" category in the near future should the causal factors continue to operate. This category includes taxa whose populations are decreasing because of overexploitation, extensive habitat destruction or other environmental disturbance; taxa with populations that have been seriously reduced and whose ultimate security has not yet been assured; and taxa with populations that are still abundant but are under threat from severe adverse factors throughout their range.

During the period 1988 to 1996 the "causal factors" (i.e., habitat loss and hunting) not only continued to operate, their rates of damage to ape populations and ape habitats accelerated (Kemf and Wilson 1997; WSPA 2000). As such, the *1996 IUCN Red List of Threatened Animals* (IUCN 1996) ranked all taxa of African ape as either "endangered" or "critically endangered" (Table 1-3).

The *1996 Red List of Threatened Animals* used a new set of criteria for assessing degree of threat (IUCN 1994, 1996). The majority of Primate Specialist Group members working with the robust chimpanzee, gracile chimpanzee, and gorilla in the wild believed that all (at that time) three recognized species should be classified as "endangered" under criterion A2. That is, based on observed or suspected changes in area of occupancy, extent of occurrence, quality of habitat, and levels of exploitation, they projected that the wild populations would decline by at least 50 percent over the next three generations. For all three species, three generations is taken to be 60 years (as defined in IUCN 1994).

The robust chimpanzee, gracile chimpanzee, and gorilla were three of six species of African primate considered to be "endangered" in 1996 (Butynski 1997; IUCN 1996). Of the subspecies of African ape, the Cross River gorilla and mountain gorilla were ranked as "critically endangered" under criterion C2. That is, their populations numbered fewer than 250 mature individuals, a continuing decline was projected in the numbers of mature individuals, and either no population contained more than 50 mature individuals (Cross River gorilla) or all individuals were in a single population (mountain gorilla). All other subspecies of African ape were rated as "endangered," because they met criterion A2 (see earlier discussion). It should be noted that five taxa were also rated as "endangered" or "critically endangered" under criteria other than A2 or C2 (IUCN 1996).

In February 2000, an IUCN/SSC Primate Specialist Group workshop reassessed the taxonomy and degree of threat of the world's primates in preparation for the *2000 IUCN Red List of Threatened Animals* (Hilton-Taylor 2000). About 25 of the world's most experienced field primatologists, primate taxonomists, and primate molecular biologists participated in this workshop. They concluded that all four species, and six of the eight subspecies, of Africa's apes are "endangered." Two subspecies, the mountain gorilla and the Cross River gorilla, are "critically endangered," as is the population of gorillas in the Bwindi–Impenetrable National Park (Table 1-3).

Numbering about 200,000 individuals, there are roughly twice as many robust chimpanzees in the wild as there are western gorillas, about five times more robust chimpanzees than gracile chimpanzees, and about 11 times more robust chimpanzees than eastern gorillas (Table 1-3). Of the four species of African ape, the biggest concern from a conservation perspective must be for the survival of those taxa with the lowest number of individuals. In this case, these are the eastern gorilla and gracile chimpanzee. Among the subspecies of African apes, conservation efforts must obviously focus on the Cross River gorilla, mountain gorilla, and Bwindi gorilla.

THREATS TO AFRICA'S APES

This section addresses the current threats to ape populations in Africa. The causes of the threats are complex, and they are often overlapping and interdependent.

Human Population Pressures

Low-density human populations have, for thousands of years, used apes and ape habitats sustainably. This changed during the 1900s, particularly since the 1950s

(Goliber 1985), as Africa's human population increased rapidly and as Western technologies capable of destroying apes and their forest habitats were widely applied.

Human population growth rates are falling on all continents except Africa, where women average 5.7 births per lifetime (Fornos 1998). Unfortunately, Africa is also the continent where food production per person continues to decline, where one third of the people are presently malnourished, where food production needs to increase 300 percent to meet demand in the year 2050, where people are 22 percent poorer than in 1975 (Conly 1998), and where civil wars and insecurity are on the increase. At a growth rate of 2.9 percent per year, Africa's population is estimated to double to more than 1,500,000,000 by the year 2025 (World Resources Institute 1994). This runaway growth means that requirements of Africans for water, food, clothing, fuel, and shelter will continue to grow rapidly. The demand for Africa's natural resources by people in Europe, Asia, and North America is also, of course, increasing rapidly. Meeting these increasing internal and external demands for natural resources is the main cause of deforestation and local extinction of species in tropical Africa (Butynski 1997; Myers 1993; Struhsaker 1996, 1997; Teleki 1989).

Today, the majority of Africans are struggling to meet their short-term survival needs. In doing so, they are fragmenting and destroying populations of chimpanzees and gorillas directly through unsustainable hunting and indirectly through habitat degradation and loss (Bowen-Jones 1998; Butynski 1997; Dupain et al. in press; Hall, Saltonstall, et al. 1998; Teleki 1989; Thompson-Handler et al. 1995). All four species of ape are increasingly confined to smaller and more isolated populations and surrounded by ever more dense human populations. Such ape populations are susceptible to extinction not only from further habitat loss and overexploitation but also from random (stochastic) genetic and demographic changes and from environmental catastrophes (e.g., disease, fire; Gippoliti and Carpaneto 1995; Hess 1994; O'Brien and Evermann 1988; Stevenson, Baker, and Foose 1992). Extirpation of the apes and of other species of large mammals from extensive areas means the loss of a natural resource that has for millennia played an important role in the diets, cultures, and economies of the indigenous peoples of tropical Africa.

Forest Loss

Destruction of forests is one of the two most serious human activities threatening the survival of chimpanzees and gorillas. The main causes of this deforestation are logging and clearing for agriculture (cash and subsistence crops and pasture). Logging disturbs the composition and functioning of the forest ecosystem (Struhsaker 1996, 1997). Furthermore, logging roads and logging trucks give people access to vast areas of remote forest, where they hunt apes and other animals and where they degrade and clear forests. Foreign logging companies, working hand in hand with international aid agencies such as the European Union and World Bank (Williams 2000b), often construct these roads. There are 100 to 130 logging companies or their "subsidiaries" active in Africa (Bowen-Jones 1998; Williams 2000a). The countries of the European Union represent the most important market for tropical timber from Africa, accounting for 87 percent of exports in 1991. The United Kingdom and France are the two largest importers of Africa's tropical timber (Ammann and Pearce 1995).

Africa is losing its forests faster than any other continent. During the 1980s, timber exportation from the 11 countries supporting populations of gorillas increased fourfold (Kemf and Wilson 1997). Only six African countries have more than 20 percent of their original forest cover remaining, and as many as 17 countries retain less than 10 percent of their original forest cover (Sayer 1992). Much of what remains is heavily degraded and fragmented (Bryant, Nielson, and Tangley 1997). About 0.6 percent of Africa's tropical moist forest is being destroyed annually. Barnes (1990) has warned that if present trends continue, 70 percent of the forests of West Africa, from Senegal to Nigeria, will disappear before the year 2040. Burundi, Rwanda, Uganda, and Kenya will lose 95 percent of their combined forest cover during this period. In Côte d'Ivoire and Nigeria, where tropical moist forest is being destroyed at an annual rate of roughly 5 percent, all moist forest could be lost by the year 2007 (World Resources Institute/International Institute for Environment and Development 1988). This loss of habitat alone will extirpate apes from these two countries.

There is no evidence that the natural tropical forests in Africa (or anywhere else) can be successfully managed for the large-scale sustainable production of timber or that such exploitation can be achieved without the loss of biodiversity (Bowen-Jones 1998; Meder 1996b; Oates 1996b; Rietbergen 1992; Struhsaker 1996, 1997). Africa is plagued by many problems, including political and economic instability, insufficient law enforcement, corruption, insecurity, and war (Bowen-Jones 1998; Bryant et al. 1997; Rietbergen 1992; WSPA 2000). As noted by Williams (2000a, 9),

> many logging companies have exploited the continuing political instability in the region, the lack of coherent forest management, effective conservation policies or enforcement capabilities. Illegal logging and corrupt practices are now prevalent in many areas.

One result is that timber extraction is almost always an unmanaged and uncontrolled process that usually represents the first step in a series of events that end in the obliteration of the forest and its apes (Ammann and Pearce 1995; Dupain et al. in press; Meder 1996b; Teleki 1989).

Hunting

Throughout their ranges, chimpanzees and gorillas are officially protected under both national and international law from being hunted, captured, or moved across international boundaries (Ammann and Pearce 1995; Bowen-Jones 1998; Lee et al. 1988). Nonetheless, during the past decade the commercial (nontraditional, nonsubsistence) hunting of apes has increased greatly in the DRC, PRC, CAR, Cameroon, Gabon, and Equatorial Guinea. In addition, European logging companies—mostly French, German, Italian, Belgian, Dutch, and Danish—continue to open up these countries to increased exploitation of timber (Ammann and Pearce 1995; Bowen-Jones 1998; Dupain et al. in press; Rose 1996; Williams 2000a; WSPA 2000; see also chapters 3 and 4, this volume).

Although all countries of West and Central Africa are signatories to CITES, these nations have not begun to meet their commitments to the Convention regarding the transborder trafficking in CITES-regulated species, particularly in the form of bushmeat. As such, violations against the Convention remain common in most of the countries where apes are found.

The organized commercial hunting of apes today is conducted at an unsustainable level in many places, especially in and around logging operations. Hunters work full-time to supply ape meat to logging company workers, to expanding agricultural communities, and to people in distant towns and cities. This trade is conducted without adequate medical or veterinary precautions, legal controls, regard for wildlife protection laws, or scientific and management information (Juste et al. 1995; Kemf and Wilson 1997; Teleki 1989). Where they occur, the chimpanzee and gorilla are often among the most sought after species (Fay and Agnagna 1992; Kano 1984; Plumptre et al. 1999; Teleki 1989; Thompson-Handler et al. 1995).

Logging roads, logging trucks, and guns enable the ape meat trade (Ape Alliance 1998; Bryant et al. 1997; Dupain et al. in press; Lahm 1996; Meder 1996b; Ngoufo 2000; Rose 1996, 1997; Teleki 1989; WSPA 1996, 2000). Logging companies, and agents acting on their behalf, assist in every aspect of this trade, often in direct contravention of the law (Ammann 2000; Ammann and Pearce 1995; Bowen-Jones 1998; Horta 1992; Kemf and Wilson 1997; Mitchell 1998; Wilkie and Carpenter 1999; Wilkie, Sidle, and Boundzanga 1992). "With the advent of modern firearms, and improved communications and transport, subsistence hunting has given way to anarchic exploitation of wildlife to supply the rapidly growing cities with game" (Juste et al. 1995, 460). In the Motaba River area of northeastern PRC,

40 percent of Aka men ate robust chimpanzee or gorilla meat in 1992. During that same year, an estimated 41 to 69 Aka and Bantu hunters killed 131 chimpanzees and 62 gorillas. Hunting in this area is for local consumption and not yet conducted commercially. Nonetheless, with an estimated 7 percent of the chimpanzees and 5 percent of the gorillas killed each year by hunters, this off-take was calculated to be unsustainable (Kano and Asato 1994).

Redmond (1989) estimated that 400 to 600 gorillas are killed each year in northern PRC. In the 4,000 sq. mi (10,000 sq. km) region of Kika, Moloundou, and Mabele, Cameroon, roughly 25 hunters with shotguns kill an estimated 800 gorillas and 400 chimpanzees each year. The number of gorillas in this region is probably not more than 3,000. This level of off-take is obviously not sustainable (Ammann and Pearce 1995; Kemf and Wilson 1997).

Rose (1997) estimated that 3,000 to 6,000 apes are killed in Africa each year for the ape meat trade. Whatever the number, the informed consensus is that this activity is out of control and unsustainable, and it continues to spread and accelerate (Ape Alliance 1998; Bowen-Jones 1998; Mitchell 1998; Redmond 1998; Robinson 1998; Yamagiwa 1999).

There are now large areas of suitable habitat where, as a result of hunting, chimpanzees and gorillas are either at low densities or no longer present (Butynski 1997; Dupain et al. in press; Fay and Agnagna 1992; Gagneux 1997; Gippoliti and Carpaneto 1995; Kano 1984; Lahm 1996; Oates 1996b; Rose 1996, 1997; Teleki 1989; Thompson-Handler et al. 1995; Tutin and Fernandez 1984). Goodall (2000, 3) argued that

> the "bushmeat" crisis in Africa, the illegal hunting of wildlife for commercial purposes, may well result in the virtual extinction of apes and other endangered species from the remaining Central and West African forests within the next 10 to 20 years.

The commercialization of bushmeat is probably a more significant and immediate threat than forest loss for all four of Africa's species of ape, as well as for many other species of primates and other large mammals (Bowen-Jones 1998; Butynski 1997; Butynski and Koster 1994; Lahm 1996; Oates 1996b; Rose 1997; Wilkie and Carpenter 1999; Wilkie et al. 1992, 1998). As Rose (1997, 8) put it,

> if the present trend in forest exploitation continues without a radical shift in our approach to conservation, most edible wildlife in the equatorial forests of Africa will be butchered before the viable habitat is torn down.

Since 1994, the World Society for the Protection of Animals (WSPA) has been leading a media campaign against the hunting of apes (Ammann and Pearce 1995;

WSPA 2000). The focus is to reduce the hunting of apes and to make governments and logging companies accountable for the consequences of their activities on wildlife and wildlife habitat. This campaign is meeting with some success. There is now greater international recognition of this problem and an increase in protests against the frequent and widespread hunting of apes. The PRC recently banned the production of shotgun cartridges used for hunting apes. A loan by the World Bank to Cameroon to improve and expand its network of logging roads was put on hold because the effects of this project on the environment were not clear (Meder 1996a). Similarly, in 1999, following strong opposition from nongovernmental organizations (NGOs), the European Union abandoned a proposal for a U.S. $50,000,000 project to support further upgrading of Cameroon's roads. The WSPA is also working for tougher controls on logging companies working in West and Central Africa, including tighter checks on the effects of European loans for road building in forest regions.

Following the lead of the WSPA, a number of other organizations have adopted the ape meat crisis as a major focus of their conservation activities. For example, the Ape Alliance, an international coalition of 34 organizations and consultants working for the conservation and welfare of apes, is now a force in the effort to halt illegal hunting of apes for meat (Ape Alliance 1998). Friends of the Earth is encouraging consumers to ensure that the timber they are buying comes from companies practicing sustainable forestry, including sustainable hunting. The Bushmeat Crisis Task Force is an international coalition with more than 100 member organizations and specialists. Its primary goal "is to facilitate the work of its members in identifying and implementing effective and appropriate controls over the commercial exploitation of endangered and threatened species" (H. Eves, personal communication).

Despite some successes over the past seven years, however, there is no evidence that illegal hunting is abating in West and Central Africa. In many regions, the illegal commercial exploitation of apes and other wildlife is greater in 2001 than in 1994. An increasing number of studies are reaching the conclusion that bushmeat hunting at present levels will continue to lead to the extirpation of wildlife species from large areas and the possible extinction of some species of large mammals within the next few decades.

Few authorities in the countries of West and Central Africa have yet to demonstrate that they take conservation matters seriously. There is a general disinterest and disregard for environmental and wildlife protection, or for the sustainable use of natural resources, at seemingly all levels of power, from government ministers, to judiciary, to local police and administrators (Ngoufo 2000; Williams 2000a; WSPA 2000).

Not only is this increase in hunting further threatening populations and taxa of apes, it also brings people and apes into closer and more frequent contact. One expected result is an increase in the rate of disease transmission between humans and apes. For example, the virus that gave rise to HIV-1 in humans may have been transmitted through blood contact during the butchering of chimpanzees for food (see later discussion in this chapter; Gao et al. 1999; Lloyd 1999). HIV/AIDS represents the most important public health threat of the new millennium. Other emerging infectious diseases, such as the Ebola virus and the Sin Nombre (hanta) virus, are major threats to global health. Medical researchers now see a strong link between the transmission of HIV-1, and other new zoonotic infections, from primates to humans with the opening up of tropical forests by logging firms and the related increased hunting of primates for human consumption.

Disease

Disease can be a major problem in efforts to conserve endangered species (Hess 1994; Karesh and Cook 1995; Leiman and Ghaffar 1996; May 1988; McCallum and Dobson 1995; Scott 1988; Thorne and Williams 1988; Wolfe et al. 1998), including the great apes (Plumptre et al. 1999; Wallis and Lee 1999; Yeager 1997). There is no doubt that disease transmission from humans to free-living apes occurs or that the impact on the apes can be catastrophic (Butynski and Kalina 1998a, 1998b; Homsy 1999; McGrew et al. 1989; Murray 1990; Teleki 1989). Nonetheless, disease as a threat to populations of free-living apes remains a relatively neglected issue (Wallis and Lee 1999; Wallis et al. 2000).

Exotic strains of pathogens have the potential to become hyperdiseases by "jumping" to hosts not previously exposed to the strain (Aguirre and Starkey 1994). This can result in 80 to 100 percent mortality, particularly in small populations. Disease can cause the rapid extinction not only of populations but also of subspecies and species (McCallum and Dobson 1995; Woodroffe 1998). Even large populations can be almost destroyed before the disease either adapts or exhausts the supply of nonimmune hosts (Macdonald 1996; O'Brien and Evermann 1988; Scott 1988; Thorne and Williams 1988; Young 1994).

Examples of the devastating impact of new diseases on susceptible populations are the high virulence and lethality of measles, smallpox, and typhoid among New World aboriginals (Cliff, Haggett, and Smallman-Raynor 1993; Panum 1940), of rinderpest on African ungulates (Percival 1918; Scott 1988), and of the extirpation of wild dogs, *Lycaon pictus,* from the Serengeti National Park as a result of rabies and canine distemper (Alexander, Richardson, and Kat 1992; Burrows 1992). Canine distemper has extirpated the black-footed ferret, *Mustela nigripes,* in the wild (May 1988; Thorne and Williams 1988) and killed more than 1,000 lions in the Serengeti National Park (Roelke-Parker et al. 1996). Pathogens brought to Hawaii by Europeans caused the extinction of nearly one half of the endemic land bird species (Warner 1968). "In our own species, epidemics have influenced the outcome of major wars, stimulated migrations, and generally have been a primary determinant in mankind's demographic history" (O'Brien and Evermann 1988, 254).

Because chimpanzees and gorillas are phylogenetically close to humans, they are highly susceptible to numerous human infectious diseases (including parasites), especially viruses (Benirschke and Adams 1980; Bonnotte 1998; Brack 1987; Homsy 1999; Kalter 1980, 1989; Ott-Joslin 1993; Wallis and Lee 1999; Wolfe et al. 1998). These include the common cold, pneumonia, whooping cough, influenza, hepatitis A and B, Epstein-Barr virus, chicken pox, smallpox, bacterial meningitis, tuberculosis, diphtheria, measles, rubella, mumps, yellow fever, yaws, paralytic poliomyelitis, encephalomyocarditis, and Ebola fever. Of particular concern are those airborne diseases such as influenza, tuberculosis, measles, mumps, and chicken pox (Holmes 1996; Homsy 1999; Rossiter 1990; Woodford, Butynski, and Karesh in press), which are readily transmitted and difficult to control. In addition, there are numerous parasitological diseases that are common to humans and apes. These include schistosomiasis, giardiasis, filariasis, strongyloidiasis, cryptosporidiosis, shigellosis, salmonellosis, *Capillaria hepatica, Entamoeba coli, Entamoeba histolytica, Endolimax nana, Ancylostoma* sp., *Oesophagostomum* sp., *Acanthocephala* sp., *Cyclospora* sp., *Chilomastix* sp., *Iodamoeba buetschlii,* and *Sarcoptes scabiei.* Toft (1986) listed 112 parasitic diseases for chimpanzees, gorillas, and orangutans. Many of these diseases are fatal or cause morbidity, with severe consequences for normal behavior and reproduction.

The risks and consequences of disease transmission between humans and apes are predicted to become more serious once stable ecosystems and large (genetically diverse) populations of apes are fragmented, reduced, and stressed by humans (Alexander et al. 1992; Hess 1994; Holmes 1996; Hudson 1992; May 1988; Rossiter 1990; Scott 1988; Teleki 1989). Small populations are likely to have diminished genetic variation, one result of which is increased vulnerability to infectious disease (O'Brien and Evermann 1988).

In the case of the apes, the stress involved with the habituation process and frequent visits by people may further challenges their well-being, compromising their ability to respond normally to disease (see later discussion). The introduction of a human-borne infection into small, stressed, genetically depressed populations of apes could lead not only to the extinction of the population but also (where the subspecies is represented by but one population) to the extinction of the subspecies (Anonymous 1986; Butynski and Kalina 1998a, 1998b; May 1988; McGrew et al. 1989).

Each year, thousands of tourists from thousands of different localities around the world step out of crowded, poorly ventilated airplanes and airports and within one or two days are close to and sometimes touching habituated apes. Tourists can carry exotic strains of pathogens while showing few clinical signs of disease. There is considerable risk that disease will be transmitted from humans to an immunologically naive ape population, triggering an epidemic (Boesch and Boesch-Achermann 1995; Homsy 1999; Nutter 1996; Sayer 1986; Wallis and Lee 1999; Wilson 1995).

Several small, already critically endangered populations of apes are being exposed to additional risk of disease through frequent close contact (less than 3.3 ft or 1 m or touching) with large numbers of tourists, guides, security personnel, researchers, and local people (Anonymous 1986; Ashford et al. 1996; Aveling 1991; Bonnotte 1998; Butynski and Kalina 1998a, 1998b; Hudson 1992; Macfie 1991; Plumptre et al. 1999; Sholley and Hastings 1989; Steklis, Gerald-Steklis, and Madry 1996–1997; Teleki 1989; Wallis and Lee 1999; Yeager 1997). All five of the gorilla tourism programs in eastern Africa are based on populations of only 240 to 340 individuals, and all seven of the robust chimpanzee tourism programs are with populations of 20 to 700 individuals.

There are only about 320 mountain gorillas remaining, and all of these are in one population located in the Virunga Volcanoes. When the security situation allows, 75 percent of the gorillas in this population, and therefore of this subspecies, are visited daily by more than 70 tourists and a similar number of guides, porters, rangers, and researchers (from three countries: Rwanda, Uganda, the DRC; Butynski and Kalina 1998a). The Morris Animal Foundation (Anonymous 1998b, 1) has argued that "the Mountain Gorilla Veterinary Center veterinarians have successfully managed several [human induced] viral outbreaks that could have eliminated the entire population." The continued daily exposure of this critically endangered population and subspecies to large numbers of people and their diseases cannot be considered either appropriate ape management or good conservation.

In 1988, in Rwanda's Volcanoes National Park, six habituated female mountain gorillas died of respiratory illness, and 27 additional cases in the habituated groups were given injections of penicillin. The illnesses occurred in three of the four tourist groups and in one of the three research groups. The epidemiology of this epidemic suggests that the disease was new to this population of gorillas, because 81 percent of the gorillas in the affected groups were sick. The serological and pathological evidence for measles in one gorilla suggested that the measles virus or a related morbillivirus was responsible. Measles is deadly for many species of nonhuman primates. As such, 65 animals in the seven habituated groups were vaccinated against human measles. No further signs of respiratory disease were seen after the initiation of the vaccination campaign, and the disease did not spread to new groups (Byers and Hastings 1991; Hastings et al. 1991; Sholley and Hastings 1989). The results of a 1989 census of the Virunga population suggests that this disease did not affect, or at least had negligible impact on, the unhabituated groups (Sholley 1991). Primates in natural habitats without human contact are thought to be free of the measles virus but are susceptible to transmission from humans (Brack 1987).

In 1990, bronchopneumonia affected 26 of 35 gorillas in a tourist group in the Volcanoes National Park. Although four of the animals were given antibiotics, two died (Macfie 1991). The position of the IUCN/SSC Veterinary Specialist Group is that people probably introduced the 1988 and 1990 respiratory diseases to the gorillas in the Virungas (J. Cooper, personal communication).

Aproximately 30 percent of adult mountain gorillas are infected by *Capillaria hepatica,* a parasite that can cause mortality in human children and that is considered a threat to gorillas. The majority of mountain gorillas infected with *C. hepatica* live close to human populations (MGVP, 2000).

In 1994, as a result of the war and genocide in Rwanda (1990 to 1995), there was a massive increase of human traffic crossing the Virunga Volcanoes, followed by a military presence. In 1996 to 1997, three intestinal parasites not previously identified in the mountain gorilla but known to infect humans were found. These parasites may have entered this population of gorillas through an increase in contact with human feces (Mudakikwa et al. 1998). In 1996, all four gorillas in a tourist group in the Bwindi–Impenetrable National Park contracted scabies, *Sarcoptes scabiei,* and one infant died. The source of this disease is thought to have been humans or domestic animals living near this forest (Macfie 1996).

During 1964, there was an epidemic of a "polio-like" disease among robust chimpanzees near Beni, the DRC. At least seven of the chimpanzees in the study community were severely handicapped by "limb paresis." This disease may have entered the chimpanzee population from the local human population (Kortlandt 1996a).

The habituated robust chimpanzees in the 21 sq. mi (52 sq. km) Gombe National Park, Tanzania, have a wider diversity and higher prevalence of parasites than unhabituated chimpanzees (F. Nutter, personal communication). About 100 people live at the Gombe Stream Research Center, a large number of people live around the park, tourists and other visitors are common, and close human proximity to the chimpanzees, including touching, occurs. In addition, chimpanzees have been provisioned with bananas since 1960, and these are often handled in unhygienic conditions. The Gombe chimpanzees also have considerable close contact with baboons, *Papio anubis,* which eat foods left by people on the beach, in and around homes, and in garbage pits (McGrew et al. 1989).

From 1966 through 1997 there were at least four disease outbreaks in the research and tourist chimpanzee (Kasekela) community at Gombe. In 1966, a paralytic disease (probably poliomyelitis) killed six chimpanzees and permanently crippled six others. Three other chimpanzees disappeared and were presumed to also be polio victims. Oral polio vaccine was given to the remaining chimpanzees in the community. It is suspected that the chimpanzees were infected by an epidemic in the local human population (Goodall 1986; Teleki 1989). In 1968, five chimpanzees are believed to have died of pneumonia (Goodall 1986; van Lawick-Goodall 1968). In 1987, at least nine (perhaps as many as 12) chimpanzees died from pneumonia (C. Stanford, cited in Hosaka 1995a; Wallis and Lee 1999). In 1997, scabies was contracted by at least 19 of 45 chimpanzees, resulting in deaths of three infants (Pusey 1998). In 1989, two intestinal parasites not previously known to occur in this chimpanzee population were found. Humans are considered to have been the most likely source of these new parasites (Murray 1990).

In 1996, a respiratory disease killed 11 of about 25 chimpanzees in Gombe's Mitumba Community (Nutter 1996; Wallis and Lee 1999). This epidemic reduced the community to about 14 chimpanzees, with only one adult male surviving. It is unlikely that this community will recover (Wallis and Lee 1999).

Based on these observations, at least 42 chimpanzees were either crippled or died from disease in Gombe over a 32-year period. During the 1980s there were about 150 chimpanzees in the Gombe National Park. Today there are fewer than 100 (Greengrass 2000; Massawe 1995). There is a strong probability that some of the diseases affecting this small population were introduced by people or their domestic animals (Adams, Sleeman, and New 1999; Wallis and Lee 1999) and that some of the decline can be attributed to these diseases.

In 1993 to 1994, at least 11 robust chimpanzees (perhaps as many as 18) died in the Mahale Mountains, Tanzania, from a flu-like illness. It is suspected that tourists or other people in contact with these chimpanzees transmitted the virus (Hosaka 1995b; Wallis and Lee 1999).

In 1994, hemorrhagic disease (Ebola virus) killed one robust chimpanzee and presumably 11 others among the 40 robust chimpanzees in the Tai Forest, Côte d'Ivoire. One student fell seriously ill (but recovered) from the same virus after participating in an autopsy (Boesch and Boesch-Achermann 1995; Le Guenno et al. 1995; Morell 1995). Two years earlier, eight chimpanzees died in two weeks from Ebola or a similar disease. These two disease outbreaks apparently killed at least 20 chimpanzees in the study community and reduced the number of adult males in this community from eight to two, greatly affecting an important 15-year field study.

There are an increasing number of robust chimpanzees in the Kibale National Park, Uganda, habituated for tourism. In 1998, Adams et al. (1999) conducted a medical survey of tourists who had close contact with chimpanzees; there were 30 Europeans, 6 Australians, 4 Americans, and 3 Africans. Among these 43 tourists, five cases of herpes virus, four cases of influenza, two cases of tuberculosis, and one case of chicken pox were considered to be infectious at the time of visitation. Forty-four percent of the tourists had never been tested for tuberculosis, and 10 percent did not know if they had ever been tested. Since their arrival in Africa, 53 percent had suffered from diarrhea, 23 percent from coughs, and 13 percent from vomiting. Seventy percent had either already visited, or intended to visit, another group of habituated chimpanzees or gorillas.

Disease transmission, particularly of novel diseases, from apes to humans is also a major concern (Brack 1987; Cooper 1996; Gagneux 1997; Murray 1990; Shellabarger 1991; Wallis et al. 2000; Woodford et al. in press). As human contact with apes increases, a corresponding increase in the emergence of new diseases in human populations is expected (Wolfe et al. 1998). Close contact activities, such as ape tourism, the keeping of apes as pets, and ape hunting must, therefore, be considered risky.

"Hunting, which involves tracking, capturing, handling, transporting, preparing, and consuming meat, may play a particularly important role in pathogen exchange" (Wolfe et al. 1998, 154). An outbreak of Ebola in 1996 in the human population at Mayibout, a village in northeast Gabon, is linked to the handling, preparation, and eating of a robust chimpanzee found dead in the forest; of the 37 reported cases of Ebola during this outbreak, 29 of the people were directly exposed to the dead chimpanzee (WHO 1996). In Cameroon, about 14 people died in the Djoum area in April 1997 of what was most likely Ebola after butchering and eating one of four gorillas found dead in the forest (K. Ammann, personal communication).

The AIDS virus HIV-1 in humans probably originated from the simian immunodeficiency virus (SIVcpz). It appears that on at least three occasions SIVcpz jumped from *P. troglodytes troglodytes* to humans and then mutated into HIV-1 (Gao et al. 1999; Lloyd, 1999; Underwood 1999). More than 14,000,000 people have died of HIV-1 and another 33,000,000 people are infected, making this the most devastating human epidemic of recent times.

There is an alpha-herpes virus that, as far as is known, is endemic to the gorillas of the Virunga Volcanoes and closely related to the human herpes simplex virus type 2 (HSV-2). A substantial portion of this gorilla population is believed to be infected, and there is some concern over the potential pathogenicity to humans. The alpha-herpes viruses currently presents the greatest problem from the standpoint of a zoonotic infectious disease (Eberle 1992). The highly pathogenic nature of certain of these herpe viruses in species other than the natural hosts provide examples of the dangers inherent in cross-species infections (Eberle 1995; Shellabarger 1991).

There are a number of published cases of gorillas biting people (e.g., Madden 1998; Sabater Pi 1966; Schaller 1963). I am aware of at least 28 cases in recent years where wild adult gorillas and chimpanzees bit people, inflicting deep puncture wounds. There were surely many more. These wounds have the potential to be extremely dangerous, because there could hardly be a better avenue for the transfer of certain diseases from one species to another. For example, monkeypox was transmitted to a girl in the DRC when a wild robust chimpanzee bit her (Mutombo, Arita, and Jezek 1983). There are numerous cases of this kind of transmission of disease between laboratory primates and laboratory workers (Wolfe et al. 1998).

Zoos, laboratories, and other facilities holding captive apes are aware of the high risk of disease transmission between people and apes and of the consequences. These facilities have strict standard operating procedures and exercise extreme caution to minimize the risks, both for the apes and for the people who come in contact with them (Adams, Muchmore, and Richardson 1995; Bonnotte 1998; Hudson 1992; Shellabarger 1991; Silberman 1993; Wallis and Lee 1999; Wallis et al. 2000). For example, captive apes are typically monitored monthly for endoparasites (via fecal examinations) and given annual physical checkups and tuberculosis tests. In addition, high levels of sanitation are maintained: Surgical masks, rubber boots, and rubber gloves are worn by staff, disinfectants are used daily; and visitors are kept several feet from the animals or behind plexiglas barriers. Veterinarians, curators, and keepers are periodically tested for diseases (e.g., tuberculosis tests every six months), and veterinarians, medicines, and treatment facilities are always close at hand. At least in some captive facilities housing apes (e.g., the National Zoo in Washington, D.C.), no person is permitted to enter the holding facility or otherwise have close contact with apes unless their recent disease history is known, they have had a recent tuberculosis test, and they are known to not be immune-compromised (B. Beck, personal communication). Indeed, captive apes are given far more protection against diseases from people than are free-living, habituated, critically endangered apes that are the subjects of tourism and field research programs (Butynski and Kalina 1998a; Wallis and Lee 1999).

Habituation, Stress, and Changes in Behavior

Habituation is a process whereby apes that are intolerant of people become tolerant of the proximity of people. The habituation process generally requires from several months to several years. For example, mountain and Grauer's gorillas usually require about one year (range 3 to 24 months) to be habituated to humans, whereas western lowland gorillas may not habituate even after several years of effort (Tutin and Fernandez 1991). The people undertaking the habituation typically visit the apes every day for several hours. Early on in the process the apes usually respond to people by fleeing or hiding in deep cover for hours. Diarrhea invariably occurs in these gorillas. Charges, other aggressive displays, and threat vocalizations are common during habituation and, in the case of gorillas, people are sometimes bitten (Anonymous 1996–1997; Butynski and Kalina 1998a; Fossey 1983; Schaller 1963; Tutin and Fernandez 1991). Gradually, the apes become tolerant, and people are able to approach to within a few feet without stimulating overt avoidance or aggressive responses.

Once the apes are habituated, the amount of time people spend with them varies from 1 hour (for tourists) to about 12 hours (for researchers) per day. The number of people visiting a group of habituated gorillas at any one time ranges from 1 to more than 32. The regulations to reduce the negative impacts on gorillas of these visits are often ignored (Amooti 1998; Anonymous 1998a; Butynski and Kalina 1998a; Plumptre et al. 1999; Schmitt 1997; Woodford et al. in press).

Habituation and frequent close visits, even by a small audience, can disrupt normal social and ranging behavior and activity patterns and can cause increased levels of stress. Stress can result in a number of pathophysiological effects leading to reduced growth, fertility, health, and survival in free-living animals (Burrows 1992; Ceballos-Lascurain 1996; Hofer and East 1998; Putman 1995; Stalmaster and Kaiser 1998; von Holst 1998; Yalden 1990), including primates (Butynski and Kalina 1998b; Fa 1992; Hudson 1992; Johns 1996; O'Leary and Fa 1993; Struhsaker and Siex 1996; Wheatley and Harya Putra 1994; Zhao and Deng 1992).

Human-induced stress probably has important negative implications for the conservation of many species, including the great apes. During the early stages of habituation, apes (particularly gorillas) experience obvious acute stress. It is reasonable to suggest that apes that are the subjects of tourism and research experience chronic stress. One concern is that these activities cause cumulative stress that might immune-compromise the apes, making them particularly susceptible to disease (Woodford et al. in press).

Kalema (1995) found that parasite burdens of habituated gorillas in the Bwindi–Impenetrable National Park are higher than those of unhabituated gorillas, and postulated that this may be a result of the stress of repeated human contact. Hudson (1992, 9) stated that

> any programme for wild gorilla tourism should bear in mind that stress may be caused by the process of habituation and certainly by over-exposure to humans, due, for example, to commercial pressures or perhaps to over enthusiastic medical intervention (e.g., "systematic monitoring") and could therefore leave a wild group more vulnerable to disease than under normal conditions.

Although there are a few studies that provide information on changes in behavior of apes as a response to habituation and human visitation (Goldsmith 2000; Johns 1996; Tutin and Fernandez 1991), there are no studies that adequately quantify and evaluate the impacts of these changes on vitality and fertility (Butynski and Kalina 1998a, 1998b; McNeilage and Thompson-Handler 1999).

Before their habituation for tourism, the gorillas of the Bwindi–Impenetrable National Park rarely slept outside of the boundaries of the national park, and there were no attacks by gorillas on local farmers (T. Butynski, personal observation, 1983–1993). In 2000, seven years after the start of the tourism program, the groups habituated for tourism spent much more time than nonhabituated groups foraging and sleeping in the densely settled farmlands (about 750 people per sq. mi; 300 people per sq. km) outside the national park. For example, one habituated group spent 35 of 36 consecutive days foraging and sleeping outside of the park (Goldsmith 2000). For an eight-month period in 1996 to 1997, the Katendengere tourist gorilla group shifted full-time to foraging on farmland (Madden 1998). While foraging in agricultural areas, these habituated gorillas have different diets and greatly reduced day ranges (Goldsmith 2000). Habituated mountain gorillas of the Virunga Volcanoes also forage on cultivated land and near human habitation (Madden 1998; McNeilage and Thompson-Handler 1999; J. Sleeman, personal communication, quoted in Woodford et al. in press).

During a ten-month period in 1996 to 1997, habituated gorillas attacked and hospitalized at least four farmers at sites outside of the Bwindi–Impenetrable National Park. Some of the farmers spent months in the hospital. Gorillas habituated for tourism have also attacked farmers off the edge of the Virunga Volcanoes Conservation Area (Madden 1998). The increase in the amount of time these gorillas spend outside these two protected areas, the shift in diet from wild plants to cultivated plants, the reduced day range, and the attacks on local people are, undoubtedly, behavioral changes resulting from habituation.

ETHICS AND THE USE AND CONSERVATION OF FREE-LIVING APES

What does the conservation status of Africa's apes and their use have to do with the ethical treatment of apes by humans? Ethical matters are an integral and legitimate part of conservation. Scientists, resource managers, politicians, loggers, hunters, tour operators, and tourists are all morally obliged to conduct their activities with respect to the viability of wild species and the integrity of natural systems.

It is increasingly recognized that not all species should be regarded as being available for use by humans (Prescott-Allen and Prescott-Allen 1996; Regan 1988). Such concerns are particularly applicable to gorillas, chimpanzees, and orangutans—endangered species that are also highly intelligent and humankind's closest living relatives. One result of this viewpoint is a growing "great apes rights crusade" to treat apes as humankind's moral equal. The Declaration on Great Apes and the Great Ape Project support the inclusion of the great apes in the human "community of equals," extending to them minimal human rights (e.g., the right to life, protection, respect, and individual liberty) on grounds of their physical, behavioral, psychological, and genetic similarities to humans (Cavalieri and Singer 1993; Diamond 1993; Teleki 1989, 1997). Nonetheless, although chimpanzees and gorillas "possess capacities that meet several commonly stated criteria of personhood" (Fox 1996, 45), they continue to be "denied the ethical and legal protection that we give to our own species" (Cavalieri and Singer 1993, 6).

Ethical concerns have been expressed over the exposure of apes to people, particularly for purposes of entertainment, research, and commercial gain (Regan 1988; Teleki 1997; Wallis et al. 2000). Is it morally acceptable to subject any wild ape to any practice, including habituation and tourism, that intrudes and exposes that ape to potential or actual harm?

Extending "rights" to the great apes necessarily creates new practical obstacles and legal difficulties for the large number of people (e.g., hunters, loggers, farmers, politicians, businesspeople, tourists, and researchers) who use, and sometimes damage, wild apes and their habitats. Must humans show more respect and compassion toward apes and more restraint in their power to dominate and exploit apes and their habitats? Should apes and their habitats be available for human use and misuse? If logging or tourism places apes under increased stress, or under increased risk of disease and death, should those who promote, fund, or participate in these activities be criminally liable? How should people who kill and eat apes be viewed and treated?

Although such questions may seem absurd, even offensive, to many people at this time, a growing number think otherwise, believing that great apes should be treated with far more respect, dignity, and kindness, and exposed to minimal harm and intervention (Dority 1998; Teleki 1997; see also chapter 16, this volume). As the chimpanzees and gorillas continue their steep decline toward extinction, and as we learn more about them, it seems likely that increasing numbers of conservationists will evoke ape rights and ethics as additional "conservation tools" in their efforts to rescue these species.

RECOMMENDATIONS TO REDUCE THE THREATS TO AFRICA'S APES

The ultimate cause of the decline of Africa's apes is the continent's rapidly expanding human population and the related poverty and insecurity. Curbing the growth of human populations and bringing them in line with the sustainable use of the natural resource base depends largely on effective action by governments, namely through the implementation of population policy (a policy that no African nation has at this time; Myers 1993; Struhsaker 1997). Major international bodies, such as the United Nations (UN), Organization of African Unity (OAU), European Union (EU), World Bank, African Development Bank (ADB), aid agencies, and NGOs all have a vital role to play in this effort by providing technical and financial support for the implementation of national population policies, as well as for related initiatives concerned with socioeconomic advancement, education, family planning, child mortality, and women's rights.

There is an urgent need for the conservation community, national governments, donors, logging companies, trade organizations, and the public to address more specifically and more intensively the two main threats to ape populations: hunting and deforestation. This can be done by promoting and supporting efforts to:

- Encourage the leaders of the governments of West and Central African countries to find the political will and courage to take the steps necessary to ensure the sustainable use of all natural resources.
- See that international aid-granting agencies (particularly the EU and the World Bank) use their considerable political and financial resources as incentives to help ensure that no "development projects" result in, or promote, the unsustainable use of natural resources, particularly wildlife and natural forests.
- Encourage international aid-granting agencies to require that all development projects undergo a rigorous and independent environmental impact assessment (EIA) before approval to ensure that forests, wildlife, indigenous people, and local communities are not adversely affected.

- Improve national protected area systems, especially through the expansion of existing national parks and the establishment of additional national parks.
- Prevent logging and hunting within protected areas, and limit logging and hunting in buffer zones around protected areas.
- Strengthen law enforcement capabilities and greatly reduce illegal, unregulated, unmanaged, commercial hunting.
- Reduce the demand for bushmeat by placing taxes on bushmeat transporters and traders so that the price of bushmeat rises relative to the price of alternative sources of meat.
- Find and develop domestic food alternatives to bushmeat.
- Increase support for environmental education and public awareness programs, particularly around protected areas.
- Change the attitudes of stakeholders toward hunting and logging, foster interest in sustainable use, and promote stakeholders' responsibility and accountability for the conservation of neighboring forests and wildlife.
- Develop and implement criteria and performance indicators for the maintenance of biodiversity and for the sustainable use of wildlife and tropical forests (including old growth forest communities).
- Require that logging companies adhere to national wildlife and forestry laws and regulations, and implement corporate codes of conduct that ensure sustainable forest use (including hunting) and the protection of biodiversity.
- Establish a strong independent international certification program for sustainable tropical forest management where bushmeat control and the maintenance of biodiversity are integral parts of the accreditation process.
- Deny import permits for uncertified timber.
- Mandate that logging companies fully finance specific programs to ameliorate both the direct and indirect damage their activities have on wildlife and on natural habitats.
- Promote enhanced global support (political, technical, and financial) for sustainable forest use and hunting practices, particularly among major decision makers, financiers, and national governments.
- Increase awareness among people in the West concerning the negative impacts that their high levels of consumption of tropical timber are having on the biodiversity, apes, and people of tropical Africa.
- Regularly survey and monitor Africa's populations of apes so that numbers, distributions, trends, and threats are far better known than at present.
- Ensure that people concerned with the disastrous consequences of unsus-

tainable logging and hunting continue to bring these issues to public attention, lobby their own governments to take action to halt these practices, and pressure the logging industry and the governments of all West and Central African countries to take responsibility for this crisis.

Major donors and conservation bodies can do more by putting into place (1) trust funds to support the required law enforcement, research, monitoring, evaluation, information, and education programs; and (2) financial incentives in the form of conditional grants and loans for logging companies, hunters, and others who exploit tropical forests. In this case, those who support the protection of national parks and reserves, and who practice sustainable forest and wildlife use, become eligible for preferential financial assistance (Struhsaker 1997).

As discussed throughout this chapter, there is an alarming lack of reliable information on the distribution and numbers of chimpanzees and gorillas. More surveys are badly needed, particularly given the rapid changes that are occurring in the face of rampant loss of habitat and commercialized hunting. It is difficult to make meaningful conservation decisions without this information. Research must also focus on the impact on ape populations of stress and disease as a result of the ever-increasing close contact with people, particularly tourists.

Field conservation activities must be "biologically appropriate." We cannot claim to be conserving and sustainably using apes if we have not undertaken the fundamental research needed to know and understand the results of our actions. In particular, much more research is needed on the benefits, impacts, and risks of habituation and tourism on small populations of apes. Greater attention needs to be given to identifying and assessing the health dangers associated with human-caused stress and disease among the great apes (Bonnotte 1998; Butynski and Kalina 1998a, 1998b; Homsy 1999; Wallis and Lee 1999). The primary goal should be to minimize human-caused stress and disease.

In addition to the strict enforcement of the regulations, every ape tourism and research program should make it mandatory that all people visiting habituated apes have a recent tuberculosis test (unless they have been vaccinated), wash their hands, and use disinfectant footbaths before entering the forest, and wear surgical masks within 33 ft (10 m) of apes (Adams et al. 1999; Wallis and Lee 1999; Woodford et al. in press). These programs should also consider requiring that all people working with or visiting the gorillas be immunized for measles, mumps, pertussis, rubella, diphteria, tetanus, poliomyelitis (killed, injectable vaccine only), hepatitis A, yellow fever, and tuberculosis (if appropriate; Bonnotte 1998; Woodford et al. in press).

Although some people view tourism of habituated, free-living apes as a conservation activity, others view this as merely another form of exploitation for entertainment and commercial gain, and as a clear sign of our failure to effectively address ape conservation issues. Can a practice that places apes at additional risk while raising serious ethical questions truly serve as an effective long-term conservation strategy? It is ironic that the conservation and ethical questions raised over ape tourism practices have also been largely ignored by the various animal welfare and animal rights advocacy groups.

Those concerned with the use and conservation of apes need to do a far better job of assessing the long-term risks and benefits of their various interventions—and their ethical consequences. They also need to be more concerned with what is good for ape conservation over the long-term rather than what is politically popular and financially rewarding over the short-term. To determine whether a proposed or ongoing intervention is appropriate, safe, and justified, it should be reviewed and evaluated by a multidisciplinary team of independent, experienced professionals (Butynski and Kalina 1998a, 1998b). At this time it is important to ask those responsible for current interventions why such reviews and evaluations are not being undertaken.

To reverse the downward trend in numbers of chimpanzees and gorillas, multidisciplinary, holistic, pragmatic approaches to their conservation are required (Bowen-Jones 1998; Bowen-Jones and Pendry 1999; Rose 1997). Indeed, all of the previously detailed activities need to be coordinated in a variety of strategies and implemented through numerous collaborative efforts. The money, workforce, motivation, political will, and policies necessary to support these integrated approaches and efforts are, however, unlikely to be forthcoming until the current crisis faced by the great apes of Africa becomes an issue of major global conservation concern. To this end, it is up to each of us to carry this issue to the public, to make clear the plight of Africa's great apes to a wide audience, and to help generate the required level of global commitment and effective conservation action.

ACKNOWLEDGMENTS

Esteban Sarmiento, John Oates, Thomas Struhsaker, Jan Kalina, Sandy Harcourt, Angela Meder, Colin Groves, Adrian Kortlandt, Jef Dupain, Jo Thompson, Delfi Messinger, Jim Moore, Caroline Tutin, and Jefferson Hall provided valuable comments and suggestions on all or part of the draft manuscript. Esteban Sarmiento, John Oates, Katherine Gonder, Ester van Adrichem, Delfi Messinger, Jo Thompson, Sally Lahm, Jef Dupain, Steve Blake, Rebecca Ham, Janis Carter, John Hart, Marc Colyn, Allard Blom, David Wilkie, Leonard Usongo, Jim Moore, Karl Ammann, Marc Languy, Rodger Fotso, Jonathan Kingdon, Marellin Agnagna, John Cooper, Jan Kalina, Jane Goodall, Gay Reinartz, Dan Warton, Ben Beck, Dietrich Schaaf, Debra Forthman,

Annette Lanjouw, Liz Williamson, Billy Karesh, Iris Weiche, Mollie Bloomsmith, Steve Gartlan, Ray Wack, Toshisada Nishida, Takayoshi Kano, Chris Hillman, Kes Smith, Fraser Smith, and Michael Boer generously provided information and unpublished data. Dennis Malewa, Pius Namachanja, Joseph Kirathe, and Mary Wandiba kindly helped compile the three ape distribution maps. Zoo Atlanta, the Disney Wildlife Conservation Fund, the Disney Institute, the National Museums of Kenya, the Kenya Institute of Primate Research, and the IUCN Eastern Africa Regional Office generously supported this project. To all these individuals and institutions I extend my deepest appreciation and thanks.

REFERENCES

Abedi-Lartey, M. 1998. *Survey of endangered forest primates in western Ghana.* Accra, Ghana: Wildlife Department. Unpublished report.

Adams, H. R., Sleeman, J., and New, J. C. 1999. A medical survey of tourists visiting Kibale National Park, Uganda, to determine the potential risk for disease transmission of chimpanzees (*Pan troglodytes*) from ecotourism. *Gorilla Gazette* 13: 32–33.

Adams, S. R., Muchmore, E. E. S., and Richardson, J. H. 1995. Biosafety Part A: General biosafety considerations; Part B: Zoonoses, biohazards, and other health risks; Part C: A model occupational health program for persons working with nonhuman primates. In B. T. Bennett, C. R. Bee, and R. Henrickson (eds.), *Nonhuman primates in biomedical research* (pp. 377–420). San Diego, CA: Academic Press.

Aguirre, A. A., and Starkey, E. E. 1994. Wildlife disease in the U.S. national parks: Historical and coevolutionary perspectives. *Conservation Biology* 8: 654–61.

Alexander, K. A., Richardson, J. D., and Kat, P. W. 1992. Disease and conservation of African wild dogs. *SWARA* 15: 13–14.

Allan, C. in press. A survey of chimpanzees *Pan troglodytes schweinfurthii* in Toro Game Reserve, Uganda. *Journal of East African Natural History* 89.

Ammann, K. 2000. Exploring the bushmeat trade. In K. Ammann (ed.), *Bushmeat: Africa's conservation crisis* (pp. 16–27). London: World Society for the Protection of Animals.

Ammann, K., and Pearce, J. 1995. *Slaughter of the apes: How the tropical timber industry is devouring Africa's great apes.* London: World Society for the Protection of Animals.

Amooti, N. 1998. Wildlife officials now gorillas' worst enemy. *The New Vision,* Nov. 24, 32.

Anderson, P. 1997. Bush refugees. *Keeping Track* (Dec.): 12–17.

Anon. 1986. Coughs and faeces spread diseases that kill mountain gorillas. *New Scientist* 109: 20.

———. 1996–1997. Learning to live with tame humans. *Digit News* 10: 3.

———. 1998a. Gorilla tourism in Uganda. *Gorilla Journal* 17: 10.

———. 1998b. *Mountain Gorilla Veterinary Project.* Englewood, CO: Morris Animal Foundation.

Ape Alliance. 1998. *The African bushmeat trade—A recipe for extinction.* London: Ape Alliance.

Ashford, R. W., Lawson, H., Butynski, T. M., and Reid, G. D. F. 1996. Patterns of intestinal parasitism in the mountain gorilla *Gorilla gorilla* in the Bwindi–Impenetrable Forest, Uganda. *Journal of Zoology* (London) 239: 507–14.

Aveling, R. 1991. *"Gorilla tourism"—Possibilities and pitfalls.* Nairobi: African Wildlife Foundation. Unpublished report.

Barnes, R. F. W. 1990. Deforestation trends in tropical Africa. *African Journal of Ecology* 28: 161–73.

———. 1997. *A brief visit to Marahoue National Park.* Washington, D.C.: Conservation International. Unpublished report.

Benirschke, K., and Adams, F. C. 1980. Gorilla diseases and causes of death. *Journal of Reproduction and Fertility* (Suppl.) 28: 139–48.

Blake, S. 1994. *A reconnaissance survey in the Kabo logging concession south of the Nouabale-Ndoki National Park.* New York: WCS/GTZ/World Bank. Unpublished report.

Blake, S., Rogers, E., Fay, J. M., Ngangoue, M., and Ebeke, G. 1995. Swamp gorillas in northern Congo. *African Journal of Ecology* 33: 285–90.

Boesch, C., and Boesch-Achermann, H. 1995. Tai chimpanzees confronted with a fatal Ebola virus. *Pan Africa News* 2: 2–3.

Bonnotte, S. 1998. *A veterinary preliminary study concerning potential risk of transmission of human diseases to the mountain gorillas* (Gorilla gorilla beringei) *colony in Bwindi Impenetrable National Park, Uganda.* Ammilly, France. Unpublished report.

Bowen-Jones, E. 1998. A review of the commercial bushmeat trade with emphasis on Central/West Africa and the great apes. *African Primates* 3: S1–S42.

Bowen-Jones, E., and Pendry, S. 1999. The threat to primates and other mammals from the bushmeat trade in Africa, and how this threat could be diminished. *Oryx* 33: 233–46.

Brack, M. 1987. *Agents transmissible from simians to man.* Berlin: Springer-Verlag.

Bryant, D., Nielsen, D., and Tangley, L. 1997. *The last frontier forests: Ecosystems and economies on the edge.* Washington, D.C.: World Resources Institute.

Burrows, R. 1992. Rabies in wild dogs. *Nature* 359: 277.

Butynski, T. M. 1997. African primate conservation—The species and the IUCN/SSC Primate Specialist Group network. *Primate Conservation* 17: 87–100.

Butynski, T. M., and Kalina, J. 1998a. Gorilla tourism: A critical review. In E. J. Milner-Gulland and R. Mace (eds.), *Conservation of biological resources* (pp. 280–300). Oxford: Blackwell.

———. 1998b. Is gorilla tourism sustainable? *Gorilla Journal* 16: 15–19.

Butynski, T. M., and Koster, S. H. 1994. Distribution and conservation status of primates in Bioko Island, Equatorial Guinea. *Biodiversity and Conservation* 3: 893–909.

Butynski, T. M., and Sarmiento, E. 1996. The gorillas of Mt. Tshiaberimu, Zaire. *Gorilla Conservation News* 10: 14–15.

Byers, A. C., and Hastings, B. 1991. Mountain gorilla mortality and climatic factors in the Parc National des Volcans, Ruhengeri Prefecture, Rwanda, 1988. *Mountain Research and Development* 2: 145–51.

Cavalieri, P., and Singer, P. 1993. A declaration of great apes. In P. Cavalieri and P. Singer (eds.), *The Great Ape Project: Equality beyond humanity* (pp. 4–7). Guildford, UK: Biddles.

Ceballos-Lascurain, H. 1996. *Tourism, ecotourism and protected areas: The state of nature-based tourism around the world and guidelines for its development.* Gland, Switzerland: International Union for the Conservation of Nature and Natural Resources.

Cliff, A., Haggett, P., and Smallman-Raynor, M. 1993. *Measles: A historical geography of a major human viral disease, from global expansion to local retreat, 1840–1990.* London: Blackwell.

Conly, S. R. 1998. Sub-saharan Africa at the turning point. *The Humanist* 58: 19–23.

Convention on International Trade in Endangered Species of Wild Fauna and Flora. 1973. Washington, D.C: Author.

Coolidge, H. J. 1929. A revision of the genus *Gorilla. Memoires of the Museum of Comparative Zoology, Harvard* 50: 291–381.

———. 1933. *Pan paniscus* (pygmy chimpanzee) from south of the Congo River. *American Journal of Physical Anthropology* 8: 1–57.

Cooper, J. E. 1996. Parasites and pathogens of non-human primates. *Veterinary Record* 139: 48.

Cousins, D. 1980. On the Koolookamba—A legendary ape. *Acta Zoologica et Pathologica Antverpiensia* 75: 79–93.

Diamond, J. 1993. The third chimpanzee. In P. Cavalieri and P. Singer (eds.), *The Great Ape Project: Equality beyond humanity* (pp. 88–101). Guildford, UK: Biddles.

Dority, B. 1998. Humanism and evolutionary humility. *The Humanist* 58: 22–24.

Dupain, J., Van Elsacker, L., and Verheyen, R. F. in press. The status of the bonobo (*Pan paniscus*). In B. Galdikas, N. Briggs, L. Sheeran, and G. Shapiro (eds.), *All apes great and small, Vol. I: Chimpanzees, Gorillas and Bonobos.* New York: Kluwer Press.

Eberle, R. 1992. Evidence for an alpha-herpesvirus indigenous to mountain gorillas. *Journal of Medical Primatology* 21: 246–51.

———. 1995. The simian herpesviruses. *Infectious Agents and Disease* 4: 55–70.

Edroma, E., Rosen, N., and Miller, P. 1997. *Population and habitat viability assessment for the chimpanzees of Uganda.* Kampala, Uganda: International Union for the Conservation of Nature and Natural Resources/SSC Conservation Breeding Specialist Group. Unpublished report.

Emlen, J. T., and Schaller, G. B. 1960. Distribution and status of the mountain gorilla (*Gorilla gorilla beringei*)—1959. *Zoologica* 45: 41–52.

Endangered Species Act. 1973. 87 Stat. 884.

Fa, J. E. 1992. Visitor-directed aggression among the Gibraltar macaques. *Zoo Biology* 11: 43–52.

Fay, J. M., and Agnagna, M. 1992. Census of gorillas in northern Republic of Congo. *American Journal of Primatology* 27: 275–84.

Fay, J. M., Agnagna, M., Moore, J., and Oko, R. 1989. Gorillas (*Gorilla gorilla gorilla*) in the Likouala swamp forests of north central Congo: Preliminary data on populations and ecology. *International Journal of Primatology* 10: 477–86.

Fleagle, J. G. 1982. Living primates as a key to human evolution. In R. A. Mittermeier and M. M. Plotkin (eds.), *Primates and the tropical forest* (pp. 37–44). Washington, D.C.: World Wildlife Fund.

Fornos, W. 1998. No vacancy. *The Humanist* 58: 15–18.

Fossey, D. 1983. *Gorillas in the mist.* Boston: Houghton-Mifflin.

Fox, M. A. 1996. Planet for the apes. *Etica & Animali* 8: 44–49.

Gagneux, P. 1997. Sampling rapidly dwindling chimpanzee populations. *Pan Africa News* 4: 12–15.

Galat-Luong, A., Galat, G., Ndiaye, I., and Keita, Y. in press. Fragmentation de la distribution et statut actuel du chimpanzé, *Pan troglodytes verus,* en limite d'aire de répartition au Sénégal. *African Primates* 4.

Gao, F., Bailes, E., Robertson, D. L., Chen, Y., Rodenburg, C. M., Michael, S. F., Cummins, L. B., Arthur, L. O., Peters, M., Shaw, G. M., Sharp, P. M., and Hahn, B. 1999. Origin of HIV-1 in the chimpanzee *Pan troglodytes troglodytes. Nature* 397: 436–41.

Garner, K. J., and Ryder, O. A. 1996. Mitochondrial DNA diversity in gorillas. *Molecular Phylogenetics and Evolution* 6: 39–48.

Gippoliti, S., and Carpaneto, G. M. 1995. The conservation of African primates: State of the art, problems and perspectives. *Rivista di Antropologia* 73: 193–216.

Gippoliti, S., and Dell'Omo, G. 1995. Status and conservation of the chimpanzee *Pan troglodytes verus* in Guinea-Bissau. *African Primates* 1: 3–5.

Goldsmith, M. L. 2000. Effects of ecotourism on the behavioral ecology of Bwindi gorillas, Uganda: Preliminary results. (Abstract). *American Journal of Physical Anthropology* (Suppl.) 30: 161.

Goliber, T. J. 1985. Sub-saharan Africa: Population pressures on development. *Population Bulletin* 40: 1–46.

Gonder, M. K., Oates, J. F., Disotell, T. R., Forstner, M. R. J., Morales, J. C., and Melnick, D. J. 1997. A new west African chimpanzee subspecies? *Nature* 338: 337.

Gonzalez-Kirchner, J. P. 1997. Census of western lowland gorilla population in Rio Muni Region, Equatorial Guinea. *Folia Zoologica* 46: 15–22.

Goodall, A. G., and Groves, C. P. 1977. The conservation of eastern gorillas. In Prince Rainier III and G. H. Bourne (eds.), *Primate conservation* (pp. 599–637). New York: Academic Press.

Goodall, J. 1986. *The chimpanzees of Gombe: Patterns of behavior.* Cambridge, MA: Harvard University Press.

———. 2000. Foreword. In K. Ammann (ed.), *Bushmeat: Africa's conservation crisis* (p. 3). London: World Society for the Protection of Animals.

Greengrass, E. 2000. The sudden decline of a community of chimpanzees at Gombe National Park. *Pan Africa News* 7: 5–7.

Groves, C. P. 1970. Population systematics of the gorilla. *Journal of Zoology* (London) 161: 287–300.

———. 1971. Distribution and place of origin of the gorilla. *Man* 6: 44–51.

———. 1996. Do we need to update the taxonomy of gorillas? *Gorilla Journal* 12: 3–4.

———. 2001. *Primate taxonomy.* Washington, D.C.: Smithsonian Institution Press.

Hall, J. S., Saltonstall, K., Inogwabini, B.-I., and Omari, I. 1998. Distribution, abundance and conservation status of Grauer's gorilla. *Oryx* 32: 122–30.

Hall, J., and Wathaut, W. M. 1992. *A preliminary survey of the eastern lowland gorilla.* New York: Wildlife Conservation Society. Unpublished report.

Hall, J. S., White, L. J. T., Inogwabini, B.-I., Omari, I., Morland, H. S., Williamson, E. A., Saltonstall, K., Walsh, P., Sikubwabo, C., Bonny, D., Kiswele, K. P., Vedder, A., and Freeman, K. 1998. Survey of Grauer's gorillas (*Gorilla gorilla graueri*) and eastern chimpanzees (*Pan troglodytes schweinfurthi*) in the Kahuzi-Biega National Park lowland sector and adjacent forest in eastern Democratic Republic of Congo. *International Journal of Primatology* 19: 207–35.

Ham, R., and Carter, J. 2000. *Population size and distribution of chimpanzees in the Republic of Guinea, West Africa.* Edinburgh, Scotland: Scottish Primate Research Report. Unpublished report.

Harcourt, A. H. 1996. Is the gorilla a threatened species? How should we judge? *Biological Conservation* 75: 165–76.

Harcourt, A. H., and Fossey, D. 1981. The Virunga gorillas: Decline of an "island" population. *African Journal of Ecology* 19: 83–97.

Harcourt, A. H., Stewart, K. J., and Inahoro, I. M. 1989. Nigeria's gorillas: A survey and recommendations. *Primate Conservation* 10: 73–76.

Hart, J. A., and Hall, J. S. 1996. Status of eastern Zaire's forest parks and reserves. *Conservation Biology* 10: 316–27.

Hart, J., and Sikubwabo, C. 1994. *Exploration of the Maiko National Park of Zaire 1989–1992.* Working Paper No. 2. New York: Wildlife Conservation Society.

Hastings, B. E., Kenny, D., Lowenstine, L. J., and Foster, J. W. 1991. Mountain gorillas and measles: Ontogeny of a wildlife vaccination program. *Proceedings of the Annual Meeting of the American Association of Zoo Veterinarians:* 198–205.

Hess, G. R. 1994. Conservation corridors and contagious disease: A cautionary note. *Conservation Biology* 8: 256–62.

Hill, W. C. O. 1967. The taxonomy of the genus *Pan.* In D. Starck, R. Schneider, and H.-J. Kuhn (eds.), *Neue Ergebnisse der Primatologie* (pp. 47–54). Stuttgart, Germany: Fischer.

———. 1969. The nomenclature, taxonomy and distribution of chimpanzees. In G. H. Bourne (ed.), *The chimpanzee. Vol. 1. Anatomy, behavior and diseases of chimpanzees* (pp. 22–49). Basel, Switzerland: Karger.

Hillman, J. C. 1982. *Wildlife information booklet: Democratic Republic of the Sudan.* New York: New York Zoological Society. Unpublished report.

Hilton-Taylor, C. 2000. *2000 IUCN red list of threatened animals.* Gland, Switzerland: International Union for the Conservation of Nature and Natural Resources.

Hofer, H., and East, M. L. 1998. Biological conservation and stress. *Advances in the Study of Behavior* 27: 405–525.

Hogarth, S. S. 1997. *The distribution and abundance of the chimpanzee* (Pan troglodytes

troglodytes) *of the lowland gallery forest of Gashaka Gumti National Park, Nigeria: A preliminary report for the Gashaka Sector.* Unpublished report.

Holmes, J. C. 1996. Parasites as threats to biodiversity in shrinking ecosystems. *Biodiversity and Conservation* 5: 975–83.

Homsy, J. 1999. *Ape tourism and human diseases: How close should we get?* Nairobi: International Gorilla Conservation Program. Unpublished report.

Horn, A. D. 1979. The taxonomic status of the bonobo chimpanzee. *American Journal of Physical Anthropology* 51: 273–82.

Horta, K. 1992. Logging in the Congo: Massive fraud threatens the forests. *World Rainforest Report* No. 24.

Hosaka, K. 1995a. Epidemics and wild chimpanzee study groups. *Pan Africa News* 2: 1.

———. 1995b. Mahale: A single flu epidemic killed at least 11 chimps. *Pan Africa News* 2: 3–4.

Hudson, H. R. 1992. The relationship between stress and disease in orphan gorillas and its significance for gorilla tourism. *Gorilla Conservation News* 6: 8–10.

International Union for the Conservation of Nature and Natural Resources. 1994. *The IUCN red list categories.* Gland, Switzerland: Author.

———. 1996. *1996 IUCN red list of threatened animals.* Gland, Switzerland: Author.

———. 2000. *The IUCN red list categories.* Gland, Switzerland: Author.

IUCN/SSC Primate Specialist Group (PSG). 2000. *Threatened African primates. IUCN/SSC PSG Workshop on Primate Taxonomy.* Washington, D.C.: Conservation International. Unpublished report.

Jensen-Seaman, M. 2000. Western and eastern gorillas: Estimates of the genetic distance. *Gorilla Journal* 20: 21–23.

Johns, B. G. 1996. Responses of chimpanzees to habituation and tourism in the Kibale Forest, Uganda. *Biological Conservation* 78: 257–62.

Jolly, C. J., Oates, J. F., and Disotell, T. R. 1995. Chimpanzee kinship. *Science* 268: 185–86.

Juste, J., Fa, J. E., Del Val, J. P., and Castroviejo, J. 1995. Market dynamics of bushmeat species in Equatorial Guinea. *Journal of Applied Ecology* 32: 454–67.

Kalema, G. 1995. Epidemiology of the intestinal parasite burden of mountain gorillas, *Gorilla gorilla beringei,* in Bwindi Impenetrable National Park, southwest Uganda. *Zebra Foundation Newsletter* (Autumn): 18–34.

Kalter, S. S. 1980. Infectious diseases of the great apes of Africa. *Journal of Reproduction and Fertility* 28: 149–59.

———. 1989. Infectious diseases of nonhuman primates in a zoo setting. *Zoo Biology* (Suppl.) 1: 61–76.

Kano, T. 1984. Distribution of pygmy chimpanzees (*Pan paniscus*) in the Central Zaire Basin. *Folia primatologica* 43: 36–52.

———. 1992. *The last ape: Pygmy chimpanzee behavior and ecology.* Stanford, CA: Stanford University Press.

Kano, T., and Asato, R. 1994. Hunting pressure on chimpanzees and gorillas in the Motaba River area, northeastern Congo. *African Study Monographs* 15: 143–62.

Karesh, W. B., and Cook, R. A. 1995. Applications of veterinary medicine to *in situ* conservation efforts. *Oryx* 29: 244–52.

Kemf, E., and Wilson, A. 1997. *Great apes in the wild.* Gland, Switzerland: World Wide Fund for Nature.

Kortlandt, A. 1976. Letters: Statements on pygmy chimpanzees. *Laboratory Primate Newsletter* 15: 15–17.

———. 1983. Marginal habitats of chimpanzees. *Journal of Human Evolution* 12: 231–78.

———. 1995. A survey of the geographical range, habitats and conservation of the pygmy chimpanzee (*Pan paniscus*): An ecological perspective. *Primate Conservation* 16: 21–36.

———. 1996a. An epidemic of limb paresis (polio?) among the chimpanzee population at Beni (Zaire) in 1964, possibly transmitted by humans. *Pan Africa News* 3: 9–10.

———. 1996b. The conservation status of *Pan paniscus. African Primates* 2: 79–80.

Lahm, S. 1996. Gabon's village hunting: Assessing its impact. *African Primates* 2: 23–24.

Lee, P. C., Thornback, J., and Bennett, E. L. 1988. *Threatened primates of Africa. The IUCN red data book.* Gland, Switzerland: International Union for the Conservation of Nature and Natural Resources.

Le Guenno, B., Formentry, P., Wyers, M., Gounon, P., Walker, F., and Boesch, C., 1995. Isolation and partial characterization of a new strain of Ebola virus. *Lancet* 345: 1271–74.

Leiman, A., and Ghaffar, N. 1996. Use, misuse and abuse of the orangutan—Exploitation as a threat or the only real salvation? In V. J. Taylor and N. Dunstone, *The exploitation of mammal populations* (pp. 345–57). New York: Chapman and Hall.

Lloyd, P. 1999. The bushmeat origin of AIDS. *Africa Environment & Wildlife* 7: 14.

Macdonald, D. W. 1996. Dangerous liaisons and disease. *Nature* 379: 400–401.

Macfie, L. 1991. The Volcanoes Veterinary Center. *Gorilla Conservation News* 5: 21.

———. 1996. Case report of scabies infection in Bwindi gorillas. *Gorilla Journal* 13: 19–20.

Madden, F. 1998. *The problem gorilla.* Nairobi: International Gorilla Conservation Programme. Unpublished report.

Marchesi, P., Marchesi, N., Fruth, B., and Boesch, C. 1995. Census and distribution of chimpanzees in Cote d'Ivoire. *Primates* 36: 591–607.

Massawe, E. T. 1992. Assessment of the status of chimpanzee populations in western Tanzania. *African Study Monographs* 13: 35–55.

———. 1995. *Present distribution of chimpanzees in Tanzania.* Kigoma, Tanzania: Mahale Wildlife Research Centre. Unpublished report.

May, R. M. 1988. Conservation and disease. *Conservation Biology* 2: 28–30.

McCallum, H., and Dobson, A. 1995. Detecting disease and parasite threats to endangered species and ecosystems. *Trends in Ecology and Evolution* 10: 190–94.

McGrew, W. C., Tutin, C. E. G., Collins, D. A., and File, S. K. 1989. Intestinal parasites of sympatric *Pan troglodytes* and *Papio* spp. at two sites: Gombe (Tanzania) and Mt. Assirik (Senegal). *American Journal of Primatology* 17: 147–55.

McNeilage, A., and Thompson-Handler, N. 1999. *The assessment of behavioral impact of tourism on gorillas. Nairobi: International Gorilla Conservation Programme.* Unpublished project proposal.

Meder, A. 1996a. Apes at risk. *Gorilla Journal* 12: 19–20.

———. 1996b. Rain forests and gorillas in Cameroon and Nigeria. *Gorilla Journal* 12: 15–19.

Mitchell, C. 1998. Countering the impacts of the bushmeat trade in Cameroon. (Abstract). In *Primate Society of Great Britain Spring Meeting 1998: Bushmeat Hunting and African Primates* (pp. 16–17). Bristol, UK: Bristol Zoo.

Morell, V. 1994. Will primate genetics split one gorilla into two? *Science* 265: 1661.

———. 1995. Chimpanzee outbreak heats up search for Ebola origin. *Science* 268: 974–75.

Morin, P. A., More, J. J., Chakraborty, R., Jin, L., Goodall, J., and Woodruff, D. S. 1994. Kin selection, social structure, gene flow, and the evolution of chimpanzees. *Science* 265: 1193–1201.

Mudakikwa, A. B., Sleeman, J., Foster, J. W., Meader, L. L., and Patton, S. 1998. An indicator of human impact: Gastrointestinal parasites of mountain gorillas (*Gorilla gorilla beringei*) from the Virunga Volcanoes Region, Central Africa. *Proceedings American Association of Zoo Veterinarians and American Association of Wildlife Veterinarians Joint Conference:* 436–37.

Murray, S. 1990. *Intestinal parasites of baboons and chimpanzees in Gombe National Park.* Senior thesis, Tufts University, MA.

Mutombo, M., Arita, I., and Jezek, Z. 1983. Human monkey pox transmitted by a chimpanzee in a tropical rain-forest area of Zaire. *Lancet* 34: 735–37.

Mwanza, N., and Yamagiwa, Y. 1989. A note on the distribution of primates between the Zaire-Lualaba River and the African Rift Valley. *Interspecies Relationships of Primates in the Tropical and Montane Forests* 1: 5–10.

Myers, N. 1993. Population, environment, and development. *Environmental Conservation* 20: 205–16.

Napier, J. R., and Napier, P. H. 1967. *A handbook of living primates.* New York: Academic Press.

Ngoufo, R. 2000. The fight against poaching in Cameroon. In K. Ammann (ed.), *Bushmeat: Africa's conservation crisis* (pp. 28–31). London: World Society for the Protection of Animals.

Nicholas, A. 1995. *A report on the results of line transect work undertaken during the dry and wet seasons in the Domaine de Chase of Garamba National Park, northeast Zaire. 2. Large mammal distribution and abundance.* Garamba, Zaire. Unpublished report.

Nishida, T. 1994. Distribution and status of chimpanzee populations in Africa. *Pan Africa News* 1: 1–10.

Nutter, F. B. 1996. Respiratory disease claims the lives of at least seven Gombe chimps. *Pan Africa News* 3: 3.

Oates, J. F. 1996a. *African primates: Status survey and conservation action plan* (rev. ed.). Gland, Switzerland: International Union for the Conservation of Nature and Natural Resources.

———. 1996b. Habitat alteration, hunting and conservation of folivorous primates in African forests. *Australian Journal of Ecology* 21: 1–9.

———. 1998. The gorilla population in the Nigeria–Cameroon border region. *Gorilla Conservation News* 12: 3–6.

Oates, J. F., McFarland, K. L., Stumpf, R. M., Fleagle, J. G., and Disotell, T. R. 1999. New findings on the distinctive gorillas of the Nigeria–Cameroon border region. (Abstract). *American Journal of Physical Anthropology* (Suppl.) 28: 213–14.

O'Brien, S. J., and Evermann, J. F. 1988. Interactive influence of infectious disease and genetic diversity in natural populations. *Trends in Ecology and Evolution* 3: 254–59.

Ogawa, H., Kanamori, M., and Mukeni, S. H. 1997. The discovery of chimpanzees in the Lwazi River area, Tanzania: A new southern distribution limit. *Pan Africa News* 4: 1–3.

O'Leary, H., and Fa, J. E. 1993. Effects of tourists on Barbary macaques at Gibraltar. *Folia Primatologica* 61: 77–91.

Omari, I., Hart, J. A., Butynski, T. M., Birashirwa, R., Upoki, A., M'Keyo, Y., Bengana, F., Bashonga, M., and Bagurubumwe, N. 1999. The Itombwe Massif, Democratic Republic of Congo: Biological surveys and conservation with an emphasis on Grauer's gorilla and birds endemic to the Albertine Rift. *Oryx* 33: 301–22.

Ott-Joslin, J. E. 1993. Zoonotic diseases of nonhuman primates. In M. E. Fowler (ed.), *Zoo and wild animal medicine* (pp. 358–73). Philadelphia: Saunders.

Panum, P. 1940. *Observations made during the epidemic of measles on the Faeroe Islands in the year 1846.* New York: American Public Health Association.

Pavy, J.-M. 1993. *Bafing Faunal Reserve. Biodiversity and human resource: Survey and recommendations.* Cheverly, MD: Author. Unpublished report.

Percival, A. B. 1918. Game and disease. *Uganda Natural History Society* 13: 302.

Plumptre, A., McNeilage, A., and Hall, J. 1999. The current status of gorillas and threats to their existence. (Abstract). *American Journal of Physical Anthropology* (Suppl.) 28: 224.

Prescott-Allen, R., and Prescott-Allen, C. 1996. Assessing the sustainability of uses of wild species: Case studies and initial assessment procedure. *Occasional Papers of the IUCN Species Survival Commission* 12. Gland, Switzerland: International Union for the Conservation of Nature and Natural Resources.

Pusey, A. 1998. Scabies in chimpanzees of Gombe National Park, Tanzania. *European Association of Zoo and Wildlife Veterinarians—Newsletter* 1: 10.

Putman, R. J. 1995. Ethical considerations and animal welfare in ecological field studies. *Biodiversity and Conservation* 4: 903–15.

Redmond, I. 1989. *Trade in gorillas and other primates in the People's Republic of Congo.* Summerville, SC: International Primate Protection League. Unpublished report.

———. 1998. Eating our relatives: Ethics, ecology and extinction. (Abstract). *Primate Society of Great Britain Spring Meeting 1998: Bushmeat Hunting and African Primates* (p. 12). Bristol, UK: Bristol Zoo.

Regan, T. 1988. *The case for animal rights.* London: Routledge.

Rietbergen, S. 1992. Forest management. In J. A. Sayer, C. S. Harcourt, and N. M. Collins (eds.), *Africa: The conservation atlas of tropical forest* (pp. 62–68). London: Macmillan.

Robinson, J. G. 1998. Limits to sustainable hunting in tropical forest. (Abstract). In *Primate Society of Great Britain Spring Meeting 1998: Bushmeat Hunting and African Primates* (p. 13). Bristol, UK: Bristol Zoo.

Roelke-Parker, M. E., Munson, L., Packer, C., Kock, R., Cleaveland, S., O'Brien, S. J., Pospischil, A., Hosmann-Lehmann, R., Lutz, H., Mwamengele, L. L. M., Mgsa, M. N., Machange, G. A., Summers, B. A., and Appel, M. J. G. 1996. A canine distemper virus epidemic in Serengeti lions (*Panthera leo). Nature* 379: 441–45.

Rose, A. L. 1996. The African great ape bushmeat crisis. *Pan Africa News* 3: 1–6.

———. 1997. Conservation becomes a global social movement in the era of bushmeat and primate kinship. *African Primates* 3: 6–12.

Rossiter, P. B. 1990. The increased risk of infectious disease with intensive utilization of wildlife. In J. G. Grootenhuis, S. G. Njuguna, and P. W. Kat (eds.), *Wildlife research for sustainable development* (pp. 124–36). Nairobi: Kenya Agricultural Research Institute/Kenya Wildlife Service/National Museums of Kenya.

Ruvolo, M., Pan, D., Zehr, S., Goldberg, T., Disotell, T. R., and von Dornum, M. 1994. Gene trees and hominid phylogeny. *Proceedings of the National Academy of Sciences USA* 91: 8900–8904.

Ryder, O. A., Garner, K. J., and Burrows, W. 1999. Non-invasive molecular genetic studies of gorillas: Evolutionary and systematic implications. (Abstract). *American Journal of Physical Anthropology* (Suppl.) 28: 238.

Sabater Pi, J. 1966. Gorilla attacks against humans in Rio Muni, West Africa. *Journal of Mammalogy* 47: 123–24.

Sarmiento, E. E., and Butynski, T. M. 1996. Present problems in gorilla taxonomy. *Gorilla Journal* 12: 5–7.

Sarmiento, E. E., Butynski, T. M., and Kalina, J. 1996. Gorillas of Bwindi-Impenetrable Forest and Virunga Volcanoes: Taxonomic implications of morphological and ecological differences. *American Journal of Primatology* 40: 1–21.

Sarmiento, E. E., and Oates, J. F. 2000. The Cross River gorillas: A distinct subspecies, *Gorilla gorilla diehli* Matschie 1904. *Novitiates* 3325.

Sayer, J. 1986. Coughs and faeces spread diseases that kill mountain gorillas. *New Scientists* 109: 20.

———. 1992. A future for Africa's tropical forests. In J. A. Sayer, C. S. Harcourt, and N. M. Collins (eds.), *Africa: The conservation atlas of tropical forest* (pp. 81–93). London: Macmillan.

Schaller, G. B. 1963. *The mountain gorilla: Ecology and behavior.* Chicago: University of Chicago Press.

Schmitt, M. 1997. Close encounter with gorillas at Bwindi. *Gorilla Journal* 14: 12–13.

Schouteden, H. 1947. De Zoogdieren van Belgischen Congo en van Ruanda-Urundi. *Annales de Musee du Congo Belga,* in-4 Deirkunde.eeks II deel III, Afl. 1-3: 1–567.

Scott, M. E. 1988. The impact of infection and disease on animal populations: Implications for conservation biology. *Conservation Biology* 2: 40–56.

Shea, B. T. 1984. Between the gorilla and the chimpanzee: A history of debate concerning the existence of the Kooloo-kamba or gorilla-like chimpanzee. *Journal of Ethnobiology* 4: 1–13.

Shellabarger, W. 1991. Overview of primate viral zoonotic diseases & their prevention. *Proceedings of the American Association of Zoo Veterinarians:* 224–34.

Sholley, C. R. 1991. Conserving gorillas in the midst of guerrillas. *Annual Conference Proceedings, American Association of Zoological Parks and Aquariums:* 30–37.

Sholley, C. R., and Hastings, B. 1989. Outbreak of illness among Rwanda's gorillas. *Gorilla Conservation News* 3: 7.

Silberman, M. S. 1993. Occupational health programs in wildlife facilities. In M. E. Fowler (ed.), *Zoo & wild animal medicine* (pp. 57–61). Philadelphia: Saunders.

Stalmaster, M. V., and Kaiser, J. L. 1998. Effects of recreational activity on wintering bald eagles. *Wildlife Monographs* 137: 1–46.

Steklis, H. D., Gerald-Steklis, N., and Madry, S. 1996/1997. The mountain gorilla: Conserving an endangered primate in conditions of extreme political instability. *Primate Conservation* 17: 145–51.

Stevens, W. K. 1997. Logging sets off an apparent chimp war. *New York Times*, May 13.

Stevenson, M., Baker, A., and Foose, T. J. 1992. *Conservation assessment and management plan for primates.* Apple Valley, MN: IUCN/SSC Captive Breeding Specialist Group (CBSG).

Struhsaker, T. T. 1996. A biologist's perspective on the role of sustainable harvest in conservation. *African Primates* 2: 72–75.

———. 1997. *Ecology of an African rain forest.* Gainesville: University Press of Florida.

———. 1998. *A survey of primates and other mammals in Marahoue National Park, Cote d'Ivoire.* Washington, D.C.: Conservation International. Unpublished report.

Struhsaker, T. T., and Siex, K. S. 1996. The Zanzibar red colobus monkey *Procolobus kirkii:* Conservation status of an endangered island endemic. *African Primates* 2: 54–61.

Stumpf, R. M., Fleagle, J. G., Jungers, W. L., Oates, J. F., and Groves, C. P. 1998. Morphological distinctiveness of Nigerian gorilla crania. (Abstract). *American Journal of Physical Anthropology* (Suppl.) 26: 213.

Sugiyama, Y., and Soumah, A. G. 1988. Preliminary survey of the distribution and population of chimpanzees in the Republic of Guinea. *Primates* 29: 569–74.

Teleki, G. 1989. Population status of wild chimpanzees (*Pan troglodytes*) and threats to survival. In P. G. Heltne and L. A. Marquardt (eds.), *Understanding chimpanzees* (pp. 312–53). Cambridge, MA: Harvard University Press.

———. 1991. *Action plan for the conservation of wild chimpanzees and protection of orphan chimpanzees in the Republic of Burundi.* Hants, UK: Jane Goodall Institute. Unpublished report.

———. 1997. Human relations with chimpanzees: A proposed code of conduct. *African Primates* 3: 35–38.

Teleki, G., and Baldwin, L. A. 1979. *Known and estimated distributions of extant chimpanzee populations* (Pan troglodytes *and* Pan paniscus) *in Equatorial Africa.* Special Report. Washington, D.C.: International Union for the Conservation of Nature and Natural Resources/SSC Primate Specialist Group.

Thompson, J. 1997. *The history, taxonomy, and ecology of the bonobo* (Pan paniscus *Schwarz, 1929) with a first description of a wild population living in a forest/savanna mosaic habitat.* Doctoral dissertation, University of Oxford.

Thompson-Handler, N., Malenky, R. K., and Reinartz, G. E. (eds.). 1995. *Action plan for* Pan paniscus. Milwaukee, WI: Zoological Society of Milwaukee County.

Thorne, E. T., and Williams, E. S. 1988. Disease and endangered species: The black-footed ferret as a recent example. *Conservation Biology* 2: 66–74.

Thys Van den Audenaerde, D. F. E. 1984. The Tervuren Museum and the pygmy chimpanzee. In R. L. Susman (ed.), *The pygmy chimpanzee: Evolutionary biology and behavior* (pp. 3–11). New York: Plenum Press.

Toft, J. D. 1986. The pathoparasitology of nonhuman primates: A review. In K. Benirschke (ed.), *Primates: The road to self-sustaining populations* (pp. 571–679). New York: Springer-Verlag.

Tutin, C. E. B., and Fernandez, M. 1984. Nationwide census of gorilla *Gorilla g. gorilla* and chimpanzee *(Pan t. troglodytes)* populations in Gabon. *American Journal of Primatology* 6: 313–36.

———. 1991. Responses of wild chimpanzees and gorillas to the arrival of primatologists: Behavior observed during habituation. In H. O. Box (ed.), *Primate responses to environmental change* (pp. 187–97). New York: Chapman and Hall.

Tuttle, R. H. 1986. *Apes of the world.* Park Ridge, NJ: Noyes.

Uchida, A. 1996. What we don't know about great ape variation. *Trends in Ecology & Evolution* 11: 163–68.

Underwood, A. 1999. How the plague began: Linking HIV-1 to chimps in west Central Africa. *Newsweek,* February 8, 55.

Vandebroek, G. 1958. Notes écologiques sur les anthropoides Africains. *Annales Societe Royale Zoologique de Belgique* 89: 203–11.

Van Krunkelsven, E., Bila-Isia, I., and Draulans, D. 2000. A survey of bonobos and other large mammals in the Salonga National Park, Democratic Republic of Congo. *Oryx* 34: 180–87.

van Lawick-Goodall, J. 1968. The behaviour of free-living chimpanzees in the Gombe Stream Reserve. *Animal Behaviour Monographs* 1: 161–311.

Vedder, A. 1987. Report from the Gorilla Advisory Committee on the status of *Gorilla gorilla. Primate Conservation* 8: 75–81.

Vedder, A., and Weber, W. 1990. The Mountain Gorilla Project (Volcanoes National Park). In A. Kiss (ed.), *Living with wildlife: Wildlife resource management with local participation in Africa* (pp. 60–83). Washington, D.C.: World Bank.

Von Holst, D. 1998. The concept of stress and its relevance for animal behavior. *Advances in the Study of Behavior* 27: 1–131.

Wallis, J., and Lee, D. R. 1999. Primate conservation: The prevention of disease transmission. *International Journal of Primatology* 20: 803–26.

Wallis, J., Woodford, M., Karesh, W., Sheeran, L., Nuller, F., and Taylor, S. 2000. American Society of Primalogy's policy statement on protecting primate health in the wild. *ASP Bulletin* 24: 9.

Warner, R. E. 1968. The role of introduced diseases in the extinction of the endemic Hawaiian avifauna. *The Condor* 70: 101–20.

Wheatley, B. P., and Harya Putra, D. K. 1994. Biting the hand that feeds you: Monkeys and tourists in Balinese monkey forests. *Tropical Biodiversity* 2: 317–27.

Wilkie, D. S., and Carpenter, J. F. 1999. Bushmeat hunting in the Congo Basin: An assessment of impacts and options for mitigation. *Biodiversity and Conservation* 8: 927–55.

Wilkie, D. S., Curran, B., Tshombe, R., and Morelli, G. A. 1998. Modeling the sustainability of subsistence farming and hunting in the Ituri Forest of Zaire. *Conservation Biology* 12: 137–47.

Wilkie, D. S., Sidle, J. G., and Boundzanga, G. C. 1992. Mechanized logging, market hunting, and a bank loan in Congo. *Conservation Biology* 6: 570–80.

Williams, J. 2000a. The lost continent: Africa's shrinking forests. In K. Ammann (ed.), *Bushmeat: Africa's conservation crisis* (pp. 8–15). London: World Society for the Protection of Animals.

———. 2000b. Trouble in the pipeline. In K. Ammann (ed.), *Bushmeat: Africa's conservation crisis* (pp. 36–39). London: World Society for the Preservation of Animals.

Wilson, M. E. 1995. Travel and the emergence of infectious diseases. *Emerging Infectious Diseases* 1: 39–46.

Wolfe, N. D., Escalante, A. A., Karesh, W. B., Kilbourn, A., Spielman, A., and Lal, A. A. 1998. Wild primate populations in emerging infectious disease research: The missing link? *Emerging Infectious Diseases* 4: 149–58.

Woodford, M., Butynski, T. M., and Karesh, W. in press. Habituating the great apes: The disease risks. *Oryx*.

Woodroffe, R. 1998. Veterinary contributions to field conservation. *Oryx* 32: 5–6.

World Health Organization (WHO). 1996. Outbreak of Ebola haemorrhagic fever in Gabon officially declared over. *Weekly Epidemiologic Record* 71: 125–26.

World Resources Institute. 1994. *World Resources 1994–95.* Oxford: Oxford University Press.

World Resources Institute/International Institute for Environment and Development. 1988. *World Resources 1988–89.* New York: Basic Books.

World Society for the Protection of Animals (WSPA). 1996. *Wildlife and timber exploitation in Gabon: A case study of the Leroy concession, Forest de Abeilles.* London: Author. Unpublished report.

———. 2000. *Bushmeat: Africa's conservation crisis.* London: Author. Unpublished report.

Yalden, D. W. 1990. Recreational disturbance of large mammals in the Peak District. *Journal of Zoology* (London) 221: 293–98.

Yamagiwa, J. 1999. Slaughter of gorillas in the Kahuzi-Biega Park. *Gorilla Journal* 19: 4–6.

Yeager, C. P. 1997. Orangutan rehabilitation in Tanjung Puting National Park, Indonesia. *Conservation Biology* 8: 410–18.

Young, T. P. 1994. Natural die-offs of large mammals: Implications for conservation. *Conservation Biology* 8: 410–18.

Zhao, Q. K., and Deng, Z. Y. 1992. Dramatic consequences of food handouts to *Macaca thibetana* at Mount Emei, China. *Folia Primatologica* 58: 24–31.

HERMAN D. RIJKSEN

2
THE ORANGUTAN AND THE CONSERVATION BATTLE IN INDONESIA

The number of orangutans confiscated by law enforcement from illegal captivity is staggering. Rehabilitation stations are filled with orphaned young apes, and primatologists studying the orangutan's ecology are finding higher densities in the wild than had been thought possible for what was thought to be a "solitary ape." Although this might appear as if there are more orangutans than had been believed, in reality such densities reflect the expansion of the modern world at the expense of wilderness.

At the turn of the twentieth century the estimated human population of Indonesia was 15,000,000. At the turn of the twenty-first century Indonesia has more than 200,000,000 people, and it is expected to expand to 300,000,000 in fewer than two decades. Because of rampant exploitation and conversion of the tropical rain forest, the last compressed populations of wild orangutans cannot avoid confrontations with people, regardless of their hiding skills. Every day, many thousands of people intrude into what used to be the apes' remote and inhospitable domain in search of commodities. In patches of habitat that were degraded because of logging amid vast areas that have been converted into plantations, some apes hang on, crowding together. Many orangutans in search of food are forced out of their pillaged forests and enter the expanding oil palm plantations or private gardens of hostile people. Once exposed, they are usually slaughtered, and their infants usually taken to be sold as pets. Even in the early 1990s more than 1,000 illegally exported infant orangutans showed up in Taiwan. Because every infant in the trade represents at least three dead apes, it was a sign of an impending disaster. Before the Taiwan scandal, Indonesian authorities seemed unaware of the illegal persecution of orangutans. Since that time the search and detection of illegal pets has been incidental. In the following years of incidental confiscation action in Borneo, more than 900 orphaned orangutan juveniles were nevertheless collected for rehabilitation. At least twice as many were probably kept and traded unnoticed.

Between 1994 and 1997, my colleagues and I conducted surveys in Kalimantan (i.e., the Indonesian part of Borneo) and Sumatra (Indonesia) to review the distribution and conservation status of the red ape (Rijksen and Meijaard 1999). Data for the Malaysian area of Borneo (Sarawak and Sabah) were extracted from recent publications (e.g., Payne 1988). The survey results show that as of early 1997, the Bornean orangutan occurred in a scatter of isolated forest fragments of an accumulated total of approximately 60,000 sq. mi (150,000 sq. km), whereas the Sumatran orangutan survived under slightly better circumstances yet occurred in a total accumulated area of only about 10,400 sq. mi (26,000 sq. km).

The figures reflect the fact that during the second half of the twentieth century, the forest cover of Borneo was reduced by about 50 percent because of encroachment for subsistence agriculture and conversion for plantations, and the remaining forest suffered massive habitat degradation as a consequence of poorly controlled timber exploitation. The prolonged drought and widespread arson by farmers and plantation speculators to claim state forest land in the period 1997 to 1998, along with accelerated forest conversion, destroyed another 30 percent of the remaining range in Borneo. In particular the forest fires caused tremendous losses of life for apes fleeing the conflagration as local people slaughtered and butchered all edible wildlife refugees. Indeed, some forest patches were set ablaze just to flush out the wildlife, for cooking and for the pet trade. It was calculated that the total population of Bornean orangutans suffered a decline from an estimated 23,000 in 1995 to about 15,400 in 1998—a reduction of about a third in less than three years (Rijksen and Meijaard 1999). If the approximately 60,000 sq. mi (150,000 sq. km) of remaining forest in Borneo had still been in its natural state, this would have meant that a total population of Bornean orangutans of more than 75,000 apes could otherwise have been expected. When we apply the same criteria to the beginning of the twentieth century, at least 230,000 orangutans must have lived in the then approximately 180,000 sq. mi (450,000 sq. km) of forest in Borneo. Thus only 7 percent of the number of 1900 orangutan population survived at the close of the twentieth century.

Forest demolition and conversion in Sumatra have been at least as extensive as in Borneo, but the surveys revealed that the survival chances for the Sumatran orangutan are slightly better. The Sumatran (sub)species survives in more rugged terrain, which has been less accessible to human encroachment, notably along the Barisan mountain range. For Sumatra it is currently believed that the total population of orangutans occurs in an accumulated total area of approximately 10,400 sq. mi (26,000 sq. km) of forest fragments (Rijksen and Meijaard 1999). Applying the estimated densities of apes in the different mosaics of habitat types in Sumatra, the current total population of surviving orangutans is approximately 12,500 individuals. If the almost 10,400 sq. mi (26,000 sq. km) of forest had not been degraded, then the population would have been more than 23,000 orangutans. At the beginning of the twentieth century the ape was probably to be found in at least 32,800 sq. mi (82,000 sq. km) of forest in Sumatra—that is, if we consider the land area north of the equator only, because very little is known of the occurrence of apes south of Lake Toba. The known population amounted to no fewer than 85,000 individuals. If one discounts the likely occurrence of the remnant populations of orangutans in several forest blocks south of the equator, this implies that Sumatra's red ape population was reduced to about 14 percent of its original size during the twentieth century.

These current total population sizes may be higher than previously believed, but the exponential decline during the twentieth century and the ongoing degradation and loss of habitat still warrant an endangered status for the ape. The prime habitat patches, namely the alluvial and flood plain (swamp) forests in which orangutan populations can attain high densities, have been or are being degraded or demolished. These areas functioned as social arenas of essential importance in the ape's reproductive strategy (Rijksen and Meijaard 1999). Without such arenas, breeding success probably will be reduced. Because the range in Borneo is fragmented into at least 61 separate forest areas of varying extent, and in Sumatra into at least 23 separate forest areas, many of the ape's regional and now isolated types are critically endangered, according to the International Union for the Conservation of Nature and Natural Resources (IUCN) criteria. Unfortunately no more than 16 percent of the current range of the orangutan in Borneo has protected status. One usually finds more optimistic figures for the coverage of protected areas in Indonesia, but such figures commonly include ever-proposed, but never-established, reserves. In reality even the formally protected status is of dismally limited practical value because the protective management of the majority of areas is absent or deficient.

The Sumatran situation is somewhat different. Because the Suharto government is accepting a new approach to nature conservation, a significant sector of the apes' range in northern Sumatra has better protection if the current authority can uphold it in the present political chaos. A large sector of state forest land in northern Sumatra was designated as the Leuser Ecosystem (9,600 sq. mi; 24,000 sq. km). It was designated a "conservation concession right" in the custody of the Leuser International Foundation. The former Gunung Leuser National Park has thereby been usurped. Its original shape, size, and location was unfit for orangutan survival, and it was virtually unprotected, in spite of a conservation staff of 130. The new protective management is to be developed by an integrated conservation and development project, called the Leuser Development Programme (LDP), sponsored jointly by the Indonesian government and the European Commission. An effectively protected Leuser Ecosystem is meant to offer survival to approximately half of the currently remaining Sumatran orangutan population (Rijksen and Griffiths 1995). All together conservation areas cover some 73 percent of the range of the orangutan in Sumatra.

DETERMINING WHY CONSERVATION IS FAILING

One of the five surviving closest relatives of humans, the orangutan, is rapidly moving toward the brink of extinction. As previously discussed, the prime cause is a growing mass of people displacing and ferociously persecuting it. On top of this is extremely poor law enforcement with respect to the use of land and natural resources. Displacement of the orangutan is not only for individual human subsistence of poor landless local people. It is primarily for a voracious world market demanding commodities of all sorts. The natural rain forest is being lost exponentially from conversion for cash crops such as oil palm, rice, maize, sugar, pulpwood, rubber, coffee, and tea. Regeneration of the pillaged natural forests under the increasing population pressure is impossible. Its remnants are being scorched.

Even without specific surveys one could have determined that the legally protected orangutan and many other unique wildlife species were facing hard times in Indonesia. Official planning documents, in spite of their inaccuracy, have spelled doom for the rain forest and its inhabitants. The official target of the Indonesian forest use and development policy for the sixth Five Year Development Programme already indicated in 1994 an annual clearing activity of about 2,744 sq. mi (6,860 sq. km), and the conversion of about 840 sq. mi (2,100 sq. km) of natural forest. However, the real clearing of forest has been at least twice as extensive. More than a decade ago, international organizations such as the United Nations Food and Agriculture Organization and the World Bank assessed 5,200 sq. mi (13,000 sq. km) of annual deforestation in Indonesia on the basis of satellite imagery comparisons.

In recent years forest conversion has even become a covert policy to remedy a deficit from scheduled logging in the large plywood industries. International banks eager to invest in oil palm estates support this policy. All forests have become vulnerable. When considering that even carefully selective timber extraction already reduces the carrying capacity of orangutan habitat by at least 50 percent (Rijksen and Meijaard 1999), it becomes evident that the official government policies for land allocation and land use put the survival of the protected orangutan at stake. In spite of a sequence of warnings since 1990, no action has been taken to redress this disastrous course.

For most human cultures, the integrity of nature and the basic interests of wildlife are subordinate to peripheral human interests dealing with wealth and welfare. Indonesia and Malaysia are no exception. Few individuals of the growing mass in a developing country of great natural wealth feel an overwhelming obligation to participate in a strenuous production process if easier cash can be found. As a consequence, many people of all walks of life seek the easiest and most economic ways of attaining welfare and wealth through unrestrained exploitation of natural resources. Authorities often lack the resolve to think in terms longer than their political careers, and thus their ecological insight is negligible. Indeed, for them the obvious historical relationship between the development of European society and the grand demise of Europe's primary forests and spectacular biodiversity allows for little restraint with respect to primeval wildlands. In their perception, the much desired development is reflected in the demolition of jungle. Thus in spite of protective laws, wildland forest areas and wildlife are seen as a relic of the dark ages. That the survival of a formally protected ape is at stake appears to make little difference.

This may perhaps be understandable for a developing country, but it is the mission of nature conservation to challenge this perspective, and whenever possible to change it. However, a similar ambiguity, giving peripheral human interest priority over protection, appears to have crept into the mind of international conservation organizations when dealing with developing nations. In 1980, a joint IUCN/UN Environment Programme, and the World Wildlife Fund (IUCN/UNEP/WWF) document called the *World Conservation Strategy* (IUCN/UNEP/WWF 1980) was professionally promoted to introduce a revolution in conservation. Several terms and concepts were redefined, and new concepts introduced, apparently with the aim of making conservation more effective and facilitating integration with development aid. However, the immediate success of the new concepts in political circles worldwide indicated that they could serve as excellent euphemisms for opening up hitherto protected areas and facilitate laissez-faire exploitation. An intellectual juggling act with the host of newspeak terms (Orwell 1976) by the world authorities in conservation gave many people the illusion that finally one could run with the hare while hunting with the hounds: Conserving wild nature was supposed to be identical with the use of wildlands and organisms, if only one added the term *sustainable.* After 80 years the political outcome of the conflict between John Muir and Gifford Pinchot in the United States was relived, but this time with global stakes and very different stakeholders. Promoted alongside considerable financial support, these newspeak concepts quickly became the guidelines for conservation authorities in many developing countries during the 1980s.

The *World Conservation Strategy* also served to spirit away the traditional focus on endangered species, although IUCN's Species Survival Commission published a pamphlet titled *Species Conservation Priorities in the Tropical Forests of Southeast Asia* in 1982 (IUCN-SSC 1982). The newly dominant "ecosystem approach" seemed to make species conservation obsolete. Like the newspeak concepts of *biodiversity, sustainability,* and *participation,* this ecosystem approach is so academic that it is extremely useful in politics but utterly useless in the real world. The ecosystem approach, assigning a strong role for human participation, indicated rural development as the focal point for funding. International "conservation" has since served primarily to boost economic development in and around protected areas. The proximate aim is to allow the local communities to participate in the "sustainable utilization" (Kramer, van Schaik, and Johnson 1997; Struhsaker 1998) of the conservation area; the ultimate aim is to make them partake of a consumer society, while hoping to save some wildlife on the side.

Was this global *Strategy* perhaps a desktop solution to redress an uneasy feeling of neocolonial interference with ex-colonial conservation areas in some Western conservation agencies? If so, then it certainly resulted in serving what it meant to redress—namely well-masked indirect neocolonialism for the voracious world market. At least in my own Southeast Asian experience over the past two decades, the undoubtedly well-meant newspeak of the international conservation corporations has powerfully contributed to the acceleration of wildland demolition and to the imminent demise of spectacular wildlife, including the orangutan. Why has it taken so long to have the effect of the newspeak policies evaluated in ecological terms? Is it perhaps "politically incorrect" and too hard to admit a blatant waste of funds and of spectacular biological diversity when poor stakeholders are involved?

Perhaps it must be realized also that for bureaucracies the conservation of nature is an unsavory issue. A bureaucracy needs tangible change and hard results of its investments for the ledger. To conserve or keep some piece of jungle successfully is hard to evaluate in administrative terms. That people around a pristine forest are supplied with a regular flow of fresh water, are provided a sustainable quantity of fish protein, breathe fresh air, and partake of a benign climate and sufficient rainfall are taken for granted, even by educated bureaucracies. Thus to placate bureaucracy, ways and means were being explored by field managers to spend some, if not most, of the annual management budgets in the construction of buildings and purchase of vehicles and the production of plans and reports, necessary or not. With the coming of the conservation revolution, however, the bureaucrat can now evaluate most of the spending in accounting terms, and the effect may even be measured in economic output such as the sale of "forest produce" and the buying power of the citizens in the market. There is still no place in the ledger for lost intangibles that were taken for granted and are of hidden macroeconomic value anyway.

The problem of forest destruction in Southeast Asia appears to be rooted in ambiguous policies, serious ideological misconceptions, bureaucratic insensitivity, opportunistic deficiencies in control, a pragmatic absence of law enforcement, and widespread graft as well as collusion through the sharing of dividends in land and resource exploitation. This exists not just among southeast Asians but also within the development banks and other international "aid" organizations for rural development and new conservation. All of this comes on top of the aspirations of a human population that has exceeded 210 million people and has booming consumer demands as well as a voracious hunger for land and resources.

In a proximate sense, this complex of factors has caused a default mismanagement of the Indonesian Ministry of Forestry (MoF) that led to the total collapse of its credibility within Indonesia and abroad. It has also facilitated the emergence of what may well be called a nationwide frenzy of natural resource looting and a sophisticated land grab during the last decade of the twentieth century. The demolition of the forests in Southeast Asia and the ongoing demise of our relative, the red ape, is perhaps the best example of a "tragedy of the commons." Another tragedy, however, is that this has been brought about by hit-and-run laissez-faire economics that were promoted by forces originating in the technologically advanced countries of the northern hemisphere, with an international "conservation" community walking in line, rather than challenging it. In a final analysis the orangutan may be one of the most spectacular victims of neocolonialism.

The international conservation-minded constituency is kept ignorant of this complex background. The public and its political representatives acquire their information on the perils of the orangutan in its rain forest primarily from the media, which is being fed with corporate propaganda and sentimental nonsense rather than information. The misled international constituency is likely to falsely blame national conservation agencies for the expanding forest degradation and the loss of unique species. Such censure on the national level adds to the confusion and frustration, and it will degrade commitment for conservation of wilderness in developing tropical countries even further. Under such conditions the future of the orangutan, which is dependent on large areas of unspoiled rain forest, looks very bleak indeed.

Such an evaluation may seem harsh for people who seem to do their best to find solutions for the problem of dwindling biodiversity, working against the odds of an exploding world population of humankind and its markets for resources. But only by knowing the weaknesses and shortcomings might one find better ways and even succeed in conservation of significant areas of unspoiled rain forest, while giving the ape a fair chance for survival. The plans have been made; the techniques are known. Even a new, improved procedure of rehabilitation has a role to play. The red ape can still be saved, if only there is a collective will to do so.

REHABILITATION: A NECESSITY AND A TRAP FOR CONSERVATION OF THE ORANGUTAN

Rehabilitation is an essential temporary emergency measure to remedy a weakness in the conservation of protected species in general and of the orangutan in particular. The apes to be rehabilitated will not themselves help to save the species from imminent extinction; as unfortunate, mentally derailed specimens of their kind they are biologically of lesser value, even if they were to return successfully to a wild existence. However, they could play an important role in reducing or even halting the persecution of their wild brethren that will soon cause their extinction. If illegal hunting and trade of protected wildlife is to be stopped, because poachers cannot readily be apprehended in the field, then confiscation of the spoils of the hunt is required. When the "spoils" is a live infant destined for the pet trade, then rehabilitation is prescribed by law to reintroduce the ape back into its natural habitat. In other words, rehabilitation is the formally required extension of law enforcement.

Corruption of the concept of ape rehabilitation is a result of media attention, to what seems to be a prevalent personal weakness in the managers, and to the lure of income for self-sufficiency or more. The media seek spectacular issues to attract public attention. Orangutan rehabilitation has it all: a mysterious jungle environment, primitive conditions, ignorant natives, and miserable yet immensely appealing orphaned apes. Moreover, there often is a Caucasian manager willing to play a heroic savior role. Many television producers genuinely believe that in their emphasis on the heroic quest of the fair-skinned savior they raise "awareness" among the public. The conservation corporations see such awareness reflected in the money that streams in to "save the orangutan."

However, rather than becoming aware, the public is being conditioned to what may be called the "Noah approach"—in other words, the approach that highlights a "savior." The coverage is not primarily about the perils of the ape, but rather about the trials and tribulations of acquiring the pitiful orphan and returning it back to the wild. Few in the audience realize that this is not what conservation is about, and under such circumstances the public will never be made properly aware. But it is difficult for the underpaid and undervalued conservationist striving to restore the independence, dignity, and wildness of apes to resist the temptation of riding such a false but lucrative wave of sensationalist publicity. Thus ape rehabilitation may be a key to the badly needed financial support for conservation, but it can also be a dangerous trap for true orangutan conservation. Even some conservationists confuse *orangutan rehabilitation* with *orangutan conservation.*

Barbara Harrisson first attempted orangutan rehabilitation, in Sarawak, during the 1960s (Harrison 1961). A government rehabilitation station was established soon thereafter at Sepilok in Sabah. Indonesia took it up a decade later. Except for the Sabah station, all other stations during the 1970s and 1980s were essentially expatriate interventions. For the first several years, rehabilitation programs appeared to be a reasonably effective means to facilitate law enforcement curbing the illegal trade in orangutans. Conditions for this initial success were that confiscation of apes was consistent, the objective of rehabilitation was focused solely on law enforcement, and the rehabilitation process was fully independent of any need to generate its own funds. An additional condition may have been the expatriate leadership and international supervision. It is perhaps telling, however, that the management of the Indonesian stations was never accorded a formal permit to receive and rehabilitate the strictly protected ape. The zealous expatriate manager in such a station was not seldom conducting the confiscation. The facilitation by a national government of such blatant neocolonialism in cases of wildlife protection is very odd indeed. Yet no such permits were provided even after the stations had been transferred into national custody during the early 1990s, implying that they had kept and handled apes illegally for at least two decades. By 1993, a total of five such stations were in operation in Borneo alone. In 1994 the first official permit for rehabilitation in Indonesia was issued exclusively to the Wanariset Samboja station in East Kalimantan, which established the new official reintroduction program (Smits, Herianto, and Ramono 1995; Smits, Susila, and Herianto 1993). None of the other stations fit the new official criteria for rehabilitation: They operate in areas where wild orangutans are still present, have bad quarantine procedures, and mix tourism with rehabilitation.

That some essential premises of the early rehabilitation technique were wrong (MacKinnon 1977; Rijksen 1978; Rijksen and Rijksen-Graatsma 1978) and that this led to the design of the new procedure and criteria in Indonesia have been widely discussed (Rijksen and Meijaard 1999; Smits et al. 1995). But less well-communicated is that the critical debates about rehabilitation among ape biologists led the international conservation corporations to lose interest in further support for orangutan conservation. Nevertheless, they did continue to exploit media coverage of rehabilitation, yielding considerable funds. One might wonder why the funds were not invested in habitat protection or ape rehabilitation. Whatever the answers, the international conservation constituency was not informed, and its donations were not applied in any meaningful way to "save the orangutan."

The media attention promoted tourism, resulting in humans seeking face-to-face contact with the wild ape (Galdikas 1991, 1995; Galdikas-Brindamour 1975). Without further foreign support, orangutan rehabilitation was obliged to become commercialized. The growing stream of visitors seemed to make that possible, but at a serious price for conservation and hidden behind a mask of awareness promotion. The lure of income from tourism helped to corrupt the concept of rehabilitation, which is essentially a delicate government affair—a follow-up of law enforcement. The exposure to visitors spoils the rehabilitation process for the apes. Yet tourism can boost local economic development, which, in the skewed ideology of the new conservation automatically leads to conservation. Thus since the late 1980s all rehabilitation projects have been obliged to bank on self-support from tourism, with only meager, often incidental, government allocations. In some periods of high tourist attention, this led to wealth for the management staff. At other times it led to scraping along by the staff. At no time, however, did the privatization yield any benefit for either the apes in custody or the wild apes in the surrounding forest. One desperate expatriate manager even began buying up the ape orphans in the illegal trade circuit, partly as an investment for another tourist circus.

In rehabilitation, the ape's nature is being transformed, and human nature is being revealed. For most people the young ape in custody is considered to be a human caricature in need of help and care. Toward the end of adolescence, it becomes strong and assertive, turning into an inherently fearsome object that can still be manipulated to facilitate a public display of human domination. Both stages appear to be attractive for spectators. It boosts a pleasant feeling of misplaced superiority over the wild humanoid beast in attendant and spectator alike and deprives the ape of privacy and self-determination. It seems to be of no consequence that the ape should be returned to the wild and that this would mean that humans should stay as much as possible out of its personal sphere.

Orphaned orangutans seem to be fully aware of their predicament and are often unwilling to be returned to an uncomfortable, if not perilous, feral existence in the jungle. As a consequence, they often cling to any opportunity to partake of the world of humans, soliciting contact and play in an active effort to forge a relationship with people. It is also a main reason why essentially feeble human attendants can still manipulate even adult orangutans. In rehabilitation the attendant should be able to resist the temptations of acting as a media hero at the expense of the ape, as well as resist the charms of clever orangutan orphans begging for affection. Yielding to the former corrupts conservation, and yielding to the latter effectively spoils the rehabilitation process—and is therefore illegal because the Indonesian law stipulates the animal must be returned to a wild state.

The result is that rehabilitation has failed for a majority of orangutans. Many were never trained to become fit for a feral existence. Some became a liability to their own survival. When an orangutan is being exposed to a regular stream of human onlookers, the ape will never learn to deal appropriately with people. Indeed, it may turn into a nuisance, readily exposing itself to fatal danger in confrontations with unfamiliar people when it becomes an assertive adult and begins to move around after it is released into the forest. Many rehabilitated orangutans probably soon meet their fate in such confrontations, as people poison or slay them after one unpleasant contact. Others could never cope with the rigors of a feral life after they lost their way in a forest or were forcefully reintroduced. Although a disappeared ape is usually accounted for as having returned to the wild, the real success of rehabilitation over the past three decades is most likely not higher than 40 percent. Perhaps that is also why the sad results of orangutan rehabilitation are usually so hard to obtain. But these results would not be too sad if rehabilitation succeeded in arresting the persecution of wild orangutans and maintaining habitat in production forests within their range.

Why could international conservation organizations not foresee the inevitable consequences of their sweeping decisions, or take heed of early warnings (e.g., Bruenig 1989; Caro 1986; Holt 1983; Ludwig, Hilborn, and Walters 1993; Wells and Brandon 1992)? One answer is expedient ideology. Whereas a rehabilitation project should operate to phase out its own existence, in reality all projects were turned into self-perpetuating circuses. The link with law enforcement was lost, and the supply of orangutans in the illegal trade circuits increased exponentially. After all, law enforcement for wildlife protection had virtually vanished because international liberalism with reference to natural resources even pervaded the conservation corporations and their "world strategies" and the invention of sustainable utilization. Thus rehabilitation of orangutans is required, but has in reality become a very serious liability to orangutan conservation.

Under the circumstances it may seem amazing that this was acknowledged in Indonesia by the Minister of Forestry. On April 23, 1991, the director general of the Indonesian Nature Conservation Department issued instructions to close the rehabilitation centers at Bohorok (North Sumatra province), Camp Leakey (Central Kalimantan province), and Teluk Kaba (East Kalimantan). It was followed by the ministerial decree of official guidelines for rehabilitation. The minister soon asked for a study of the status of the ape and invited a proposal for an integrated Orangutan Survival Programme. He also initiated the establishment of the Wanariset Samboja reintroduction facilities, according to the new guidelines, through the intervention of the International MoF–Tropenbos research program (Smits et al. 1993). These concrete measures earned the minister, Ir Jamaludin Suryohadikusumo, a well-deserved high international conservation award, Commander in the Order of the Golden Ark, awarded by His Royal Highness Bernhard of the Netherlands.

Unfortunately, these measures also soon revealed the weakness of Indonesian authority in forestry and nature conservation. The official instructions have had no effect. Many forestry officials flatly ignored them or even challenged them. The outdated stations still operate, and law enforcement is still lax. Little has changed. Perhaps one major reason is that the instructions were not backed by financial support. Perhaps the decree and measures were meant in the first place to invite fresh international financial support and guidance and assistance for the new, official Orangutan Survival Programme, including the improved reintroduction procedure (PHPA 1995; Rijksen and Meyaard 1999; Soemarna et al. 1995). Following annual campaigns to "save the orangutan" in the European media, all concerned may have expected that the international conservation corporations would react to this opportunity. The Great Apes Conference, organized by scientist Biruté M. F. Galdikas in 1991, seemed to enforce the expectations among the Indonesian authorities that international support was imminent. Perhaps the conservation authority was loath to enforce dissolution of the old stations and transfer their staffs before having the facilities to receive the swell of illegal ape captives to come. In any case, the expected support did not come, in spite of buzzing regional offices of all major international conservation corporations, fully informed on the issue, and a string of international conferences. Indeed, the intellectual discussions and uninformed opinions on whether expensive rehabilitation is less important than so-called ecosystem conservation, and the fruitless search for consensus on actions among experts, continue to rage. Perhaps more serious, in view of the modern emphasis on participation, is that no sponsor seems prepared to accept the official Indonesian proposals in the Orangutan Survival Programme and begin negotiations for implementation. Funds are accumulating in banks or are squandered in growing numbers of bureaucratic staff and corporate propaganda. Rather than setting the aim at protecting wildlife and its habitat, the international conservation corporations feel obliged to be more than politically correct. Rather than promoting effective birth control and good governance, they have sought to include the boundless desires and aspirations of an exponentially growing mass of poor local "stakeholders" in their considerations. As a consequence they are committed to integrate local participation and socioeconomic development into their "ecosystem approach," even within protected areas. But unanswered is the question of why exploiters with a vested interest should be expected to govern and constrain themselves for the common good of the nation (Holt 1983).

To save the orangutan from imminent extinction it is imperative that a united world conservation community agrees on the objective and begins serious negotiations for quid pro quo arrangements to implement the Orangutan Survival Programme (see Soemarna et al. 1995). The opportunities are still available, but it will cost money and unwavering commitment to hardcore protection to save the ape and many other unique rain forest organisms. Because the Ministry of Forestry and Plantations of the former Suharto regime has lost its credibility to corruption and inefficiency, it is questionable whether this ministry and its subservient Directorate General for Nature Conservation can seriously conduct such negotiations and arrive at meaningful arrangements. Reorganizing this outdated ministry and according it a mission focused on forest conservation and regeneration is immensely important for Indonesia's future forest production capacity and its nature conservation. Such a ministry could restore confidence, attracting much more international support for the effective sustainable use and conservation of living resources than has hitherto been thought possible. In the currently changing political atmosphere, some Indonesian nongovernmental organizations in the field of forest conservation (e.g., Skephi) have already issued proposals in this direction. Their constructive efforts should be supported by the international conservation community.

REFERENCES

Bruenig, E. F. 1989. Use and misuse of tropical rainforests. In H. Lieth and M. J. A. Werger (eds.), *Tropical rainforest ecosystems* (pp. 611–36). New York: Elsevier.

Caro, T. M. 1986. The many paths to wildlife conservation in Africa. *Oryx* 20(4): 221–29.

Frey, R. 1976. *Orangutan rehabilitation centre Bohorok. World Wildlife Fund Annual Report.* Washington, D.C.: World Wildlife Fund.

Galdikas, B. M. F. 1991. *Protection of wild orangutans and habitat vis a vis rehabilitation.* Proceedings of The Great Apes Conference, Ministry of Forestry, Jakarta, December.

———. 1995. *Reflections of Eden; My life with the orangutans of Borneo.* London: Gollancz.

Galdikas-Brindamour, B. M. F. 1975. Orangutans, Indonesia's "people of the forest." *National Geographic Monograph* 148(4): 444–72.

Harrison, B. 1961. Orangutan: What chances of survival? *The Sarawak Museum Journal* 10: 238–61.

Holt, S. 1983. Who really threatens whales and seals? *Oryx* 17(2): 68–77.

IUCN/UNDP/WWF. 1980. *The world conservation strategy.* Gland, Switzerland: Author.

Kramer, R., van Schaik, C., and Johnson, J. (eds.). 1997. *Last stand: Protected areas and the defence of tropical biodiversity.* New York: Oxford University Press.

Ludwig, D., Hilborn, R., and Walters, C. 1993. Uncertainty, resource exploitation and conservation: Lessons from history. *Science* 260: 17–36.

MacKinnon, J. R. 1977. Pet orangutans: Should they return to the forest? *New Scientist* 74: 697–99.

Payne, J. 1988. *Orangutan conservation in Sabah.* Kuala Lampur, Malaysia: World Wide Fund for Nature.

Orwell, G. 1976. *1984.* New York: Penguin Books.

Perlindungan dan Pengawetan Alam (PHPA). 1995. *Orangutan survival strategy.* Internal publication.

Rijksen, H. D. 1978. *A field study on Sumatran Orangutans (*Pongo pygmaeus abelli *Lesson, 1827): Ecology, behaviour and conservation.* Wageningen, the Netherlands: H. Veenman and Zonen.

Rijksen, H. D., and Griffiths, M. 1995. *Masterplan for the Leuser Development Programme.* Wageningen, the Netherlands: H. Veenman and Zonen.

Rijksen, H. D., and Meijaard, E., 1999. *Our vanishing relative: The status of wild orangutans at the close of the twentieth century.* Amsterdam: Kluwer.

Rijksen, H. D., and Rijksen-Graatsma, A. G. 1978. Rehabilitation: A new approach is needed. *Tigerpaper* 6(1): 16–18.

Smits, W. T. M., Herianto, and Ramono, W. S. 1995. A new method for rehabilitation of Orangutans in Indonesia: A first overview. In R. Nadler, B. F. M. Galdikas, and L. K. Sheeran (eds.), *The neglected ape* (pp. 69–77). New York: Plenum Press.

Smits, W. T. M., Susilo, A., and Herianto, 1993. *Year report 1992: Orangutan reintroduction at the Wanariset 1 Station, Samboja, East Kalimantan.* Jakarta: Ministry of Forestry. Unpublished document.

Soemarna, K., Ramono, W. S., Poniran, S., van Schaik, C. P., Rijksen, H. D., Leighton, M., Dahjuti, D., Lelana, A., Karesh, W., Griffiths, M., Seal, U. S., Traylor-Holzer, K., and Tilson, R., 1995. Conservation action plan for Orangutans in Indonesia. In R. Nadler, B. F. M. Galdikas, L. K. Sheeran (eds.), *The neglected ape* (pp. 123–28). New York: Plenum Press.

Struhsaker, T. T. 1998. A biologist's perspective on the role of sustainable harvest in conservation. *Conservation Biology* 12(4): 930–32.

Wells, M., and Brandon, K. 1992. *People and parks: Linking protected area management with local communities.* Washington D.C.: World Bank, World Wildlife Fund, and USAID.

KARL AMMANN

3
BUSHMEAT HUNTING AND THE GREAT APES

Propped against a dozen bananas, the head of a female gorilla lay soaking in her own blood. Her left eye, still open, stares at the floor of the small hut. I walked into this scene at Mambalele Road Junction, southeast Cameroon, while investigating the causes and effects of the bushmeat trade in Central Africa.

The story behind this image is in many ways representative of what is happening on a daily basis in many parts of Central Africa. The hunter who shot this gorilla explained to me that he did so at the request of the police chief of Mouloundou, a major administrative center in southeast Cameroon. The police chief sent a gun on the daily bush taxi. The hunter bagged the gorilla the next day and sent the carcass and gun back, minus the head and one arm, which he was allowed to keep for his "contribution."

This chapter is a collection of such anecdotes, which I hope will illustrate that the commercialization of the bushmeat trade has now reached crisis level and represents a very real risk to the survival of our closest relatives in most parts of Central Africa.

For the past ten years I have regularly traveled the region. Most of my visits concentrated on the Democratic Republic of Congo (DRC), the Congo Republic, and Cameroon, frontline countries as far as the bushmeat trade is concerned. I found that the results were most representative when traveling in bush taxis and sleeping on bamboo mats in hunting camps. Once I started traveling with television crews and in shiny four-wheel-drive vehicles, it became more difficult to establish contact with the guy who pulls the trigger.

I chose to focus my efforts on documenting the impact of the bushmeat trade on gorilla and chimpanzee populations. I believe the apes above all other species have the potential to be ambassadors in the fight to save other endangered wildlife. If we cannot do anything for the great apes, what hope is there for the giant pangolin or the forests these creatures live in?

THE COMMERCIALIZATION OF THE BUSHMEAT TRADE IN CENTRAL AFRICA

I am trying to illustrate, with anecdotal country-by-country data, that we are not talking about isolated cases but that the butchering of great apes and other protected wildlife, and the corrupt practices associated with such butchering, is a daily scenario. I appreciate that in scientific terms, anecdotal data can be questioned. However, one wonders how much more anecdotal material is needed for the scientific community to accept that the commercialization of bushmeat has reached crisis level as far as the off-take of most hunted species is concerned.

Democratic Republic of Congo (formerly Zaire)

I first became interested in the bushmeat issue in 1988 while traveling on one of the legendary Zaire riverboats (Ammann 1991, 1996b). Villagers in pirogues would bring goods to trade with the merchants based on the boat. This included thousands of primates, mostly smoked and tied together in packages of four or five. Some traders also arrived with fresh carcasses, which went directly into the many onboard freezers. By the time these boats reach their destination, Kinshasa or Kisangani, they resemble floating butcher shops. On this trip three orphaned chimpanzee babies also came on board. However, I did not see the meat of their mothers. When I asked, a hunter told me that the orders from the captain were that no heads and no hands or feet of great apes should be seen since Western tourists traveling on the steamers might get the "wrong impression."

I again traveled to the DRC in 1994, this time with a World Society for the Protection of Animals (WSPA) representative. At Kinshasa we filmed massive quantities of ivory in the main tourist market. Selling or possessing ivory was illegal, because there had been no legal hunting of elephants since the late 1980s. In an up-market restaurant we photographed a bushmeat menu, advertising elephant and primate specialties, including chimpanzee.

In 1998, a group of us, at the request of Reinhard Behrend of Rettet den Regenwald (a German forest protection organization), visited a logging concession on the southern bank of the Congo River, owned and operated by the Société Industrielle et Forestière Congo Allemand (SIFORZAL, now SIFORCO), a subsidiary of the German-based Danzer, the biggest logging company in the country. Our objective was to establish the conservation status of the bonobo, *Pan paniscus,* and the influence of logging on it. We traveled for two weeks along large sections of the Lopori, Yekokora, and Lotondo Rivers. We interviewed dozens of villagers, village chiefs, hunters, former and present employees of SIFORZAL/ Danzer, missionaries, and logging company executives from a neighboring concession. To validate the interviews, we cross-checked all information with different sources. We confronted Danzer with our major findings, which follow (Ammann 1998d).

1. Most of the 12-gauge shotguns used in this area were produced in the workshop/garage at this concession. Hunters said they ordered them from SIFORZAL workers, and delivery time was generally about two months. Our informants assumed that SIFORZAL tools and materials were used to produce these guns.
2. Hunters explained that the SIFORZAL boats were used to supply practically all the cartridges used in the region. Independent traders traveled on the boats with ammunition, but to a large extent the trade was reportedly controlled by the boat captains and the Danzer personnel manager, Mr. Lobilo. The cartridges were sold in shops at the SIFORZAL port, at the different market sites, and at Lobilo's residence.
3. When cutting crews go out in the morning, as many as six hunters travel on the logging lorry. In general, one or two company employees bring their own guns and cartridges. They are then excused from logging duty by the team chief and spend the day hunting, returning on the same vehicle in the afternoon and distributing part of their bounty. The professional hunters are deposited en route or at the road head, where they follow the same routine.
4. In the past, most of the procured meat had been eaten by the more than 200 SIFORZAL employees. But since 1998 most of it reportedly had been exported on SIFORZAL boats and floated to Kinshasa. It would appear that this coincided with the company lifting the ban on passengers traveling on these timber floats, and employees and their wives were now offered free passage. As a result, many wives of employees had become bushmeat traders, buying meat from hunters along the various rivers and at the bimonthly market at Bompindo, then transporting the meat to the capital.
5. During this investigative trip up the Yekokora River we saw and filmed seven smoked bonobo carcasses, even though there was a serious shortage of cartridges.

6. A man had, on at least two occasions, bought orphaned bonobos to be sold as pets or to a medical research laboratory in Kinshasa for NZ 800,000 (U.S. $10 in 1998) at this same market, and transported them on the SIFORZAL boats to Kinshasa.
7. In many villages and at the market, we found heavy steel cables used to tie the log floats together. They were for sale in what appeared to be standard lengths of about 5 ft (1.5 m). Hunters explained that these are then unwound to make snares, and it was confirmed that all these cables came from—possibly stolen from—SIFORZAL.
8. According to hunters who worked in areas previously logged by SIFORZAL, when the base moves on, there is practically no game left, which led us to question if there remained any genetically viable populations of bonobos to repopulate logged-out areas.

This relationship between large-scale mechanized logging and the commercialization of bushmeat seems very similar throughout most of Central Africa.

Congo-Brazzaville

I started visiting the Congo regularly in the early 1990s, mainly to document the operations of the three great ape sanctuaries. All of their orphans are a by-product of the bushmeat trade and thus are a pretty clear indicator of the level of great ape hunting. Some of the facts I compiled on these trips follow (Ammann 1994).

1. Bushmeat from a wide variety of species was available for sale in all the major markets, regardless of whether the hunting season was closed or open.
2. Although the meat of protected species was disguised in some markets, it was openly on display in others.
3. For a while, frozen and vacuum-packed elephant steaks were on sale in the capital's most up-market supermarket chain. The French manager told me it had been imported from Chad, and he thus thought it was legal.
4. The prime minister went on television during the closed hunting season to encourage all schoolchildren to spend their holidays hunting and fishing (Aliette Jamart, personal communication).
5. When some concerned individuals in the West responded to the initial publicity on the magnitude of the bushmeat problem by writing letters

to the Congo Embassy in Washington, they received a reply from Ambassador Daniel Mouellet stating, "There is no poaching problem in the Congo" (personal communication to Julie Kaye, April 5, 1996).

6. At the Conkouati Wildlife Reserve, we filmed a lorry being loaded with bushmeat, right next to an International Union for the Conservation of Nature and Natural Resources (IUCN) vehicle. When we asked one of the traders why the cost of the meat doubled by the time it reached the coastal town of Pointe-Noire, we were told that the government rangers staffing the roadblocks would need to be "paid." When we asked how much, we were told the more protected the species, the higher the price.
7. On our first and only evening in Ouesso, the gateway to the renowned Nouabale Ndoki National Park, we filmed a lorry carrying tons of bushmeat, including the carcass of a silverback gorilla that was openly for sale.
8. The next day, the police chief ordered us out of town because we had no permit out of Brazzaville permitting us to go through the CIB concession or onward to the Nouabale Ndoki National Park, and told us to charter a pirogue to take us to neighboring Cameroon. He gave us an armed escort. We assumed this was for our own protection. In the first village out of town our pirogue stopped to load a large bag of ivory, which was to be escorted to Cameroon. Two years later, an ABC crew filmed an elephant graveyard halfway between the Nouabale Ndoki National Park and the Odzala National Park. They counted 280 carcasses.
9. The Réserve de la Chasse de la Lefini is the largest protected reserve in the Congo. It is also the site where the first of several groups of orphaned gorillas was rehabilitated. I visited twice, and walked for hours in the savanna and forest without seeing any trace of wildlife. The local trackers informed me that there were two hippos left and that the last chimpanzees and gorillas had been shot in the 1960s. I could not find any evidence that population pressure or habitat loss were the reasons behind the disappearance of wildlife. In my opinion, market hunting for the capital Brazzaville, some two and a half hours away, had resulted in the wildlife being wiped out. With regular flights from Ouesso carrying bags of meat as a primary cargo, it is easy to guess what supply and demand will do to wildlife in the long term, even in the more remote parks and reserves such as Odzala and Ndoki.
10. When I asked an official if I could see records that any poacher had

been prosecuted, he told me there were none. Poachers do not get prosecuted or even arrested in Central Africa.

Cameroon

I have visited Cameroon more than any other Central African country. Based on interviews with loggers, hunters, government officials, and correspondence with various organizations and individuals, I feel Cameroon represents the very front-line of logging and the unsustainable trade in bushmeat. It is a drastic example of how things get out of hand, and it is an indicator of what the Congo River Basin might look like in a decade if no drastic measures are taken. Some of my findings from these trips follow.

1. The daily express train from the north of Cameroon now regularly stops en route, in the middle of the bush, to allow bushmeat traders to load their meat. When authorities tried to stop the unloading of the meat in the capital's main railway station, the traders forced the train engineer to stop at a bridge a few miles away. That is now where tons of meat are unloaded every day. Some of it is later on display in a nearby bushmeat market, which is open year-round despite the hunting season being closed.
2. Government officials have gone on record stating that both the army and police are deeply involved in the commercial bushmeat trade (Carte Blanche TV 1996). Officials from the Ministry of Environment and Forests (MINEF) reportedly rent out their guns for commercial poaching.
3. There is a MINEF employee who owns a restaurant in the same concession in which he is supposed to control and oversee a French logging company. The hunter who hunts for the MINEF official stated that he rents the official's gun and that three weeks earlier he had killed three gorillas with it. The hunter explained that half the meat went to the MINEF official's restaurant, the other half to the hunter. This same official owned a truck used by the official's wife for bushmeat transport and trading.
4. In the town of Mouloundou we stayed at a Catholic mission, where a Polish missionary told us that he ate gorilla and chimpanzee meat on a weekly basis. He had just dispatched a team of hunters to arrange the meat for a first-communion feast, and he was sure that it would include at least one gorilla.
5. While traveling by bush taxi the following day, we were stopped at

Mambalele Junction, where the police pulled us from the car and forced us to wait 24 hours for the next taxi to arrive. It was here that we found the gorilla head propped against the bananas described at the beginning of the chapter. We had time to waste, so we decided to interview some of the traders operating the shops at the junction. The traders offered us "elephant bullets," as they called them, to be used in "army guns." They also offered us a variety of other shotgun cartridges. They explained that although officially one would need various permits to acquire such ammunition, nobody cared in the forest.

6. We also spoke with a lorry driver coming from neighboring People's Republic of Congo (PRC), whose vehicle had broken down in the middle of the junction. He told us that he had bought a smoked gorilla down the road on the Congo side and had hidden it in the engine compartment to cross the border. We later discovered the meat was a bunch of chimpanzee arms and legs.
7. We then trekked to a nearby logging camp, rumored to be holding a live baby gorilla, probably to be sold as a pet. Unfortunately, the infant had died that morning, and by the time we got there, it was little more than a scrawny, limp body.
8. We returned to Yaounde to ensure safe dispatch of our videotape. Once in the capital we added a few more images to our collection, including a baby chimpanzee kept in a freezer near the railway station. At another market, we found a variety of fetish items openly on display, by-products of the bushmeat trade. Many of the items, such as gorilla skulls, elephant tails, and chimpanzee hands and feet, were from protected species and could not have been legally hunted under any circumstances.
9. In August 1998 I visited the same hunting camp that I had visited several times previously. The village chief explained to me how upset he was about the amount of commercial hunting going on in his area. He had prepared a letter to an official in Lomie that stated that if hunting continued at the present rate, he expected there would be no wildlife left in two years.
10. We trekked with a hunter who told me he had shot three gorillas that very morning. When we arrived at the site, there were actually three dead females with two dead babies. It was the worst massacre I had ever seen (Ammann 1998a).
11. In October 1998, I returned to Cameroon with a German film crew to document the trade. In the Pallisco logging concession we met two

women and a hunter holding a baby chimpanzee. At their feet were the butchered carcasses of four adult chimpanzees in three carrying crates (Ammann 1998c).

12. We have pictures of an ex-police chief from Kika with six guns he reportedly rents out. Some of these are said by locals to have been confiscated by MINEF.
13. The MINEF director of wildlife, Yadje Bello, went on record stating, "We don't want to create other political or social problems because of bushmeat" (Ammann 1998b). This implies that the Cameroon government has already written off its wildlife. The trade has now gone so commercial, with so many parties depending on it, that there will be no way to enforce any aspect of the law without causing some hardship. When asked why no poachers have ever been jailed, Bello responded,

> This is an internal problem related to the collaboration of two administrations. You should go to the magistrate and ask them why they don't want to put poachers in jail. . . . All the matters which have been handed to the court end sometimes with very joking judiciary measures. (Ammann 1998b)

14. Creating protected areas in Cameroon has not resulted in protecting forests and wildlife. According to a local IUCN official, the Dja Reserve, a World Heritage Site in southeast Cameroon, has hunters penetrating at least 31 mi (50 km) inside the park, taking out endangered wildlife, including gorillas, chimpanzees, and elephants, at an unsustainable rate. The reserve is almost completely surrounded by timber concessions, most of which are controlled by multinational French companies. Encouraged by the improved road access to Lomie, which was supported by the European Union, more timber companies have started work in the area (Van der Wal 1998). A major hunting camp within the Dja Reserve supplies 330 lb (150 kg) of bushmeat every week, and there are an estimated 100 such semipermanent hunting camps inside the Reserve, supplying a large part of the meat consumed or sold at the Pallisco concession (Van der Wal 1998).
15. At the end of 1989 a Cameroon general named Aos was filmed moving his logging equipment on army low loaders to begin logging in a concession in the Dja buffer zone, even though local officials report that his concession should have been canceled under the stipulations of the World Bank structural adjustment loan (World Bank, personal communication by confidential documents).

16. Cameroon has an official bushmeat traders' association.
17. Transparency International recently rated Cameroon as the most corrupt country in the world (Transparency International 1998).

Gabon

Gabon is the only country where I have seen chimpanzee carcasses openly on sale in the markets of the capital. It is also the most prosperous country in the region, with a per-capita annual income of U.S. $4,010, compared to U.S. $610 in Cameroon. Steel (1994), working on behalf of the World Wide Fund for Nature, found that the annual trade and consumption of bushmeat in Gabon is a U.S. $47,000,000 industry for a human population of only one million. Thus it seems the bushmeat trade is a serious problem, regardless of economic status. Observations from travel in Gabon follow.

1. Gabonese newspapers advertised baby chimpanzees for sale (Libreville newspaper advertisement 1994).
2. A French logging company was operating in the heart of the private presidential reserve of Omar Bongo. They reputedly paid a royalty of U.S. $60,000 per month directly to the president.
3. Iseroy, a multinational French logging company, had started operations in the Lope Reserve.
4. In 1996 there was another outbreak of Ebola on the upper reaches of the Ivindo River. The minister of health issued a statement in a local newspaper in 1996 saying that villagers in the region should not eat any great apes they killed or found dead in the forest that behaved strangely or looked sick. By implication, he approved hunting healthy apes despite their being protected under national law.
5. Patrick Houben of Vétérinaires sans Frontièrs, who ran the cane rat breeding program in Libreville, attended a government workshop and banquet on poaching, where only bushmeat, including that of three protected species, was served (P. Houben, personal communication).

LOGGING

The logging industry has almost single-handedly brought the bushmeat trade to the unsustainable, commercial level we see today. Funding for infrastructure projects by the World Bank as well as Western governments, particularly the European Union, has greatly enhanced the profitability of logging in the region. From supplying vehicles and equipment to funding logging roads, the international community has indirectly contributed greatly to the slaughtering of apes (ACP-EU Joint Assembly 1996; Ammann and Pearce 1995; Behrend 1997; Bowen-Jones 1998; Horta 1997).

Logging trucks illegally transport meat and hunters. Steel cables, normally used in logging operations, are used to make illegal snares. Hunters use new and abandoned logging roads to gain access into once undisturbed forest. Logging employees represent a captive audience for the camp-based commercial hunters. In some concessions workshop staff earn extra income by making guns for resident hunters.

Some of the German logging companies we investigated requested of the independent timber transport company that the drivers should no longer be allowed to carry bushmeat. The drivers went on strike and regained permission to carry the illegal commodity.

One of the most prominent loggers in Cameroon went on record stating that it was much easier to operate illegally than legally. No government authority wanted him to operate legally, because that would reduce the corruption potential (J. Liboz, personal communication). None of the logging firms in Central Africa, local or foreign, had been awarded a recognized sustainable forestry certificate.

Interministerial meeting minutes on logging in the DRC state that the government plans to put 280,000,000 cu. ft (8,000,000 cu. m) of timber annually on the international market. This calculation is based on Cameroon producing 70,000,000 cu. ft (2,000,000 cu. m) a year and having only one quarter of the forest of the DRC (personal communication). Minutes of a meeting of the executive officers of a World Bank-initiated logging company state that IUCN considers logging to be totally out of control in the Cameroon (Meeting Minutes of Working Group 3a 1998).

CORRUPTION AND POLITICAL WILL

The level of official corruption is probably highest in the bushmeat sector of the economy in these countries. Aron Lima, the former MINEF assistant director for wildlife, stated on camera: "One of the main problems is the involvement of the army and police in the trade. Few conservation projects are geared to deal with these realities. The sooner the conservation community accepts these facts the better" (Carte Blanche TV 1996). The commercialization of the bushmeat trade has grown to be a major economic factor in the past ten years in many of the countries concerned, despite a wide range of laws restricting hunting (Bowen-Jones 1998). Although (1) the guns used by 99 percent of hunters are illegal because they have no permit; (2) the cartridges used are illegal for the same reason; (3) transport of bushmeat and hunters on logging trucks is illegal; (4) snares made from steel cable are illegal; (5) commercial bushmeat hunting in most countries is restricted by law; (6) hunters illegally settle in zones meant for logging rather than human settlement; (7) licensing of bushmeat during the closed hunting season (6 months per year) is illegal; and (8) the hunting and trading of endangered species such as gorilla, chimpanzee, and elephant is illegal, I have never seen any proof that a hunter, logger, or trader was arrested, prosecuted, and jailed for disregarding these laws.

I asked in a meeting with World Bank officials why the recruitment of a foreign director of wildlife for Cameroon was not an option to be negotiated. Uganda had recently placed an advertisement in *The East African* newspaper calling for applications for executive director of Uganda Wildlife Authority to "ensure a more effective, efficient, and commercial organization" (Ugandan Wildlife Authority 1998). The World Bank officials replied that the difference was that there is no political will in Cameroon (Cynthia Cook, personal communication).

The WWF's National Elephant Management Plan for Cameroon states that, "Implementation and enforcement of basic trade controls, particularly CITES regulations, is inadequate because of a lack of enabling legislation [and] an absence of political will and motivation" (Tchamba, Barnes, and Ndiang 1997, 31). A Conference on the Impacts of Forest Exploitation on the Wildlife was held in Bertoua, Cameroon, in April 1996. It was attended by representatives from local nongovernmental organizations (NGOs), members of parliament, MINEF officials, mayors, teachers, and foreign conservation organizations. The following are some of the recommendations and action steps that were agreed on (WSPA 1996): (1) government strengthening of conservation policy by ensuring the implementation of existing laws or establishing more appropriate ones; (2) creation of check points; (3) transformation of abandoned logging concessions into sustainable hunting zones with guides and rangers; (4) destruction of logging roads and bridges after the logging operations have ended; (5) banning settlement of local people in abandoned logging concessions; (6) providing alternative means of income to local populations around logging concessions; and (7) converting poachers by offering alternative employment—for example, as farmers, ecoguards, and cattle breeders. Three years later there was no sign that any of these resolutions had been implemented. T'Sas has blamed the Cameroon economic crisis of the 1980s and 1990s not on the fall of commodity prices but rather on the fact that "the crisis had been fueled by years of staggering corruption, nepotism and other malpractice in government circles" (1997, 33).

CONSERVATION POLITICS AND SOLUTIONS

There is little hope for conservation in Central Africa when poachers are not arrested, loggers are allowed and encouraged to operate illegally, police chiefs rent out guns to poachers, and the wives of wardens become bushmeat traders. However, can we put all the blame on the doorsteps of African governments and the European logging companies? To what extent has the "quiet diplomatic approach" (Jane Goodall, personal letter, Sept. 21, 1995) advocated by most conservation organizations active in Central Africa contributed to the present state of affairs?

In my opinion, a letter from the Congo ambassador in Washington in April 1996 (to Julie Kaye) provides the answer: He wrote, "We would like you to know that organism composed of Congolese and International experts (Japanese, British and Americans) and sponsored by the World Bank and the Planning Ministry in collaboration with the Ministry of Forests take care of and protect the Congolese Wildlife against any danger." This is the same Congo where a year later 280 elephant carcasses were found in one forest clearing. Hiding behind the presence of foreign conservation NGOs is the order of the day in Central Africa.

Any real solution, at this stage, will require some hard choices. The governments concerned are not ready to make them. Recall the quote of the MINEF director of wildlife: "We can not risk any social or economic problems because of bushmeat." Based on letters I have received from conservation organizations and discussions with researchers and conservation executives, a confrontational stance and "boat rocking" are not considered as options (Gartlan 1999; Wildlife Conservation Society [WCS] 1994; Wilhelm 1998). When I offered the editor of the WCS in-house publication a feature article on bushmeat, I got the following response:

> The chief drawback, of course, was our firm conviction that publishing your article with your compelling photographs would have wide repercussions that certainly would adversely impact our scientists in Africa. An essential and exhaustive part of their job is to maintain good relations with the governments and indigenous people so the society's conservation projects will be permitted to continue. (WCS 1994)

Yet the African wildlife and forest situation worsens every year. Timber companies continue unsustainable and often illegal timber extraction (Ammann 1996a; Ammann and Pearce 1995; Behrend 1997; Horta 1997; J. Liboz personal communication; Verhagen 1993). The past ten years has seen a dramatic increase in the bushmeat trade, particularly in great ape meat (Ammann and Pearce 1995). In a 1998 public lecture in Nairobi at the National Museum of Kenya, Jane Goodall announced, "The bushmeat trade is the greatest danger to apes in Central and West Africa today."

Although conservation organizations have won some minor battles (such as establishing very remote reserves where the lack of infrastructure has prevented establishment of the commercial bushmeat trade), the war and most of the main battles are being lost. I am convinced that conservation organizations that are not willing to stand up, speak out, and risk having their offices closed are more of a liability than a benefit. The International Monetary Fund (IMF) and other donor organizations regularly pull out of countries or suspend loans, especially if there is widespread corruption and no political will. In 1997, the IMF halted a U.S. $205 million structural adjustment loan to Kenya on the grounds of rampant government corruption. This caused Kenya to slip into a deep recession (Redfern 1999). The IMF and the World Bank reinstated funding to Kenya in 2000 and suspended it again in 2001 after Kenya did not live up to some of the commitments that had been made.

I have had the opportunity to discuss bushmeat and related topics with many conservation executives. Coming from a business background, what has surprised me more than anything else is the lack of criteria for measuring conservation performance and results. The result appears to be to "sell" the minor success stories and ignore that the war is being lost on a very wide front. As long as the conservation community arrives at the negotiating table with "peanuts" compared to the logging industry, I see little hope for creating the necessary political will. We will need much bigger carrots and sticks to be taken seriously and the leverage to enforce agreements and promises. No grassroots projects will have a real impact as long as the villagers respond to arguments of not killing this gorilla or cutting that tree by pointing out that the minister eats bushmeat at every official function and the army generals all have become loggers. Mr. Liboz, a prominent French logger in Cameroon, went on camera stating that what was happening was "total destruction" and that there was no point in counting on the government, the loggers, or the conservation community to effect any real change. He felt only a major international outcry would make a difference (J. Liboz, personal communication).

The problem seems to be getting those involved to admit that the bushmeat trade has reached crisis level, has become a major economic factor in several of the countries concerned, and that any solution will require some hard choices that the governments in question are no longer willing or able to make. These admissions would allow looking for appropriate solutions, such as mobilizing the bilateral donor community to introduce more environmental conditionalities. These solutions might currently be beyond the scope of the conservation community. Richard Leakey, director of Kenya Wildlife Services, made the following point in a 1998 speech:

> Kenya cannot eradicate polio without international money, we cannot deal with the problems of children without international money, we cannot deal with the problems of public health without international support. We can take millions of dollars from WHO [World Health Organization] to attempt to eradicate polio and other diseases. . . . So why can we not find a way to support the cost of conservation measures in Kenya and other African countries through international funds? . . . Why not set up a decade of support for wildlife management programs to pay the project implementation costs, not for theories or experts but for the guys on the ground? Why not structure it in such a way that if we did not deliver in terms of audit we would not get any more money and not only would we not get any more money for wildlife but we would not get any more money for polio and roads and other things?

The first step in a drastically new big-picture approach will be to admit that the traditional approach has failed or is failing. That, however, seems to be as much of a problem as getting the hunters to stop pulling the triggers.

REFERENCES

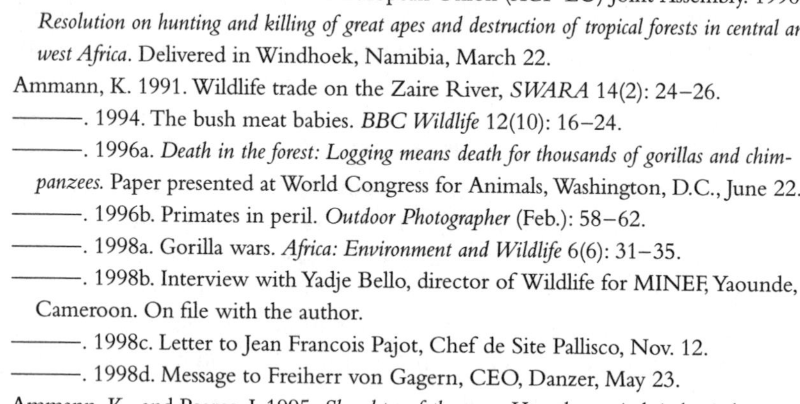

African Caribbean Pacific Nations/European Union (ACP-EU) Joint Assembly. 1996. *Resolution on hunting and killing of great apes and destruction of tropical forests in central and west Africa.* Delivered in Windhoek, Namibia, March 22.

Ammann, K. 1991. Wildlife trade on the Zaire River, *SWARA* 14(2): 24–26.

———. 1994. The bush meat babies. *BBC Wildlife* 12(10): 16–24.

———. 1996a. *Death in the forest: Logging means death for thousands of gorillas and chimpanzees.* Paper presented at World Congress for Animals, Washington, D.C., June 22.

———. 1996b. Primates in peril. *Outdoor Photographer* (Feb.): 58–62.

———. 1998a. Gorilla wars. *Africa: Environment and Wildlife* 6(6): 31–35.

———. 1998b. Interview with Yadje Bello, director of Wildlife for MINEF, Yaounde, Cameroon. On file with the author.

———. 1998c. Letter to Jean Francois Pajot, Chef de Site Pallisco, Nov. 12.

———. 1998d. Message to Freiherr von Gagern, CEO, Danzer, May 23.

Ammann, K., and Pearce, J. 1995. *Slaughter of the apes: How the tropical timber industry is devouring Africa's great apes.* London: World Society for the Protection of Animals.

Behrend, R. 1997. Affenmord. *Regenwald Report* 2: 2.

Bowen-Jones, E. 1998. *The African bushmeat trade: A recipe for extinction.* London: Ape Alliance.

Carte Blanche/M-Net South Africa TV. 1996. *The bushmeat orphans.* TV Documentary.

Gartlan, S. 1999. E-mail letter to American Zoo and Aquarium bushmeat discussion list, March 5.

Horta, K., 1997. Letter to Bernd von Droste, director of Ecological Sciences, UNESCO.

Leakey, R. 1998. Insight. *SWARA* 21(3): 43.

Libreville newspaper advertisement of chimpanzees for sale. 1994. *Zoom* 87: 11–24.

Meeting Minutes of Working Group 3a. 1998. Meeting held at Zurich Koten Hilton Hotel on Nov. 30.

Redfern, P. 1999. Kenya—IMF aid deal may return. *Nation* (Feb.): 2.

Steel, E. 1994. Study of the value and volume of bushmeat commerce in Gabon. *Report for World Wide Fund for Nature Program for Gabon*. Libreville, Gabon: World Wide Fund for Nature Program Pour le Gabon.

Tchamba, M., Barnes, R. F. W., and Ndiang, I. N. 1997. *National elephant management plan for Cameroon*. Yasunde, Cameroon: WWF and Ministry of Environment and Forestry. Unpublished document.

Transparency International. 1998. *Corruption perceptions index*. Website www.transparency.de.

T'Sas, V. 1997. A right royal president. *BBC Focus on Africa* 8(3): 32–33.

Ugandan Wildlife Authority. 1998. Advertisement for executive director of Ugandan Wildlife Authority. *East African,* July 13–19, p. 15.

Van der Wal, M. 1998. *Large mammals of the Dja, Conservation and management*. Paper presented at the Regional workshop on the sustainable exploitation of wildlife in the southeast of Cameroon, WWF Cameroon, Bertoua, April 14–15.

Verhagen, P. 1993. *Logging and conflicts in the rainforests of Cameroon*. Tilburg, Netherlands: International Union for the Conservation of Nature and Natural Resources. Unpublished document.

Wildlife Conservation Society (WCS). 1994. Letter from WCS to Karl Ammann, January 31.

Wilhelm, P. 1998. Letter to Parador Television Productions, June 5.

World Society for the Protection of Animals (WSPA). 1996. *General report of the conference on the impacts of forest exploitation on the wildlife*. Bertoua, Cameroon: Author.

DAVID S. WILKIE

4
BUSHMEAT TRADE IN THE CONGO BASIN

Bushmeat hunting for subsistence and as a source of income is a common component of household economies in the Congo Basin and more generally throughout sub-Saharan Africa (Anadu, Elamah, and Oates 1988; Asibey 1977; Geist 1988; Juste et al. 1995; King 1994; ma Mbalele 1978; Martin 1983). Some ethnic groups such as the Mvae, Yassa, and Kola of Cameroon eat more meat, 161 lb per capita per year (73 kg/capita/year—primarily bushmeat), than the average person in France (Chardonnet et al. 1995) or elsewhere in the industrialized world (66 lb/capita/year or 30 kg/capita/year). Human population across the region is likely to have at least doubled since the 1920s (Hochschild 1998), and given average growth rates of 2.7 percent (range 1.5 to 3.3 percent) the population is expected to double again in 25 to 30 years. If demand continues to grow as expected, and consumers do not or cannot switch to eating meat of domestic livestock, then hunting of wild game will increase in the future. Can wildlife in the forests of the Congo Basin sustain present and projected levels of hunting, and are the great apes particularly at risk from the bushmeat trade?

GREAT APES AND HUNTING

The Congo Basin supports the greatest diversity and number of great apes anywhere in the world. Yet nowhere in the region are great apes truly abundant. Fewer than 120,000 gorillas, 100,000 chimpanzees, and 30,000 bonobos are believed to exist in the forests of the Congo Basin (Oates 1996). Ape density within the forests of the Congo Basin, on average, is fewer than .34 per sq. mi (0.13 per sq. km; assuming that Congo Basin forests cover approximately 720,000 sq. mi or 1,800,000 sq. km). That is 1/10th to 1/200th of the density of the human populations that inhabit the same forest.

Most information on hunting of apes for bushmeat is anecdotal or, worse, apocryphal. No bushmeat study has focused on apes, possibly because they are relatively uncommon in markets (Bowen-Jones 1998). Apes are most often hunted opportunistically and tend to constitute the "by-catch" of hunters seeking the more abundant and, in absolute terms, more lucrative duikers (Eves 1995; Wilkie and Carpenter 1999). This is not surprising, because gorillas are "considered to be the most dangerous species to hunt" (Pearce and Ammann 1998, 8). On occasion gorillas may be targeted explicitly by hunters, because gorilla hands are considered a delicacy by some consumers (Bowen-Jones 1998). However, because of apes' absolute scarcity and low reproductive rates, though they rarely constitute more than 1 percent of the carcasses brought into markets, even present levels of hunting may threaten the long-term survival of all ape populations within all range states in the Congo Basin (Oates and Davies 1986).

Paucity of quantitative data on ape abundance, distribution, and off-take by hunters throughout the region leaves us with little direct evidence to assess just how much danger apes face from the commercial bushmeat trade. We can argue, however, that if empirical data suggest that commercial hunting adversely affects wildlife species such as duikers, which because of their relative abundance and life-history characteristics makes them less at risk from overhunting, then by proxy we can estimate the likely risk to great apes as a result of the commercial trade in bushmeat.

Given the importance of bushmeat to rural and urban consumers in Congo Basin countries, and rising concerns about the impact of the bushmeat trade on apes and other forest wildlife (chapters 1 and 3, this volume), it is critical that we impartially assess what we know about the scale and causes of bushmeat hunting. Without this information we will be unable to propose approaches to managing the use of bushmeat in the region that balance local realities with global concerns for biodiversity conservation.

Twenty years ago it would have been impossible to evaluate, other than anecdotally, the scale and impact of the bushmeat trade on wildlife in the Congo Basin. However, since the early 1990s growing concern about bushmeat exploitation in the region has prompted research that has substantially filled the gaps in our knowledge. These studies can be divided into (1) largely anecdotal accounts that focus world attention on the issue (Klemens and Thorbjarnarson 1995; ma Mbalele 1978; McRae 1997; Pearce 1996a; Pearce and Ammann 1998); (2) more quantitative assessments of bushmeat consumption and the economic value of bushmeat at national and household levels (Anstey 1991; Asibey 1974b; Aunger

1992; Auzel 1996; Bailey and Peacock 1988; Bennett Hennessey 1995; Chardonnet et al. 1995; Colyn, Dudu, and ma Mbalele 1987; de Garine 1993; Delvingt 1997; Dethier 1995; Eves 1995; Heymans and Maurice 1973; Hladik, Bahuchet, and de Garine 1990; Hladik and Hladik 1990; Koppert and Hladik 1990; Koppert et al. 1996; Lahm 1993a; Laurent 1992; Mares and Ojeda 1984; Njiforti 1996; Noss 1995; Pierret 1975; Steel 1994; Takeda and Sato 1993; Wilson and Wilson 1989, 1991); and (3) quantitative surveys of wildlife population densities, production rates, and hunting impacts (Auzel and Wilkie 2000; Bennett Hennessey 1995; Delvingt 1997; Dethier 1995; Dubost 1980; Fa et al. 1995; Feer 1993; Gally and Jeanmart 1996; Hart 1978; Ichikawa 1983; Infield 1988; Koster and Hart 1988; Lahm 1993b, 1994; Malonga 1996; Ngnegueu and Fotso 1996; Noss 1995; Payne 1992; Steel 1994; WCS 1996; White 1994; Wilkie 1989; Wilkie and Curran 1991; Wilkie and Finn 1990; Wilkie et al. 1998a).

Do we now know enough to assess the scale and impact of market hunting and to propose approaches to manage the harvesting, sale, and consumption of bushmeat in the Congo Basin? To answer this, we need to review the results of these recent studies to ask if hunting to meet present and future demand for bushmeat is sustainable, and how important bushmeat is to the diet and income of families. We also need to determine if there are substitutes for bushmeat.

DETERMINING IF HUNTING FOR BUSHMEAT IS SUSTAINABLE

Ideally to assess the impact of hunting on forest wildlife we would like to know the present biomass, the harvesting intensity, and the annual production of all common bushmeat species across a stratified random sample of forest blocks within the Congo Basin using standardized methods. However, even measuring the density of the most common species is difficult and the results often uncertain. The small size of most common bushmeat species, the discrete behavior of the animals, and dense vegetation in the habitat make systematic and reliable visual censuses extremely difficult (White 1994) within a single forest area. Differences in plant species composition and understory density between sites make cross-site bushmeat density estimates difficult even when the same researchers use the same methods. In the Congo Basin, not only have researchers used very different methods to conduct censuses in different areas, the reliability of index measures of density (e.g., dung or track counts) is poorly understood. Similarly, production estimates are usually derived from an equation reported in Robinson and Redford (1994) that relies on uncertain biomass estimates and scarce data on life history characteristics (East 1995; Estes 1991; Haltenorth and Diller 1980; Kingdon 1997) that are often based on a very small sample of captive animals. Moreover, the Robinson and Redford equation assumes no mortality of juveniles and adults up to the age of last parturition. Slade, Gomulkiewicz, and Alexander (1998) suggested that this results in overestimates of production, which risks concluding that bushmeat species are sustainably exploited when they may not be. Comparing results across studies is, therefore, fraught with difficulty. Yet, though results of individual studies are uncertain, if most converge or triangulate on the same conclusion we can feel more confident in the results and more certain about the true impact of bushmeat hunting.

From market studies we now know that (1) residents of Congo Basin countries eat as much if not more meat than residents of northern industrial countries (average of 103 lb/person/year or 47 kg/person/year versus 66 lb/person/year or 30 kg/person/year); (2) although urban families eat less bushmeat than rural families, urban demand for bushmeat may, given population distributions, exceed rural demand; (3) bushmeat, not the meat of domesticated animals, constitutes the primary source of meat for most residents of the Congo Basin; and (4) the gross quantity of bushmeat consumed in forest and urban areas across the Congo Basin may exceed 1,100,000 short t (1,000,000 metric t) per year (Wilkie and Carpenter 1999).

Though we can infer, in gross terms, the scale of bushmeat consumption from this information, we are unable say whether or not hunting is sustainable, because we do not know if the bushmeat sold in a market came from 2,470 a (1,000 ha) or 400 sq. mi (1,000 sq. km) of forest. To assess the impact of the bushmeat trade on wildlife populations and appraise the sustainability of hunting we need to review the information available on the intensity of bushmeat hunting (i.e., lb bushmeat/sq. mi/year) and the quantity of bushmeat produced by the forest each year.

Duikers, bush pigs, primates, and rodents are the most commonly hunted groups of animals in the forest, with duikers both numerically (> 75 percent) and in terms of biomass being the most important bushmeat species group (Auzel and Wilkie 2000; Bennett Hennessey 1995; Delvingt 1997; Dethier 1995; Fa et al. 1995; Gally and Jeanmart 1996; Hart 1978; Ichikawa 1983; Lahm 1994; Malonga 1996; Ngnegueu and Fotso 1996; Noss 1995; Steel 1994; Wilkie 1989; Wilkie and Curran 1991). Primates rarely constitute more than 20 percent of the animals sold in bushmeat markets, and apes fewer than 0.5 percent. Rodents gain in importance in long-established markets, presumably because slow reproducing primates and large duikers have been depleted in accessible forests (Fa et al. 1995; Steel 1994). In fact, the ratio of duiker-to-rodent biomass found in markets or the presence or absence of primates in markets may provide a rough index of bushmeat overexploitation or diminishing hunter access to dense forests (e.g., the ratio in rural Ekom, Cameroon, is 14 to 1, whereas in urban Libreville, Gabon, the ratio is as low as 1 to 1.25), and could provide us with a fairly simple tool to monitor the impact of bushmeat hunting over time.

Though primates and particularly apes are likely to be more susceptible to overhunting than are the more numerous, faster reproducing, and less visible duikers and rodents, and should have been a focus of bushmeat studies, information on primate harvest rates is largely anecdotal (Bowen-Jones 1997; Harcourt 1980; Kalivesse 1991; McRae 1997; Pearce 1996a, 1996b; Pearce and Ammann 1998). Quantitative data from across the Congo Basin (Dethier 1995; Fa et al. 1995; Feer 1993; Gally and Jeanmart 1996; Infield 1988; Ngnegueu and Fotso 1996; Noss 1995; WCS 1996; Wilkie et al. 1998a) do exist for duikers, and show enormous variability in harvest rates (182 to 1541 lb/sq. mi; 32 to 270 kg/sq. km with a median of 445 lb/sq. mi or 78 kg/sq. km). Similarly, duiker biomass estimates from one site to another across the Congo Basin vary from 575 to 8,515 lb/sq. mi (101 to 1,493 kg/sq. km), depending on the methods used (Dethier 1995; Dubost 1980; Feer 1993; Koster and Hart 1988; Lahm 1993a; Payne 1992; White 1994; Wilkie and Finn 1990). Visual counts using either encounter transects or drive counts generate more comparable numbers ranging from 576 to 1,143 lb/sq. mi (101 to 201 kg/sq. km), although even these still differ by almost 100 percent.

Comparison of average harvest rates (553 lb/sq. mi; 97 kg/sq. km/year) with average production rates (970 lb/sq. mi; 170 kg/sq. km/year) suggests that duikers are being overharvested across much of the Congo Basin, assuming that, as Robinson and Redford suggested (1994), relatively short-lived animals should not be harvested at a rate that exceeds 40 percent of annual production (i.e., 387 lb/sq. mi/year; 68 kg/sq. km/year).

Lahm's (1994) research on game densities in hunted and nonhunted patches of forest near Makokou, Gabon, has confirmed that commercial hunting can deplete wildlife populations. Assuming that habitat, vegetation density, and visibility are comparable in the hunted and nonhunted sites, her data suggest that hunting resulted in a decline in game densities of 43 to 100 percent. Primates and large-bodied species were most severely affected. Six of 14 species of wildlife were effectively extirpated from hunted areas, four of which were primates (gorilla, chimpanzee, mangabey, and colobus). Gray-cheeked mangabey, *Cercocebus albigena*, densities were reduced from more than 128 individuals per sq. mi (51 individuals per sq. km) to fewer than 8 individuals per sq. mi (3 individuals/sq. km), a reduction in species biomass of more than 1,600 lb/sq. mi (280 kg/sq. km).

Ngnegueu and Fotso (1996) have shown that snare trapping by hunters in settlements located along the northern edge of the Dja Reserve is depleting wildlife populations within 6 mi (10 km) of villages. Dethier (1995) also showed that around the Dja Reserve red duikers constituted a smaller proportion of all duiker captures within 3 mi (5 km) of settlements, compared with captures in more isolated forest locations (51 percent and 82 percent, respectively). Fimbel and Curran (WCS 1996) working in the forests surrounding the Lobéké Reserve in southeastern Cameroon show that red duiker populations (e.g., *Cephalophus dorsalis, C. nigrifrons, C. leucogaster*) have been depleted close to villages and that both blue, *C. monticola,* and red duikers were being overexploited, with harvest exceeding production by 338 to 2,707 percent, in all zones outside the reserve.

Noss (1995) evaluated the sustainability of snare and net hunting in the Dzanga-Sangha Special Reserve in the Central African Republic (CAR). His data suggest that hunters using traditional net-hunting techniques harvest more animals per unit area than hunters using illegal "modern" snares. Net hunts that can be conducted from hunters' permanent villages overexploit all duiker species. Only the brush-tailed porcupine, *Atherurus africanus,* appears to be harvested at sustainable levels by net hunters. In contrast, snare hunters appear only to be overexploiting the larger duikers (*C. callipygus* and *C. dorsalis*). Noss has warned, however, that the combined impacts of net and snare hunting are likely to overexploit *all* duiker species and that total harvest rates are in reality even higher because the impact of shotgun hunting was not included in the study. Combining data gathered by Infield (1988) and Payne (1992) shows that harvesting of blue duikers and Ogilby's duiker, *C. ogilbyi,* around Korup National Park in Cameroon was 1.3 to 13.2 times greater than was sustainable.

The impact of snare hunting is also likely to be more severe than shown by capture studies because 25 percent or more of animals trapped rot or are lost to scavengers. Delvingt (1997) reported that losses account for 4 to 36 percent of all animals trapped, whereas Dethier (1995) showed that in distant forest 26 to 39 percent of all trapped animals were left to rot in the traps and that some hunters left traps set and unchecked for as long as 29 to 77 days. Village traps were checked more frequently, and waste was estimated to be 11 percent (Dethier 1995).

Though these studies used different methods, in different forest areas, over different time periods, convergence of results from studies across the Congo Basin leaves little doubt that the bushmeat trade is clearly having an adverse impact on forest wildlife throughout the region, and commercial hunting of wildlife is likely to be unsustainable for primates, all large-bodied duikers, and in many areas all mammals larger than rodents.

DETERMINING HOW IMPORTANT BUSHMEAT IS TO THE DIET AND INCOME OF FAMILIES

Plausible approaches to mitigating the adverse impacts of the bushmeat trade on wildlife populations requires assessing the importance of bushmeat to rural and urban households. We need to know (1) how much income is generated by hunting per capita relative to alternative sources of money; (2) the relative contribution of bushmeat to an average individual's diet; and (3) the availability of affordable substitutes for bushmeat, such as the meat of domesticated animals.

Hunting typically contributes between 30 to 80 percent of protein consumed by forest-dwelling families in the Congo Basin (Koppert et al. 1996). Agricultural crops provide most of the calories. Thus unless Congo Basin families have access to true substitutes for bushmeat, any attempt to curtail bushmeat production may result in children suffering the consequences of protein deficiency: slowed growth and learning delays. Fish and domestic animals are the most obvious substitutes for bushmeat as a source of protein.

Why people eat bushmeat is controversial. Some argue that bushmeat is a cultural preference and cite consumers' willingness to pay a price premium over domestic meat for the privilege of eating bushmeat (Chardonnet et al. 1995; ma Mbalele 1978). Steel (1994) found in Libreville, Gabon, that the average price for the most popular bushmeat species was U.S. $1.68 per lb (U.S. $3.70/kg), more than 1.6 times the price of the most popular cut of beef. More recent evidence suggests simply that bushmeat is often the only source of animal protein available and tends to be cheaper than domestic substitutes. Gally and Jeanmart (1996) found that the price of bushmeat per pound was .10 to .25 times the price of available substitutes in three markets in Cameroon, Congo, and the CAR. In Bayanga, CAR, beef prices are two to three times the price of bushmeat (Noss 1998). Similarly, a pound of bushmeat in various towns near the Ngotto forest in the CAR ranged from $0.14 to $.34 per lb ($0.32–0.75/kg), whereas goat was $0.80 per lb ($1.75/kg), chickens were $1.60 per lb ($3.52/kg), and caterpillars were $1.66 per lb ($3.65/kg; Delvingt 1997).

Most consumers probably eat bushmeat because it has few less expensive substitutes and is an open-access resource available to anyone willing to go hunting. Urban elites, however, may view bushmeat, and particularly the meat of apes, as a cultural heritage luxury item and thus may be willing to pay a premium price. If population growth rates continue at present levels, per capita demand remains constant, effective substitutes remain unavailable, and bushmeat continues to be an open-access resource, it is highly likely that bushmeat species will be extirpated from all areas of forest proximal to population centers (i.e., sources of most demand). Even when bushmeat scarcity causes prices to rise and substitutes to be more competitive, hunting will continue in areas where bushmeat capture and transport costs remain comparable to the costs of livestock rearing or where there is a demand for bushmeat as a luxury.

Though numerous studies document bushmeat entering markets, few have documented the economic value of bushmeat to the hunter and trader (Ambrose-Oji

1997). Noss (1998) reported that snare hunters trapping within the Dzanga-Sangha Special Forest Reserve earn between U.S. $400 to $700 per year. Hunters earn more than the CAR's official minimum wage and an amount comparable to guards employed by the park ($450 to $625 per year). In the Congolaise Industrielle du Bois (CIB) logging concession in northern Congo, residents of the logging camp and a nearby village that had access to markets for bushmeat on average sold between 36 to 52 percent of all bushmeat captured and generated income of approximately $300 per household per year. Because logging concession employees earn about $4 to 12 per day, bushmeat sales contribute between 6 to 40 percent of all households' daily income (Wilkie et al. in press).

Gally and Jeanmart (1996) demonstrated the benefits to hunters, traders, and restaurant owners who sell bushmeat by tracing the sale of three monkeys killed with a shotgun. In this case the hunter netted $6.30 (30 percent profit) from the sale of the monkeys, the trader made $10.20 (19 percent profit), and the restaurateur made $20.60 (21 percent profit). These authors also reported that the economic returns to six hunters in Cameroon generated an annual income from hunting that ranged from $330 to $1,058, an amount well above the national average. In Congo, Dethier (1995) showed that hunters generated between $250 to $1,050 per year from selling bushmeat. Near the Dja Reserve in Cameroon, Ngnegueu and Fotso (1996) showed that individual hunters could generate as much as $650 per year from selling bushmeat. In the six months of their study 30 hunters generated more than $9,500 in income from bushmeat sales. Bushmeat is clearly an important dietary item, with few if any cheaper substitutes, and the bushmeat trade is an important economic option for rural and urban households that may have few, if any, alternatives.

APPROACHES TO MANAGING BUSHMEAT HUNTING

Given the importance of hunting to household economies, and that the market for meat is primarily urban centers, strategies to reduce incentives for, and the impact of, market hunting will have to address both economic and law enforcement issues. The importance of bushmeat to local economies is likely to be the single most important barrier to mitigating overexploitation, because producers and consumers will resist attempts to change their behavior, and governments have very little incentive to impose restrictions on bushmeat use that would further lower the welfare of their already poor citizens. To move toward sustainable use of bushmeat in the Congo Basin the supply of and demand for bushmeat need to be brought into balance. Consumer demand must be reduced, or supply available to hunters increased.

Demand-Side Mitigation Options

Consumer demand for bushmeat can be altered by changing the availability and relative price of substitutes, the price of bushmeat relative to substitutes, and consumer tastes or preferences. The first two options rest on the assumption that bushmeat is a normal good and that demand is elastic—in other words, an increase in the price of bushmeat relative to substitutes will result in a reduction in demand (Wilkie and Godoy 2000). Given that bushmeat is the primary source of animal protein for most Congo Basin families, unless substitutes are available, bushmeat demand may be inelastic and thus will not decline significantly with increasing price.

CHANGING CONSUMER PREFERENCES Consumers, in response to social marketing and education efforts, may be willing to change their preferences for luxury goods such as ivory (O'Connell and Sutton 1990) or for goods that contribute little to individual diets or household income (e.g., gorillas and chimpanzees). However, changing consumer preference is unlikely to be effective when the good being consumed satisfies some basic human need, for which substitutes are not available (Freese 1997). Given the importance of bushmeat to local diets and the absence of alternative sources of protein, changing consumer preferences through environmental education and social marketing is only likely to be effective for bushmeat consumed as a luxury item. In this case, capital city elites and restaurants should be targeted, as should expatriate consumers living abroad. Primates, given their assumed role in the spread of Ebola-like haemorrhagic diseases, may also be an effective focus for social marketing efforts to alter consumer tastes for certain types of bushmeat.

INCREASING THE AVAILABILITY OF SUBSTITUTES Though preliminary data from Bolivia and Honduras suggest that demand for bushmeat is elastic—a 1 percent decrease in the price of beef results in an 8 percent decrease in the consumption of bushmeat (Wilkie and Godoy in press)—it might be prudent to assume that it is in fact inelastic and that any attempts to increase the scarcity and price of bushmeat would not result in a reduction in demand. In fact, without available substitutes, increasing the price of bushmeat, given that it is an open-access resource, will provide economic incentives for hunters to intensify their harvest of wildlife and will encourage more individuals to enter the bushmeat trade in search of profits. Thus seeking ways to strengthen markets for nonwild sources of animal protein are critical to address the unsustainability of bushmeat hunting in the Congo Basin.

Domestic animals in Central Africa are primarily viewed as savings and insurance rather than as sources of protein. Furthermore, tsetse flies and trypanosomiasis, rather than absence of pasture, severely limit cattle raising in the region. Considerable attention has been paid, therefore, to nontraditional livestock rearing and the potential for raising bushmeat species that would provide direct substitutes for wild-harvested individuals.

Cane rat, *Thryonomys* spp., and giant rat, *Cricetomys* spp., production is possible using domestic food scraps and agricultural waste (Asibey 1974a; Jori, Mensah, and Adjamohoun 1995; Jori and Noel 1996; Tewe and Ajaji 1982). However, low production rates and handling difficulties of recently domesticated cane rats may limit their utility as livestock, at least in the short-term (D. Messinger, personal communication). Promoting small livestock production (Branckaert 1995; Hardouin 1995; NRC 1991) such as rabbit raising has proven effective in Cameroon in areas where bushmeat is already scarce (HPI 1996). Several pilot projects are underway in Gabon to raise cane rat (Jori and Noel 1996), brush-tailed porcupine, and bush pig–domestic pig hybrids to reduce demand for bushmeat in cities (Steel 1994). Small-game raising activities are also part of a United Nations Development Program/Global Environment Facility (UNDP/GEF) project in Gabon that focuses on commercial use of forest flora and fauna (Steel 1994). Raising small domesticated animals such as rabbits is attractive in that methods of husbandry and veterinary care are well-known. Feer (1993) argued that the meat productivity of pigs exceeds that of zebu cattle, which exceeds that of cane rats, which exceeds that of duikers. As a consequence, it makes more sense to promote pig production, which is well-understood, than to attempt to raise cane rats or duikers.

Pig or rabbit rearing as an alternative to bushmeat hunting is only likely to be successful, however, when the labor and capital costs of production are less than the costs of bushmeat hunting and marketing (i.e., when game becomes too scarce to be worth searching for and livestock transportation costs are not prohibitive). Of course, if domestic production of meat only becomes economically viable after wild game has become so scarce as to be unprofitable to hunt, the strategy is clearly ineffective as a conservation measure. In addition, the use of domesticated wildlife species as livestock risks the infiltration of markets with wild-caught individuals sold as "raised" meat and may promote habitat loss as forests are converted to produce fodder or pasture for livestock.

Small animal raising is viable in suburban areas that are close to sources of demand and where proximal bushmeat species populations have already been depleted (Lamarque 1995). Promotion of small livestock raising in suburban areas will of course disrupt the flow of economic benefits of urban consumption from hunters to poor rural producers and may, perversely, encourage intensification of bushmeat hunting to maximize profits before prices drop as more domestic substitutes enter the market.

Supply-Side Mitigation Options

A reduction in the scale of the commercial trade in bushmeat can also be attempted by controlling the supply of wildlife that enters the market.

CONTROLLING DOMESTIC HUNTING Controlling hunting for domestic consumption is likely to be untenable given the size of the area to be policed and the importance of bushmeat to the nutrition and economies of forest-dwelling families throughout the Congo Basin. Any attempt at legal control of household bushmeat consumption will likely fail, because households depend on bushmeat as a nutritional staple and as a source of income and are unlikely to relinquish this without considerable pressure or access to dietary and economic substitutes and because sufficient repression would require large numbers of trustworthy law enforcers. One guard per village of 50 people, paid at least U.S. $1 per day, with an additional U.S. $1 per day for equipment and supplies, would cost more than U.S. $46,720,000 per year for the Basin, assuming that 30 percent of the population is rural and 20 percent of the rural population lives in the forest. No national agency can afford this, nor is an international donor likely to finance it. Additional history shows clearly that any law that hurts the majority or the politically powerful is highly unlikely ever to be effectively implemented and policed. So whether "hunting" is illegal may be moot, if the majority of residents believe that the law has no standing or legitimacy and if they know that they are unlikely to be caught and even less likely to be punished even if they flout their activities.

Thus banning or substantially curbing bushmeat hunting for domestic consumption without providing an acceptable substitute is unrealistic from a cultural, practical, and financial viewpoint. Furthermore, stopping domestic bushmeat hunting will confirm rural communities' fears of national infringement on traditional resource use rights, which may fuel resentment toward the government and may result in retaliatory hunting of rare and endemic species and increased elephant poaching.

In a few rare cases within the Congo Basin, blocks of forest are sufficiently isolated from human settlements that they presently experience minimal human impact. If demand for bushmeat were not to increase and roads were not being built into these isolated blocks of forest, this de facto protection is, for the short to midterm, likely to be as effective as legal protection. There are currently insufficient resources and incentives for the rest of the Congo Basin to curb or halt the domestic consumption of bushmeat.

CONTROLLING MARKET SUPPLY Unlike hunting for domestic consumption, market hunting is more amenable to command-and-control measures, because as bushmeat is transferred from individual hunters to individual consumers, it is concentrated temporarily by traders who transport the meat from the forest and trade it in central sales locations. Control of market hunting can therefore ignore the numerous hunters and consumers and focus only on the far fewer bushmeat traders. Police need only to conduct random inspections along roads, railways, and at airports, or raid marketplaces on random occasions to enforce bushmeat market regulations. Guards could confiscate the bushmeat and fine the bushmeat traders. These interventions still require substantial finances to support a large, incorruptible (i.e., well-paid) corps of law enforcers, and assume that alternative distribution and marketing systems will not emerge, neither of which is likely in the Congo Basin at present.

Bushmeat does not need to be confiscated at roadblocks; police could merely charge a market tax. Taxation will raise the effective price of bushmeat and, if demand is elastic, will drive down demand. Not confiscating bushmeat avoids the need to dispose of the game in a way that does not encourage corruption (assuming that the police are not going to purloin the taxes) and prevents the sale of confiscated game at reduced prices, thus fueling demand. As traders' costs increase with taxation (even if the police steal the tax moneys), profits will fall as they attempt to keep rising prices from driving down demand. As demand and profits fall, the price that traders are willing to pay hunters will decline and the income-generating incentive for hunting will decline. Setting the bushmeat tax sufficiently high might mitigate any cultural preferences for bushmeat and consumer willingness to pay a price premium for bushmeat (Steel 1994). Market prices of bushmeat and domesticated alternatives should be monitored regularly so that the level of taxation can be maintained high enough to curb consumer demand for bushmeat. Manipulating the price of bushmeat is unlikely to have a significant impact on curbing luxury demand for bushmeat. That can only be addressed through effective law enforcement and environmental education.

Though taxation of bushmeat would appear to be an option, in the short-term the importance of bushmeat to the diet and income of forest-dwelling families, the huge areas of forest involved, the shortage of well-paid (i.e., less corruptible) and trained forestry officers, and little government interest in regulating the bushmeat trade are likely to preclude the use of command-and-control measures to limit market hunting outside more circumscribed areas such as logging concessions and protected areas.

THE SPECIAL CASE OF LOGGING CONCESSIONS Command-and-control measures may work within the confines of logging concessions, because logging companies could be required to pay for sufficient numbers of incorruptible law enforcers and to provide to them the transportation and equipment necessary for monitoring hunting. Wildlife law enforcers should not be paid directly by the logging concessions. Instead the companies should be required to post a conservation bond, paid to the appropriate government ministry, for an amount indexed to the area of forest to be exploited that year. These monies would be earmarked for natural resource conservation within logging concessions, and thus could only be used to support forestry and wildlife law enforcers and plant and animal surveyors stationed in logging concessions. Repayment of the bond to the logging concession could be indexed to the ratio of pre- and postlogging game survey figures, with the highest rebates occurring at parity. If the bond was set high enough, logging companies might comply with recommendations that wildlife and firearms laws of the country be respected by personnel of logging companies and agree that vehicles, roads, facilities, and company time should not be used in support of illegal bushmeat hunting. Using a logging company bond-financed fund, earmarked for natural resource conservation within logging concessions, would allow the development of wildlife management plans and a regulated harvest of forest protein. A conservation bond would also help strengthen national capacity and institutions to enforce wildlife protection, as Verschuren (1989) urged. This approach will only work if (1) the price of the conservation bond does not make logging uneconomic; (2) the logging companies do not attempt to bribe forestry and wildlife officers; (3) the forestry ministry establishes and enforces wildlife conservation bond legislation and uses the earmarked funds appropriately; and (4) assessment of pre- and postlogging wildlife populations is cost-effective and perceived by all parties as legitimate.

CURBING TRANSPORTATION Another approach is to control the shipment of bushmeat from the concessions to the point of sale. This directly affects the profitability of market hunting, which is largely determined by access to, and cost of, transportation. When the CIB started transporting logs to Douala from the Sangha river port at Sucambo (near Ouesso), bushmeat from Cameroon soon made up more than 13 percent of the game sold in Ouesso markets (Bennett Hennessey 1995). Yet a dispute between the trucking company and the concession in August 1995 halted traffic from Congo through Cameroon, resulting in the temporary collapse of the bushmeat market and the closure of hunting camps that border the roads (Pearce and Ammann 1998). Bushmeat marketing is, thus, a risky business. If the truck does not arrive to ship meat to the market, the meat may rot and become worthless. The key to reducing market hunting is, therefore, curbing transportation of bushmeat on logging vehicles owned by concessions and by transport companies. This could be accomplished by using roadblocks for bushmeat, assuming willingness and the capacity on the part of the national governments to enforce wildlife laws. The Wildlife Conservation Society is already working with the CIB in northern Congo on a joint project to curb the export of bushmeat from the concession. Results of this experiment will be a useful test of the effectiveness of an nongovernmental organization (NGO)/private sector collaboration to implement national wildlife laws and reduce or halt the illegal bushmeat trade within logging concessions.

Controlling Access

Another option for constraining the supply of bushmeat is to change the open-access nature of the bushmeat trade by restricting who has the right to hunt, and limiting where hunting can occur.

COMMUNITY RESOURCE MANAGEMENT An often discussed approach to game conservation in developing countries is community resource management (Bissonette and Krausman 1995; Hannah 1992; Kiss 1990; Wells, Brandon, and Hannah 1992). Though all forest resources within the Ituri forest region of the northeastern Democratic Republic of Congo (DRC) are under legal control by the government, local households have de facto management authority. Direct local ownership of game is a fact throughout much of the Congo Basin. However, for this form of management to result in wildlife conservation, communities have to be relatively small and stable, be able to defend their resource from free riders, and must not discount the future at a high rate (Becker and Ostrom 1995). Unless communities exhibit these characteristics, externalities will continue to exist, lowering the true value of forest resources and resulting in their irrational overexploitation. Poverty's "have-to-eat-today" principle (Bodmer 1994) and the absence of effective political or cooperative institutions above the household or clan level in most Congo Basin forest communities make it highly unlikely that these prerequisites for community-based resource conservation could be met in the near future.

Few governments in the Congo Basin appear ready to devolve ownership and management rights of forest resources to local communities. Furthermore, the prerequisites for common-property resource management may not evolve within forest-dwelling communities before bushmeat consumption seriously affects forest animal populations. As local communities presently have de facto control over forest resources in most of the region, it is exceedingly important that local communities are involved in developing and implementing all policies associated with sustainable management of wildlife populations. Unless local communities are advocates for bushmeat management, no command-and-control measure is likely to work, and demand-side approaches may be unacceptable or not considered worth adopting. Ignoring the human factor in the sustainable management of bushmeat is a clear recipe for failure (Stephensen and Newby 1997).

PRIVATIZATION In theory, the production of bushmeat could be increased in its native habitat by increasing the production of species that constitute the food of bushmeat species or controlling bushmeat predators and other competitors. Even if we understood enough about the biology of bushmeat species, their food sources, and their competitors to develop appropriate interventions, the open-access nature of the bushmeat trade and the problem of free-riders make investment in bushmeat production by hunters unlikely. Moreover, the costs of increasing bushmeat production in the forest would probably match or exceed the costs of livestock rearing, making it an unlikely option even if control of the forest was privatized.

Privately owned wildlife ranches, reserves, and conservancies have been able to expand biodiversity conservation outside protected areas in southern Africa. However, at present they are economically viable not from the sale of meat but through the sale of live animals to restock more recently established conservancies (Bojo 1996; Crowe et al. 1997; Kreuter and Workman 1994) and through revenue generated from trophy hunting (Leader-Williams, Kayera, and Overton 1996). Even the Hopcraft game ranch in Athi, 25 mi (40 km) from Nairobi, Kenya, and with a well-equipped slaughterhouse, could not make a profit selling bushmeat to local consumers at prices that were competitive with beef and chicken (Stelfox et al. 1983). Bushmeat ranching at Athi was only economically viable when the meat was sold to tourist hotels and restaurants at a considerable premium.

Given transportation costs, the present price structure of the bushmeat trade, and production rates of commonly exploited forest bushmeat species, it is questionable whether private bushmeat "ranchers" could afford to pay the costs of managing a ranch—for example, excluding "poachers," paying rent to local communities to cover the opportunity costs of not hunting—and yet keep harvest levels sufficiently low to be sustainable.

SPATIAL CONTROL OF HARVEST LEVELS A paper by McCullough (1996) proposed use of a changing mosaic of hunted and unhunted areas as an alternative to harvest quotas to control hunting intensity and harvest rate within a wildlife management area. The approach argues that using quotas to control hunting requires baseline data on species numbers and productivity to set quotas at sustainable levels and monitoring and law enforcement to ensure that hunters do not exceed their assigned quotas, both of which are likely to be prohibitively expensive in the

Congo Basin context. As an alternative, McCullough suggested that rather than trying to set and monitor harvest quotas it would be easier, from a management cost-effectiveness perspective, to allow hunters to take as many animals as they want but to constrain where they hunt. If hunters are allowed to hunt only within certain zones, dispersal of surplus animals produced in adjacent unhunted zones could compensate for the individuals harvested by hunters. By increasing or decreasing the relative proportion of hunted to unhunted areas in response to time-series data on harvest levels the manager can establish the maximum area that can be hunted without resulting in declining harvest levels.

Spatial control of hunting certainly requires fewer data to maintain harvests at sustainable levels. However, in the Congo Basin context, where travel is the primary cost associated with hunting, convincing hunters to bypass reserve areas to hunt in more distant harvest zones is unlikely unless law enforcement is ubiquitous and strict and penalties a sufficient deterrent. Most hunters are central-place foragers, and hunting intensity declines with distance from their home (Wilkie 1989; Wilkie and Curran 1991). Establishing buffer zones around settlements that approximate the average distance that hunters travel may be a viable alternative to spatially distributed hunting zones if protected areas bordering the hunting zones are sufficiently large to serve as dispersal reservoirs. The width of the buffer zone could be increased or decreased in response to harvest returns monitored over time. Buffer zones may be more practical in terms of enforcement. However, few areas in the Congo Basin are likely to be large enough to leave core (unhunted) areas of sufficient size to restock hunted areas that meet even current demand for bushmeat. In the Ituri forest of northeastern DRC, even the 3,211,000 a (1,300,000 ha) Okapi Wildlife Reserve appears too small to provide a sustainable supply of bushmeat to meet domestic demand from the area's 30,000 inhabitants (Wilkie et al. 1998a, 1998b). Simply stated, in most forests of the Congo Basin zoning is only likely to result in sustainable hunting if off-take is lower than the present level of demand. Lowering demand for bushmeat is therefore the key to wildlife conservation in the region.

CONCLUSION

This review shows that uncertainty still exists in our ability to quantify bushmeat consumption, off-take rates, and production, and that the evidence is overwhelming that present bushmeat hunting is unsustainable for most primates and large-bodied forest duikers and may only be sustainable for highly productive animals such as rodents.

The first message suggests that further study is needed. Yet much of the uncertainty that exists in the data is less a result of inadequate effort and more a statement of the difficulties associated with studying tropical forest animals. Though scientists are wont to delay taking action until they have enough information or reliable figures, the effects of environmental uncertainties and measurement errors ensure that correct or exact numbers are rarely obtainable. As a consequence, the substantial increase in effort necessary to enhance our confidence in the data is probably not worth the investment, particularly as it is unlikely to alter the second message—that bushmeat hunting is probably unsustainable if present trends in population growth and forest access continue.

Thus it makes most sense to spend scarce resources on mitigation rather than further study to assess the impacts of hunting. There are three categories of options for mitigation: (1) interdiction, (2) bushmeat price manipulation through fines and taxation, and (3) development of substitutes. No single solution is likely to be effective in all contexts; rather, the relative importance of each approach is likely to change with land use and population density. For example, whereas interdiction may predominate in sources of supply such as protected areas and logging concessions, taxation and provision of substitutes may be more effective near urban sources of demand.

The importance of bushmeat in the diet and economies of Congo Basin families, the high demand for bushmeat, the lack of effective substitutes, and political resistance to controlling bushmeat hunting make command-and-control measures such as interdiction, fines, and taxation unlikely to be implemented effectively. As a consequence, if we are concerned about conservation of a globally scarce resource that is still relatively abundant locally, it is essential that (1) we develop a better understanding of the elasticity of bushmeat demand, (2) that pilot bushmeat substitution projects are supported and their impact on demand evaluated, and (3) social marketing activities are put in place to attempt to deflect consumer preferences for animal protein away from bushmeat species that are particularly susceptible to overexploitation.

The few options available to mitigate bushmeat hunting reemphasize the importance of protected areas where, unlike the majority of forested areas, biodiversity conservation is the primary objective in land use. Thus strategic and sufficient financing of protected areas is going to be critical in ensuring that a representative sample of forest wildlife continues to inhabit the Congo Basin in the future.

Given the role that timber exploitation plays in facilitating intensive market hunting in the farthest reaches of the Congo Basin, donors and international NGOs must seek ways to work with concessions to minimize the wildlife impacts of logging. Lobbying, green conditionality (environmental conditions) on loans from bilateral or multilateral donors, green labeling (labels indicating that a product was produced using environmentally sustainable methods), and consumer preference may be effective in encouraging logging companies, particularly those with European home offices, to manage wildlife populations sustainably within their concessions.

ACKNOWLEDGMENTS

Thanks to the U.S. World Wildlife Fund, the Wildlife Conservation Society, Ecosystèmes Forestiers d'Afrique Centrale (ECOFAC), and the Avenir des Peuples des Forêts Tropicales (APFT) for access to unpublished reports. This study was supported by funding from the U.S. Agency for International Development Central African Regional Program for the Environment (CARPE) project, and by National Science Foundation Grant SBR-940-2034.

REFERENCES

Ambrose-Oji, B. 1997. Valuing forest products from Mount Cameroon. In S. Doolan (ed.), *African rainforests and the conservation of biodiversity: Proceedings of the Limbe Conference* (pp. 140–50). Oxford: Earthwatch Europe.

Anadu, P. A., Elamah, P. O., and Oates, J. F. 1988. The bushmeat trade in southwestern Nigeria: A case study. *Human Ecology* 16: 199–208.

Anstey, S. 1991. *Wildlife utilization in Liberia.* Gland, Switzerland: World Wildlife Fund International.

Asibey, E. O. A. 1974a. The grasscutter. *Thyronomys swinderianus,* Temmick, in Ghana. *Symposium of the Zoological Society of London* 34: 161–70.

———. 1974b. Wildlife as a source of protein in Africa south of the Sahara. *Biological Conservation* 6: 32–39.

———. 1977. Expected effects of land-use patterns on future supplies of bushmeat in Africa south of the Sahara. *Environmental Conservation* 4: 43–49.

Aunger, R. 1992. *An ethnography of variation: Food avoidance among horticulturists and foragers in the Ituri forest, Zaire.* Doctoral dissertation, University of California, Los Angeles.

Auzel, P. 1996. *Agriculture/extractivisme et exploitation forestière. Étude de la dynamique des modes d'exploitation du milieu dans le nord de lUFA de Pokola, nord Congo.* Bomassa, Republic of Congo: Wildlife Conservation Society/GEF Congo.

Auzel, P., and Wilkie, D. S. 2000. Wildlife use in northern Congo: Hunting in a commercial logging concession. In J. G. Robinson and E. L. Bennett (eds.), *Evaluating the sustainability of hunting in tropical forests* (pp. 414–26). New Haven, CT: Yale University Press.

Bailey, R. C., and Peacock, N. R. 1988. Efe pygmies of northeast Zaire: Subsistence strategies in the Ituri forest. In I. de Garine and G. A. Harrison (eds.), *Uncertainty in the food supply* (pp. 88–117). Cambridge: Cambridge University Press.

Becker, D. S., and Ostrom, E. 1995. Human ecology and resource sustainability: The importance of institutional diversity. *Annual Review of Ecology and Systematics* 26: 113–33.

Bennett Hennessey, A. 1995. *A study of the meat trade in Ouesso, Republic of Congo.* Bronx, NY: Wildlife Conservation Society.

Bissonette, J. A., and Krausman, P. R. 1995. *Integrating people and wildlife for a sustainable future.* Bethesda, MD: Wildlife Society.

Bodmer, R. E. 1994. Managing wildlife with local communities in the Peruvian Amazon: The case of the Reserva Comunal Tamshiyacu-Tahuayo. In D. Western, R. M. Wright, and S. C. Strum (eds.), *Natural connections: Perspectives in community-based conservation* (pp. 113–24). Washington, D.C.: Island Press.

Bojo, J. 1996. *The economics of wildlife: Case studies from Ghana, Kenya, Namibia, and Zimbabwe.* Washington, D.C.: World Bank.

Bowen-Jones, E. 1997. *A review of the current depth of knowledge on the commercial bushmeat trade (with emphasis on central/west Africa and the great apes).* Cambridge: Ape Alliance/Flora and Fauna International.

———. 1998. *The African bushmeat trade—A recipe for extinction.* Cambridge: Ape Alliance/Fauna and Flora International.

Branckaert, R. D. 1995. Minilivestock: Sustainable animal resource for food security. *Biodiversity and Conservation* 4: 336–38.

Chardonnet, P., Fritz, H., Zorzi, N., and Feron, E. 1995. Current importance of traditional hunting and major contrasts in wild meat consumption in sub-Saharan Africa. In J. A. Bissonette and P. R. Krausman (eds.), *Integrating people and wildlife for a sustainable future* (pp. 304–307). Bethesda, MD: Wildlife Society.

Colyn, M. M., Dudu, A., and ma Mbalele, M. 1987. Data on small and medium scale game utilization in the rain forest of Zaire. In WWF, *Wildlife management in sub-Saharan Africa: Sustainable economic benefits and contribution towards rural development* (pp. 109–45). Harare, Zimbabwe: World Wide Fund for Nature.

Crowe, T. M., Smith, B. S., Little, R. M., and High, S. H. 1997. Sustainable utilization of game at Rooipoort estate, northern Cape Province, South Africa. In C. H. Freese (ed.), *Harvesting wild species: Implications for biodiversity conservation* (pp. 359–92). Baltimore: Johns Hopkins University Press.

de Garine, I. 1993. Food resources and preferences in the Cameroonian forest. In C. M. Hladik, A. Hladik, O. F. Linares, H. Pagezy, A. Semple, and M. Hadley (eds.), *Tropical forests, people and food: Biocultural interactions and applications to development* (pp. 561–74). Paris: UNESCO.

Delvingt, W. 1997. *La chasse villageoise: Synthèse régionale des études réalisées durant la première phase du Programme ECOFAC au Cameroun, au Congo, et en République Centrafricaine.* Gembloux, Belgium: ECOFAC AGRECO-CTFT, Faculté Universitaire des Sciences Agronimiques des Gembloux.

Dethier, M. 1995. *Étude chasse.* Yaounde, Cameroon: ECOFAC.

Dubost, G. 1980. L'ecologie et la vie sociale du Cephalophe bleu (*Cephalophus monticola,* Thunberg), petit ruminant forestier africain. *Zeitschrift für Tierpsychologie* 54: 205–66.

East, E. 1995. *Antelopes—Global survey and regional action plan: Part 3—West and Central Africa.* Gland, Switzerland: International Union for the Conservation of Nature and Natural Resources.

Estes, R. D. 1991. *The behavior guide to African mammals: Including hoofed mammals, carnivores, primates.* Berkeley: University of California Press.

Eves, H. E. 1995. *Socioeconomics of natural resource utilization in the Kabo logging concession northern Congo.* New York: Wildlife Conservation Society.

Fa, J. E., Juste, J., Perez del Val, J., and Castroviejo, J. 1995. Impact of market hunting on mammal species in Equatorial Guinea. *Conservation Biology* 9: 1107–15.

Feer, F. 1993. The potential for sustainable hunting and rearing of game in tropical forests. In C. M. Hladik, A. Hladik, O. F. Linares, H. Pagezy, A. Semple, and M. Hadley (eds.), *Tropical forests, people and food: Biocultural interactions and applications to development* (pp. 691–708). Paris: UNESCO.

Freese, C. H. 1997. *Harvesting wild species: Implications for biodiversity conservation.* Baltimore: Johns Hopkins University Press.

Gally, M., and Jeanmart, P. 1996. *Étude de la chasse villageoise en forêt dense humide d'Afrique centrale. Travail de fin d'études.* Gembloux, Belgium: Faculté Universitaire des Sciences Agronomiques de Gembloux.

Geist, V. 1988. How markets for wildlife meat and parts, and the sale of hunting privileges, jeopardize wildlife conservation. *Conservation Biology* 2: 15–26.

Haltenorth, T., and Diller, H. 1980. *A field guide to the mammals of Africa including Madagascar.* London: Collins.

Hannah, L. 1992. *African people, African parks: An evaluation of development initiatives as a means of improving protected area conservation in Africa.* Washington, D.C.: Conservation International.

Harcourt, A. H. 1980. Gorilla-eaters of Gabon. *Oryx* 15: 248–51.

Hardouin, J. 1995. Minilivestock: From gathering to controlled production. *Biodiversity and Conservation* 4: 220–32.

Hart, J. A. 1978. From subsistence to market: A case study of the Mbuti net hunters. *Human Ecology* 6: 32–53.

Heymans, J. C., and Maurice, J. S. 1973. Introduction a l'exploitation de la faune comme ressource alimentaire en Republique du Zaire. *Forum Universitaire* 2: 6–12.

Hladik, C. M., Bahuchet, S., and de Garine, I. 1990. *Food and nutrition in the African rain forest.* Paris: UNESCO-MAB.

Hladik, C. M., and Hladik, A. 1990. Food resources of the rain forest. In C. M. Hladik, S. Bahuchet, and I. de Garine (eds.), *Food and nutrition in the African rain forest* (pp. 14–18). Paris: UNESCO-MAB.

Hochschild, A. 1998. *King Leopold's ghost: A story of greed, terror, and heroism in colonial Africa.* New York: Houghton-Mifflin.

Heifer Project International (HPI). 1996. *Boyo rural integrated farmer's alliance, Cameroon: Project summary.* Little Rock, AK: Author.

Ichikawa, M. 1983. An examination of the hunting dependent life of the Mbuti pygmies, eastern Zaire. *African Studies Monographs* 4: 55–76.

Infield, M. 1988. *Hunting, trapping and fishing in villages within and on the periphery of the Korup National Park.* Gland, Switzerland: World Wildlife Fund.

Jori, F., Mensah, G. H., and Adjamohoun, E. 1995. Grasscutter production: An example of rational exploitation of wildlife. *Biodiversity and Conservation* 4: 257–65.

Jori, F., and Noel, J. M. 1996. *Guide pratigue d'élevage d'aulacodes au Gabon.* Lyon, France: Vétérinaires Sans Frontières.

Juste, J., Fa, J. E., Perez del Val, J., and Castroviejo, J. 1995. Market dynamics of bushmeat species in Equatorial Guinea. *Journal of Applied Ecology* 32: 454–67.

Kalivesse, A. 1991. Supplying the local Bangui markets with bushmeat. *Nature et Faune* 7: 39–42.

King, S. 1994. Utilisation of wildlife in Bakossiland, West Cameroon with particular reference to primates. *TRAFFIC Bulletin* 14: 63–73.

Kingdon, J. 1997. *The Kingdon field guide to African mammals.* San Diego, CA: Academic Press.

Kiss, A. 1990. *Living with wildlife: Wildlife resource management with local participation in Africa.* Washington, D.C.: World Bank. Technical Paper No. 130.

Klemens, M. W., and Thorbjarnarson, J. B. 1995. Reptiles as a food resource. *Biodiversity and Conservation* 4: 281–98.

Koppert, G. J. A., Dounias, E., Froment, A., and Pasquet, P. 1996. Consommation alimentaire dans trois populations forestières de la région côtière du Cameroun: Yassa, Mvae et Bakola. In C. M. Hladik, A. Hladik, and H. Pagezy (eds.), *L'Alimentation en forêt tropicale. Interactions bioculturelles et perspectives de développement* (pp. 477–96). Paris: Orstom.

Koppert, G. J. A., and Hladik, C. M. 1990. Measuring food consumption. In C. M. Hladik, S. Bahuchet and I. de Garine (eds.), *Food and nutrition in the African rain forest* (pp. 59–61). Paris: UNESCO-MAB.

Koster, S. H., and Hart, J. A. 1988. Methods of estimating ungulate populations in tropical forests. *African Journal of Ecology* 26(2): 117–27.

Kreuter, U. P., and Workman, J. P. 1994. Costs of overstocking on cattle and wildlife ranches in Zimbabwe. *Ecological Economics* 11: 237–48.

Lahm, S. A. 1993a. *Ecology and economics of human/wildlife interaction in northeastern Gabon.* Doctoral dissertation, New York University, New York.

———. 1993b. Utilization of forest resources and local variation of wildlife populations in northeastern Gabon. In C. M. Hladik, A. Hladik, O. F. Linares, H. Pagezy, A. Semple, and M. Hadley (eds.), *Tropical forests, people and food: Biocultural interactions and applications to development* (pp. 213–26). Paris: UNESCO.

———. 1994. *Hunting and wildlife in northeastern Gabon: Why conservation should extend beyond protected areas.* Makokou, Gabon: Institut de Recherche en Ecologie Tropicale.

Lamarque, F. A. 1995. The French co-operation's strategy in the field of African wildlife. In J. A. Bissonette and P. R. Krausman (eds.), *Integrating people and wildlife for a sustainable future* (pp. 267–70). Bethesda, MD: Wildlife Society.

Laurent, E. 1992. *Wildlife utilization survey of villages surrounding the Rumpi Hills Forest Reserve.* Mundemba, Cameroon: Deutsche Gesellschaft für Technische Zusammenarbeit.

Leader-Williams, N., Kayera, J. A., and Overton, G. L. 1996. *Tourism hunting in Tanzania.* Gland, Switzerland: International Union for the Conservation of Nature and Natural Resources.

ma Mbalele, M. 1978. Part of African culture. *Unasylva* 29: 16–17.

Malonga, R. 1996. *Dynamique socio-economique du circuit commerical de viande de chasse à Brazzaville.* Bronx, NY: Wildlife Conservation Society.

Mares, M. A., and Ojeda, R. A. 1984. Faunal commercialization and conservation in South America. *Bioscience* 34: 580–84.

Martin, G. H. G. 1983. Bushmeat in Nigeria as a natural resource with environmental implications. *Environmental Conservation* 10: 125–34.

McCullough, D. R. 1996. Spatially structured populations and harvest theory. *Journal of Wildlife Management* 60: 1–9.

McRae, M. 1997. Road kill in Cameroon. *Natural History* 2: 36–47, 74–75.

Ngnegueu, P. R., and Fotso, R. C. 1996. *Chasse villageoise et conséquences pour la conservation de la biodiversité dans la réserve de biosphère du Dja.* Yaounde, Cameroon: ECOFAC.

Njiforti, H. L. 1996. Preferences and present demand for bushmeat in northern Cameroon: Some implications for wildlife conservation. *Environmental Conservation* 23: 149–55.

Noss, A. J. 1995. *Duikers, cables and nets: A cultural ecology of hunting in a central African forest.* Doctoral dissertation, University of Florida, Gainesville.

———. 1998. Cable snares and bushmeat markets in a Central African forest. *Environmental Conservation* 25: 228–33.

National Research Council (NRC). 1991. *Microlivestock: Little-known small animals with a promising economic future.* Washington, D.C.: National Academy Press.

Oates, J. F. 1996. *African primates: Status survey and conservation action plan.* Gland, Switzerland: International Union for the Conservation of Nature and Natural Resources.

Oates, J. F., and Davies, A. G. 1986. Primate conservation in West Africa. *Primate Eye* 29: 20–24.

O'Connell, M. A., and Sutton, M. 1990. *The effect of trade on international commerce in Africa elephant ivory.* Washington, D.C.: World Wildlife Fund.

Payne, J. C. 1992. *A field study of techniques for estimating densities of duikers in Korup National Park, Cameroon.* Master's thesis, University of Florida, Gainesville.

Pearce, J. 1996a. A bridge too far. *Animals International* 53: 18–20.

———. 1996b. *Wildlife and timber exploitation in Gabon: A case study of the Leroy concession, Forest des Abeilles.* London: World Society for the Protection of Animals.

Pearce, J., and Ammann, K. 1998. *Slaughter of the apes: How the tropical timber industry is devouring Africa's great apes.* London: World Society for the Protection of Animals.

Pierret, P. V. 1975. *La place de la faune dans le relèvement du niveau de vie rurale au Zaire.* Kinshasa, DRC: Institut Zairois pour la Conservation de la Nature.

Robinson, J. G., and Redford, K. H. 1994. Measuring the sustainability of hunting in tropical forests. *Oryx* 28: 249–56.

Slade, N. A., Gomulkiewicz, R., and Alexander, H. M. 1998. Alternatives to Robinson and Redford's method for assessing overharvest from incomplete demographic data. *Conservation Biology* 12: 148–55.

Steel, E. A. 1994. *Study of the value and volume of bushmeat commerce in Gabon.* Libreville, Gabon: World Wildlife Fund.

Stelfox, J. B., Sisler, D. G., Hudson, R. J., and Hopcraft, D. 1983. *A comparison of wildlife and cattle ranching on the Athi Plains, Kenya.* Mimeo.

Stephensen, P. J., and Newby, J. E. 1997. Conservation of the Okapi Wildlife Reserve, Zaire. *Oryx* 31: 49–58.

Takeda, J., and Sato, H. 1993. Multiple subsistence strategies and protein resources of horticulturists in the Zaire basin: The Ngandu and Boyela. In C. M. Hladik, A. Hladik, O. F. Linares, H. Pagezy, A. Semple, and M. Hadley (eds.), *Tropical forests, people and food: Biocultural interactions and applications to development* (pp. 497–504). Paris: UNESCO.

Tewe, G. O., and Ajaji, S. S. 1982. Performance and nutritional utilization by the African giant rat (*Cricetomys gambianus,* W.) on household waste of local foodstuffs. *African Journal of Ecology* 20: 37–41.

Verschuren, J. 1989. Habitats mammals and conservation in the Congo. *Bulletin de l'Institut Royal des Sciences Naturelles de Belgique Biologie* 59: 169–80.

Wells, M. P., Brandon, K., and Hannah, L. 1992. *People and parks: Linking protected area management with local communities.* Washington, D.C.: World Bank.

White, L. J. T. 1994. Biomass of rain forest mammals in the Lope Reserve, Gabon. *Journal of Animal Ecology* 63: 499–512.

Wildlife Conservation Society (WCS). 1996. *The Lobéké forest southeast Cameroon: Summary of activities—1988–1995.* Bronx, NY: Author.

Wilkie, D. S. 1989. Impact of roadside agriculture on subsistence hunting in the Ituri forest of northeastern Zaire. *American Journal of Physical Anthropology* 78: 485–94.

Wilkie, D. S., and Carpenter, J. F. 1999. Bushmeat hunting in the Congo Basin: An assessment of impacts and options for mitigation. *Biodiversity and Conservation* 8: 927–55.

Wilkie, D. S., and Curran, B. 1991. Why do Mbuti hunters use nets? Ungulate hunting efficiency of bows and nets in the Ituri rain forest. *American Anthropologist* 93: 680–89.

Wilkie, D. S., Curran, B., Tshombe, R., and Morelli, G. A. 1998a. Managing bushmeat hunting in the Okapi Wildlife Reserve, Democratic Republic of Congo. *Oryx* 32: 131–44.

———. 1998b. Modeling the sustainability of subsistence farming and hunting in the Ituri Forest of Zaire. *Conservation Biology* 12: 137–47.

Wilkie, D. S., and Finn, J. T. 1990. Slash–burn cultivation and mammal abundance in the Ituri forest, Zaire. *Biotropica* 22: 90–99.

Wilkie, D. S., and Godoy, R. A. 2000. Economics of bushmeat. *Science* 287: 973.

———. in press. Can consumer demand for bushmeat be reduced? Income and price elasticities of demand in lowland Amerindian societies. *Conservation Biology.*

Wilkie, D. S., Sidle, J. G., Boundzanga, G. C., Blake, S., and Auzel, P. in press. Defaunation or deforestation: Commercial logging and market hunting in northern Congo. In R. Fimbel, A. Grajal, and J. G. Robinson (eds.), *The impacts of commercial logging on wildlife in tropical forests.* New York: Columbia University Press.

Wilson, V. J., and Wilson, B. L. P. 1989. *A bushmeat market and traditional hunting survey in south-west Congo.* Bulawayo, Zimbabwe: Chipangali Wildlife Trust.

———. 1991. La chasse traditionelle et commerciale dans le sud-oest du Congo. *Tauraco Research Report* 4: 279–88.

Section 2

GREAT APES IN CAPTIVITY

TARA S. STOINSKI, JACQUELINE J. OGDEN,
KENNETH C. GOLD, AND TERRY L. MAPLE

5
CAPTIVE APES AND ZOO EDUCATION

Current estimates of the impact of bushmeat hunting and loss of rain forest habitat suggest that some species of African apes may be extinct before the middle of the new century (e.g., Conway 1999). Given these and other threats, we believe that zoos have a mandate to educate guests about great apes in general and about the complex and emotive issues of ape conservation. Although all member institutions in the American Zoo and Aquarium Association (AZA) are committed to such an educational responsibility (IUDZG and IUCN/SSC 1993), their effectiveness in accomplishing this priority is unclear. Our goal in this chapter is to examine the educational efficacy (defined as the power to produce an effect or change) of zoos and discuss needed improvements and future directions. Because the focus of this volume is on great apes, the most desirable approach would be to focus on the educational efficacy of this taxonomic group. Unfortunately, few of the data available address apes. Thus we will discuss zoo education programs more generally and add available information on the great apes when possible.

THE ROLE OF ZOOS AS EDUCATORS

In our expanding urban societies, zoos (defined as in Regan 1995, 38, as "A professionally managed zoological institution accredited by the AZA and having a collection of live animals used for conservation, scientific studies, public education, and public display") are increasingly important for contact with wild creatures (Conway 1969; Maple, McManamon, and Stevens 1995). The need for such contact remains strong (Kreger and Mench 1995a, 1995b), and the effect of such connections should not be undervalued. Desmond Morris (1968, 78), in describing his first visit to a zoo, summarizes perfectly the role of zoos in modern society:

> That visit did more for my later interest in animals than a hundred films or a thousand books. The animals were real and near. . . . If zoos disappear, I fear that our vast urban population will become so physically remote from animal life that they will eventually cease to care about it.

Not everyone, however, supports the idea of zoos as effective educational institutions. For example, Robert Sommer disagreed about the role of zoos in education:

> Unless one's goal is to learn about the effects of confinement, the educational value of the zoo is probably more negative than positive. Despite excellent intentions, it is likely that even the best public zoos are creating stereotypes about animal behavior that are not only incorrect but work against the interests of wildlife preservation. (1974, 67)

In the almost 25 years since Sommer made that statement, prominent zoos and aquariums have changed dramatically. In fact, in a presentation given at the 1999 Environmental Design Research Association meeting (Sommer 1999), Sommer acknowledged these significant changes and commended zoos for their efforts. However, actual evidence of changes in educational impact may be harder to come by, especially because zoos are places of informal learning, which is difficult to assess (Kellert and Dunlap 1989). Because of this, we will look beyond traditional cognitive measures of learning in this chapter and also examine how zoos' influence attitudes/affect toward wildlife and conservation-related attitudes and behaviors.

THE EDUCATIONAL EFFICACY OF MODERN ZOOS FOR THE PUBLIC

Surveys have clearly demonstrated that people view environmental education as one of the primary roles of zoos and aquariums (Birney 1986, cited in Tunnicliffe 1996; Kellert 1980, 1999; Kreger and Mench 1995a, 1995b; but see Rosenfeld 1980). While at the zoo, visitors tend to prefer educational opportunities that involve interactive experiences with both people and animals (Chicago Zoological Society and Lincoln Park Zoological Society 1993; Gold and Benveniste 1995; People, Places, and Design 1992; Wolf and Tymitz 1979). Zoological parks provide a variety of these experiences, including immersion in naturalistic exhibits; outreach programs; live animal demonstrations; participatory displays such as interactive graphics, biofacts, and artifacts; formal education programs; technology-assisted programs; and interactions with staff and docents. The question then becomes, "Are these effective in changing the knowledge, affect/attitudes, and behavior of zoo visitors?"

Cognitive Learning

Research evaluating the cognitive impact of education programs in zoos and other informal learning centers, such as museums or nature centers, has demonstrated several ways such institutions achieve cognitive gain. For example, just the presence of live animals can produce positive cognitive gains that are superior to those obtained with animal alternatives such as biofacts and mounted specimens (MacMillen 1994; Sherwood, Rallis, and Stone 1989). In addition, the way animals are displayed in zoos affects cognitive learning. Whereas zoos used to house animals taxonomically in sterile environments that bore no relationship to their natural habitat, naturalistic exhibits displaying animals in species-typical associations are becoming more common (Polakowski 1987). Data on such exhibits suggest they can increase staying time at an exhibit (Ogden, Capanzano, and Maple 1994; Price, Ashmore, and McGivern 1994); improve the understanding of relationships between natural habitats and their inhabitants (Burton 1990; Ford and Burton 1991); and increase knowledge of animal behavior (Burton 1990; Ford and Burton 1991; Price et al. 1994) compared with traditional zoo exhibits. In addition, most zoos include interactive or participatory activities in exhibits, such as graphics, games, and demonstrations. These activities significantly increase both learning (Bielick and Doering 1997; Derwin and Piper 1988; Heinrich and Birney 1991; Yerke and Burns 1991) and learning-related behaviors, such as asking or answering questions, reading exhibit text, or explaining the exhibit to others (Borun et al. 1997). Finally, formal zoo education programs that are integrated into school curricula or include previsit preparation also enhance cognitive gains (Gutierrez de White and Jacobson 1994).

Although evaluations of formal aspects of zoo visits, such as education programs and live animal demonstrations, demonstrate cognitive change, most studies of informal zoo visits show little or no increase in wildlife knowledge (Gutierrez de White and Jacobson 1994; Kellert 1984; Kellert and Dunlap 1989; Koran and Baker 1979; Parsons 1999; Wolf and Tymitz 1979). Clearly, more data are needed to identify variables associated with positive cognitive gain. For example, what is the effect of structured versus unstructured zoo visits? How does the overall effect of a zoo visit compare with the effect of specific exhibits or aspects of a zoo visit? Are there differences in the way different types of visitor groups (i.e., family groups compared to school groups) learn? Educational evaluation in zoos is just beginning to address such points. For example, results from Bielick and Karns (1998) suggest that cognitive gain in adults may be influenced by the presence of children during their visit. Such information could lead to exhibit design that maximizes educational efficacy for a variety of visitor types.

Affect/Attitudes toward Animals

The affective aspect of zoo learning may be more important than cognitive learning in changing peoples' perception of animals and their desire to conserve them (Gutierrez de White and Jacobson 1994). Unfortunately, as with cognitive gains, there are limited data examining change in affect as a result of zoo visits. We do know that the presence of live animals sparks positive affective responses among children (Birney 1996; Tunnicliffe 1996) and can produce significant affective change (Davison et al. 1993; Morgan and Gramann 1989; Yerke and Burns 1993). Such results have not been found with animal alternatives (Sherwood et al. 1989). In addition, the way zoos display animals can change affect/attitudes. Naturalistic exhibits increase positive feelings toward the exhibit (Bitgood, Ellingsen, and Patterson 1990; Finlay, James, and Maple 1988; Ford and Burton 1991; Shettel-Neuber 1988), the animals (Finlay et al. 1988; Ford and Burton 1991; Gold 1998; Rhoads and Goldsworthy 1979), and the welfare of the animals (Price et al. 1994) compared with traditional exhibits. Naturalistic exhibits may also positively affect perceptions of animals, such as viewing an animal as wild rather than tame (Gold in press; Wolf and Tymitz 1979). However, naturalistic exhibits are not required for affective change to occur. A majority of visitors to "Think Tank," an exhibit at the National Zoo in Washington, D.C., that focuses on the science of animal thinking and contains little aesthetic naturalism, reported the exhibit positively influenced the way they thought about animals (Bielick and Karns 1998).

In contrast to the studies mentioned, Kellert and Dunlap (1989) found that informal visits to a zoo produced no clear change in attitudes toward animals. The relationships between zoo experiences and attitudes/affect toward animals clearly are not simple, and, as with cognitive learning, more research is needed to address these questions.

Attitudes toward Conservation and Conservation-Related Behavior

The rapid loss of habitats and species makes it imperative that zoos influence conservation-related attitudes and behavior, in addition to the more traditional cognition and affect. However, we have even less information on the effect of zoo experiences on conservation-related areas than on cognitive and affective changes. In terms of visitor interest in conservation, the data are mixed. Studies by Tunnicliffe (1994, 1995a, 1995b) reported that few visitors spontaneously discuss conservation issues while at the zoo, but Gold and Benveniste (1995) found that interest by zoo visitors in ape conservation was second only to interest in ape intelligence. The public's view of zoos as conservation centers is perhaps more positive. Although Birney (1986, cited in Tunnicliffe 1996) reported that few visitors described zoos as serving a conservation purpose, a decade later Reade and Waran (1996) found that conservation was considered to be the main role of the zoo.

With respect to how zoo visits affect attitudes toward conservation, we do know that, as with cognitive and affective learning, the manner in which zoos display their animals can influence conservation interest. Ogden and her colleagues (1994) found a self-reported increase in conservation interest following a visit to a naturalistic gorilla exhibit. Similarly, Ford and Burton (1991) found that concern for conservation was higher at naturalistic gorilla exhibits than at traditional exhibits, with greater understanding by viewers of the need for habitat preservation. Finally, we also know that specific aspects of zoo visits can improve attitudes toward conservation. Both Yerke and Burns (1991) and Swanagan (1993) found a significant, proconservation attitude shift after exposure to live animal demonstrations.

Unfortunately, few studies have gone the next step to examine how interest in conservation affects conservation-related behavior. A study by the Chicago and Lincoln Park Zoological Societies found that zoo visitors had higher conservation knowledge scores than nonvisitors and that conservation knowledge was positively correlated with environmentally responsible consumer behavior, voting behavior, and participation in conservation-oriented volunteer activities (Chicago Zoological Society and Lincoln Park Zoological Society 1993). Similarly, Swanagan (1993) found that frequent zoo visitors, particularly those with a participatory experience, such as viewing an animal demonstration or having animal contact, were more likely to engage in conservation-related action than nonfrequent visitors or visitors with only a passive experience of viewing animals at a distance. Finally, the Brookfield Zoo found that visitors who reported spending a substantial amount of time in the zoo's wetland exhibit showed significantly higher interest in participating in wetland conservation behaviors than visitors who reported spending little or no time in the exhibit (Rabb and Saunders 1999). Like the data examining cognitive and affective change, these data also suggest that although some aspects of zoo visits may influence conservation-related behaviors, zoos could be doing more to ensure that conservation-related behavioral change occurs in all visitors.

ZOOS AS VEHICLES FOR TEACHING, RESEARCH, AND CONSERVATION

Although zoo and aquarium educators recognize the visiting public as their largest educational audience (AZA 1999; Coe 1995; Gold 1997), there are educational opportunities beyond those available to the general visitor. For example, more than nine million kindergarten through 12th-grade students participate annually in on-site, formal zoo education programs (AZA 1999). Zoos also serve as laboratories for high schools and for college courses in conservation biology, comparative psychology, experimental analysis of behavior, environmental design, population biology, and zoo biology and management (e.g., Carter 1998; Greenburg 1987; Lukas, Marr, and Maple 1998; Maple and Hoff 1980; Stoinski, Lukas, and Maple 1998a; Wineman and Stoinski 1999). Finally, zoos provide professional training opportunities for college students through on-site research programs, many of which provide funding for these students (Stoinski et al. 1998b).

Another educational opportunity provided by zoos is scientific advancement though basic research. Although much zoo research is applied, basic or theoretical research questions, such as those pertaining to animal learning and cognition and proximate mechanisms of behavior, can also be explored in a zoo setting (Glatston and Hosey 1997; Hosey 1997). Unfortunately, zoos appear to be underused by scientists for basic research (Hosey 1997), despite the vast majority of zoos permitting outside scientists to conduct research (Stoinski et al. 1998a). Thus zoos should work together with the academic community to "promote research which is of common interest but whose goal goes beyond just the better maintenance of animals in captivity" (Hosey 1997, 206).

Finally, as zoo research can extend beyond basic scientific knowledge to the conservation of wild animals, zoos provide educational opportunities to further our understanding of conservation methodologies and strategies. Although some scientists argue that the artificiality of zoos prevents meaningful conservation research, we agree with Glatston and Hosey (1997) that quality, conservation-relevant research can be conducted in zoos. For example, concepts and techniques developed in captivity can also help with conservation in the field, especially in the areas of satellite telemetry, health, and small population management. Such work provides two educational opportunities. First, scientists increase their knowledge about effective conservation techniques. Second, as more zoos become involved in conservation (especially in situ conservation), the stories of these partnerships can be shared with zoo visitors through exhibit design and interpretive displays (Gold 1997), which in turn can generate funding for conservation (Conway 1999).

THE FUTURE OF EDUCATION IN ZOOS

Having discussed the efficacy of educational opportunities at modern zoos, we would now like to address how zoo education programs might change to become more effective in the future.

Content of Educational Messages

Zoos are places of informal learning; each visitor brings his or her own set of experiences and perceptions and chooses what parts of the zoo to see. Thus both the content and delivery of zoo messages need to be carefully developed to ensure that all visitors receive the intended information. It has been argued that zoos have not adopted such an approach. For example, Tunnicliffe (1995a, 1995b) claimed that zoo education messages, particularly those related to conservation, are often beyond the scientific understanding of the visitor. She maintained zoos should focus their educational messages at the level of visitor interest, including the categorization and attributes of animals. Similarly, we feel that there are topics in which the public may be interested that zoo educational messages should focus on, including animal welfare and related aspects of zoo management and the reality of life in the wild.

In terms of animal welfare, most zoo visitors are unaware of the high level of care that zoo animals receive, yet are very interested in learning about it (Kreger and Mench 1995b). Local welfare issues pertaining to captive animals outside the zoo, such as biomedical animal research, roadside zoos, primates as pets, and circuses could also be addressed in educational messages.

Zoos also need to educate the public honestly about their animal management, acquisition, and disposition policies (Norton et al. 1995). For example, zoos often cite their involvement in AZA Species Survival Programs (SSPs), which are designed to manage threatened or endangered species "collectively for genetic variability, demographic stability, behavioral compatibility, and long-term viability" (Norton et al. 1995, 330). However, how animals were acquired before establishing successful captive breeding programs (and in some cases still are, although this is not the case with apes) and issues related to animals surplus to the breeding program are rarely mentioned. Although zoos have traditionally imported very few wild animals compared to the laboratory, pet, and animal product trades (Conway 1969), zoos should openly discuss their past and present role in removing animals from the wild and the issues involved with such practices. By excluding such information from educational messages, zoos may appear to be avoiding negative issues directly related to their role in animal conservation and welfare, thus undermining their desired perception as conservationists.

Finally, when addressing the reality of life in the wild, most zoos tend to purposefully exclude information that may be distressing to the public. As Polakowski (1987, 3) stated,

> The "blood and guts" of the predator–prey relationship and the slow and sickening aspect of death by starvation through overpopulation of the species may be an unwelcome educational message to the visitors who have come to the zoo for entertainment and en-

joyment. They are anticipating a good time and are ready to have fun, daydream about "wild" fantasies and create mental illusions caused by the presence of wondrous zoo animals. . . . The risk of offending the public by presenting images of reality may be too great and costly for the zoo management to present.

However, we believe avoiding such messages weakens the educational opportunities of zoos. Bushmeat is a prime example; as mentioned earlier in this volume, bushmeat may be the greatest threat facing many species of fauna in West and Central Africa, including three ape species, yet it remains very rarely addressed in zoo educational messages. Pilot survey data at Zoo Atlanta found that only 7 percent of exiting visitors could define bushmeat, despite the fact that western lowland gorillas and drills, both hunted for their meat, are housed at the zoo (Stoinski and Maple unpublished data). How can zoos expect the public to understand and care about the complex problems facing animals in the wild if we fail to inform them?

To assist with the challenges all zoos have faced with providing clear, consistent messages, the Conservation Education Committee of the AZA has developed a recommended template of priority conservation messages, with the intent of assisting zoos and aquariums in developing and reviewing their educational messages.

Delivery of Educational Messages

A number of factors are critical when reviewing the delivery of educational messages in institutions. First, messages must be consistent throughout the facility; that is, both exhibit design and facility operation must be compatible with our messages. For example, describing the importance of recycling in graphics without providing recycling bins on zoo grounds sends an inconsistent message of the importance of this conservation strategy to the public.

Second, messages must be relevant to the visiting public. Visitors may feel too distant from the complex issues facing wildlife in Africa or Asia to believe they can make a difference. By presenting conservation issues that are closer to home, or relating them to our visitors' everyday lives, zoos may increase their potential to produce behavioral change. Alternatively, local issues may be politically charged and may open zoos to public criticism. For example, zoo visitors in the Pacific Northwest are unlikely to take issue with an exhibit on bushmeat in Africa but may do so if the exhibit criticized logging practices in nearby forests. As institutions that depend on public support, zoos should be aware of the potential for controversy but not avoid addressing complex issues. Finally, zoos should ensure that their messages are effectively presented to visitors and are not "purveyed through detailed labels written in a didactic manner and which do not attract the attention of the general visitor" (Tunnicliffe 1995b, 190). Tunnicliffe goes on to suggest that zoos develop conservation education techniques that introduce visitors to key concepts at their interest and educational levels.

The Need for Future Research

We hope that as zoos move away from the Noah's ark paradigm, which was hailed as the *raison d'être* for zoos in the 1970s and 1980s, and toward an educational focus, more research to determine the educational efficacy of programs and the best methods for producing cognitive, affective, and behavioral change will occur. Currently, fewer than 50 percent of AZA institutions conduct such evaluations, and even fewer publish their findings in peer-reviewed journals (Stoinski et al. 1998b). In addition, there is little published information on educational techniques that do not work, which would serve to prevent institutions from repeating mistakes.

When evaluating education, we recommend several additional areas of focus. First, zoo and aquarium professionals are most interested in seeing improvement in conservation-related attitudes and behaviors, which may be more valuable than those related to factual gain or affect. As different educational methods appear to produce different types of learning (e.g., Stronk 1983), zoos may want to further research the most effective methods for ensuring changes in conservation-related attitudes and behaviors.

Second, as was mentioned earlier, zoos should evaluate the most effective educational methods for different groups of zoo visitors (i.e., family groups, teachers, adults, children, school groups). For example, Wineman, Piper, and Maple (1996) found that although significant changes to facilitate conservation education have occurred in the design of many zoo exhibits, the design of children's zoos has remained static. The authors conclude "these visitors are being inadequately served and less than maximally involved in conservation education" (Wineman et al. 1996, 95).

Third, some species may provide greater educational opportunities than others. For example, Gold and Benveniste (1995) found that primates (and great apes in particular) were among the most popular animals among visitors. In addition, Stoinski and Maple (unpublished data) found students who studied nonhuman primates in a formal education program at Zoo Atlanta were significantly more likely to report satisfaction with the species they studied than students who studied nonprimates. These data demonstrate that nonhuman primates may offer a powerful opportunity for educating the public and provide some evidence arguing against the suggestion that apes be eliminated from zoo collections (chapters 10 and 16, this volume).

To facilitate additional educational research, the Conservation Education Committee of the AZA has initiated a multi-institutional project assessing the impact of an overall visit to a zoo or aquarium, including effects on conservation-related knowledge, affect, attitudes, and behavior. The goal of this project is to answer some of these remaining questions, as well as to provide tools to zoos and aquariums to address more specific research issues.

Educational Objectives

In summary, when developing or refining educational messages and objectives, zoos and aquarium professionals should consider taking the following steps:

- Focus on messages based on inspiring conservation-related attitudes and behavior. As an example, behavioral change opportunities can be incorporated into the zoo experience (e.g., having "Save a Rainforest" parking meters; see also the Congo exhibit description later in the chapter).
- Focus on current issues (e.g., bushmeat), realizing these may be controversial.
- Ensure that messages are consistent throughout the park.
- Focus on messages related to animal welfare and zoo management.
- Strive to bring high levels of cutting-edge scientific knowledge into education and exhibitry (Kleiman 1996) by keeping abreast of the current field literature, conducting scientific studies, and aligning with a research institution or field researcher.
- Incorporate the latest in educational research into interpretation and exhibitry (e.g., technology-based teaching, different learning modalities, and inquiry-based learning) to ensure state-of-the-art educational opportunities, to reach a variety of audiences (e.g., families, children, youth), and to reach underserved populations.
- Provide a hierarchy of information to ensure that visitors with various levels of knowledge will be able to learn something from their visit. This hierarchy is designed to allow frequent visitors to focus on deeper messages, while providing opportunities for first-time visitors with less knowledge to simply increase their awareness.
- Provide attention to detail to create a natural exhibit that ensures the exhibit's place within the overall message of the zoo. Including in-house education, research, and keeper staff as well as the standard outside consultants in the design process will help integrate the exhibit with the prevailing institutional philosophy, ensure that educational messages are appropriate for visitors, and facilitate staff use of the exhibit.
- Focus on all aspects of the visitor experience. According to Maslow's hierarchy of needs (Maslow 1968), visitors' basic physiological needs and

safety concerns must be met before they will be receptive to education and behavior change (Rabb and Saunders 1999). This could include providing visit schedules and maps when possible to prevent the adjustment to a novel environment from interfering with on-site learning (Falk, Martin, and Balling 1978).

- Avoid the assumption that visitors and zoos share the same interests (Tunnicliffe 1995a, 1995b). Instead, zoos ought to gather information on public interest with formative evaluations and use the results to assist in exhibit design. For example, a survey of visitors to the great ape house at Lincoln Park Zoo found that visitors ranked ape intelligence as the topic of most interest (Gold and Benveniste 1995), yet to our knowledge only one American zoo exhibit, "Think Tank" at the National Zoo, specifically addresses ape intelligence. Similarly, the public repeatedly demonstrates interest in the similarities between humans and animals, particularly the great apes, in their spontaneous anthropomorphic comments and responses on exit surveys (Gold and Benveniste 1995; Tunnicliffe 1995a). However, most zoos actively avoid anthropomorphism in their educational messages. It might be more effective to use similarities between humans and animals to capture peoples' attention, and then use this opportunity to educate them further about the unique qualities of the animals.
- Refurbish older exhibits, when funds are unavailable for large renovations, to achieve educational objectives. For example, using data on visitor interests and preconceptions, Lincoln Park Zoo was able to redesign interpretive programs within an older ape exhibit to provide effective learning opportunities (Gold and Benveniste 1995).
- Scientifically evaluate and document educational achievements and failures to provide a greater understanding of how zoos affect knowledge, attitudes, and behavior change. These costs can be incorporated into the overall price of an exhibit from the outset or into the operating budget. Collaboration with universities may provide an inexpensive way to staff these studies; zoos can also consult with the Evaluation Task Force of the AZA Conservation Education Committee when designing evaluation projects.

CASE STUDIES

Many institutions are striving to meet the educational objectives outlined earlier in the chapter. Next we discuss five case studies of institutions working to meet these objectives.

Zoo Atlanta: Merging Technology and Research

By integrating a traditional research approach with computer-based telecommunications technology, Zoo Atlanta strives to bring science into the realm of education for staff, collaborating scientists, and the public. Zoo Atlanta's partnership with the Georgia Institute of Technology and The Dian Fossey Gorilla Fund International (DFGFI) has resulted in the formation of the DFGFI Institute for Conservation Science and Technology. Housed in part in the zoo's new Conservation Action Resource Center (Conservation ARC), the institute applies technological, scientific, and conservation leadership and direction to educate American and African scientists and to affect field conservation. This is a multidisciplinary approach, combining Geographic Information Systems (GIS) and Global Positioning System (GPS) technologies with traditional field census techniques to create an understanding of vegetation zones, their use by nonhuman primates, and human impact on habitat and biodiversity in East Africa. These data, along with satellite and video images of gorillas sent from the field, will also be incorporated with data collected at the zoo by research staff and used by zoo educators to provide up-to-date information on both wild and captive gorillas.

Another type of technology used in education is virtual reality. In 1995, Zoo Atlanta and the Georgia Institute of Technology Graphics, Visualization, and Usability Center (GVU) began work on a Virtual Reality (VR) Gorilla Exhibit "to explore issues in the use of VR in education and to create an example of a virtual environment that provides meaningful educational experiences for children" (Allison et al. 1996, 1). The VR exhibit allows visitors to virtually step into one of the Zoo Atlanta gorilla habitats as a juvenile gorilla and interact with members of the group. Inappropriate behavior, such as approaching a dominant animal too quickly or direct eye contact, results in species-typical aggressive responses from the other group members and an eventual "time-out" for bad behavior. Evaluations of the effectiveness of this type of learning are currently underway.

Finally, Zoo Atlanta uses distance learning technology to provide off-site educational opportunities. With videos or remote cameras at exhibits, zoo staff guide students as they observe and collect data on the animals at the zoo. Preliminary assessment shows students rate live shots of gorillas a full Likert-scale point (7-point scale) higher than previously taped videos, suggesting a benefit to seeing the animal behavior through a live feed rather than on tape (Stoinski and Maple unpublished data).

"Think Tank" at the National Zoological Park: Integrating Science and Exhibitry

In 1995 a revolutionary exhibit called "Think Tank" opened at the National Zoo. Designed to educate visitors about animal thinking, the exhibit includes demonstrations, texts, and graphics focused on three areas of scientific inquiry: tools, language, and social behavior. The exhibit emphasizes "concepts and relationships instead of traditional zoological exhibits which focus on animals and species. Visitors may examine animal cognition and the way in which humans negotiate, construct, and define their relationship to other animals" (Bielick and Karns 1998, 4). By integrating research and exhibitry, "Think Tank" provides an excellent opportunity for both scientific and public education: Research fellowships enable scientists to conduct studies on-site, and these studies are accessible to the public through live demonstrations by the scientists. Unlike demonstrations from the early 1900s, which dressed animals in human clothes for tea parties to educate visitors about "the mentality of animals and their wonderful likeness to man" (Hornaday 1906, cited in Mench and Kreger 1995, 379), demonstrations at "Think Tank" are designed to "instill a new respect for animals by showing how they think with the ultimate goal of raising concern for conservation" (*Think Tank Training Manual* 1995, 1).

Assessments of "Think Tank" reveal that the exhibit improves visitors' knowledge about animal thinking and increases their scientific understanding of how scientists study this unobservable phenomenon. Follow-up telephone surveys conducted a year after a "Think Tank" visit found that half of the visitors still remembered learning something new from their visit and that the exhibit positively affected their attitudes toward animals (Bielick and Karns 1998).

Congo Forest at Wildlife Conservation Society: An Experiment in Participatory Conservation

Opening in 1999, the Congo Forest exhibit at the Bronx Zoo has spawned a new era in zoo exhibitry. Described as an "experiment in participatory conservation" (Conway 1999), this exhibit took ten years to bring from concept to completion. Formative evaluations dictated the educational focus on issues where visitors were lacking in knowledge: the extent of rain forest habitat loss, the role science plays in protecting rain forests, and how individuals can act to save rain forests. Using landscape immersion, interactives, drama, and live animals, the series of animal exhibits and galleries challenges visitors to act as scientists ("discover the facts"), field workers ("involve local people"), and conservationists ("conserve wildlife and habitats"; Conway 1999). The merging of education and conservation culminates in the final gallery of the exhibit by allowing visitors to use their newly gained knowledge to take action for conservation by choosing one of more than 300 conservation projects to receive the $3 initial admission charge. This "experiment in participatory conservation" contributes between $750,000 and $1,000,000 to conservation each year (Conway 1999). Summative evaluations are currently being conducted to determine how well the exhibit helps visitors discover science as a conservation tool, promotes care for African rain forests, and encourages visitors to take action for conservation outside of the exhibit setting.

Disney's Animal Kingdom: Integrating Science, Conservation, and Entertainment

Disney's Animal Kingdom (DAK) opened in 1998 as a "new species of theme park." DAK education mission includes three elements: to celebrate animals; to reconnect guests to nature; and, perhaps most important, to inspire conservation action. The goal is to provide conservation messages through a variety of experiences, including animals displayed in naturalistic settings; rides such as "Kilimanjaro Safari," "Countdown to Extinction," and "Kali River Rapids" that have conservation messages and ecological themes; and live animal shows.

The only great ape species exhibited at DAK is the western lowland gorilla, *Gorilla gorilla gorilla,* with both a bachelor and a family group. The area is designed to replicate a conservation school and research center dedicated to studying and preserving the lives of endangered wildlife. All interpretation follows the theme of research, including two structures simulating field study stations. Staff members are trained to initiate conversations with guests on topics ranging from poaching, to the wide variety of plants gorillas eat, to gorilla demography and genetics.

DAK most strongly emphasizes its conservation messages in its Conservation Station. At Conservation Station there are a variety of interactive ways to learn about conservation, including specific issues related to great apes, as well as a behind the scenes experience focusing on animal welfare messages such as veterinary care and husbandry. Guests may select and view short videos on "eco-heroes," including Jane Goodall, and learn how to get involved in conservation efforts on the Eco Web, a program that allows guests to type in questions on a computer about organizations in their community that are active in conservation, including efforts related to great apes.

One of the primary goals at DAK is to assess the effectiveness of delivering conservation messages. The goal is to develop a technique that assesses effectiveness in achieving cognitive, affective, and behavioral goals and that can be continually reused as changes are made in the visitor experiences. DAK is now working to focus on its key conservation messages.

The "Quest" at Brookfield Zoo: Demonstrating the Way to Behavioral Change

Chicago's Brookfield Zoo has taken the novel approach of developing an exhibit solely focused on influencing lifestyle changes. This exhibit, called "Quest to Save the Earth," is located just outside of Tropic World, an indoor facility housing gorillas as well as other tropical rain forest species from South America, Asia, and Africa. Tropic World focuses on choices nonhuman primates make in their environments. The "Quest" continues this theme, focusing on the choices of human primates and their effect on the health of our environment. The "Quest" follows a pathway, and consists of four primary activities. In the first, "The Bog of Habits," guests must cross a bog by stepping on various stones. Each stone is inscribed with a different habit or behavior, and guests learn about the positive and negative environmental impact of these behaviors. The second is "Doors of Doubt." One door looks hard to open, but when visitors know where to push, it opens easily, which is often the case with conservation action. Visitors must team up to open another door, reinforcing the importance of collaborating for conservation. Third is the "Path of Temptation," where visitors who take the "easy way," end up with all the things they have thrown away in a landfill. Those who take the "harder way" and get past all their excuses for not taking action find themselves in a beautiful garden featuring reused objects. The last challenge is the "Tower of Balance," where visitors build a tower of blocks, with each block representing recreational choices. Blocks representing recreations with more negative impacts on the environment are harder to stack than are those with little or no impact. The trick is to balance the recreational choices to have a lighter impact on the Earth. The finale features an interactive globe, where visitors must pledge to perform environmentally friendly behaviors. The more pledges that are made, the prettier and greener the globe becomes. Throughout the exhibit, visitors are applauded (literally in some cases) for the positive actions they are already doing and given examples of what else they might do. "Quest" was developed with a great deal of formative evaluation, and a summative evaluation is being conducted to determine its actual effects.

CONCLUSION

As more becomes known about how people learn, zoo and aquarium exhibits are being better designed. At the very least, a visit to a zoo should leave even the most (cognitively) unorganized visitor with an appreciation for the intrinsic beauty of nature. A focused visitor should begin to gain a sense of the interdependence of all living things, and a frequent or truly engaged visitor should begin to incorporate these messages into a personal strategy for involvement with conservation of the environment.

With respect to great apes, zoos are perfectly positioned to educate guests about the complex issues surrounding these animals in both captivity and the wild, and, as discussed earlier, many institutions have developed innovative ways of communicating these messages. There are limited data on the effectiveness of aspects of our educational programs that suggest that we do accomplish some of our educational goals. Unfortunately, we do not have definitive data addressing the overall educational efficacy of zoos. There is a clear mission on the part of the AZA to address these questions, and the AZA Conservation Education Committee has a task force devoted to evaluation issues. In addition, the educational initiatives proposed by individual institutions may help to address these larger questions. With these efforts in place, we hope that we will soon have the data to show how zoos influence the knowledge, attitudes, and behaviors of our visitors regarding the importance of conserving nature and to determine how we can improve our educational efficacy in the future.

ACKNOWLEDGMENTS

The authors would like to thank Cynthia Vernon, vice president of Communications at the Chicago Zoological Society, and Carol Saunders, director of Communications Research, for their assistance in providing background materials on the "Quest to Save the Earth" exhibit at the Brookfield Zoo, as well as Sue Dale Tunnicliffe, Michael Kreger, Kathy Lehnhardt, Debra Forthman, and Nancy Pratt for their thorough reviews and helpful comments on this chapter.

REFERENCES

Allison, D., Wills, B., Hodges, L. F., and Wineman, J. 1996. *Gorillas in the bits.* Atlanta, GA: Graphics, Visualization and Usability Center, Georgia Institute of Technology.

American Zoo and Aquarium Association (AZA). 1999. *The collective impact of America's zoos and aquariums.* Bethesda, MD: Author.

Bielick, S., and Doering, Z. D. 1997. *An assessment of the "Think Tank" exhibition at the National Zoological Park.* Washington, D.C.: Smithsonian Institution. Report No. 97-1.

Bielick, S., and Karns, D. A. 1998. *Still thinking about thinking.* Report No. 98-5. Washington, D.C.: Smithsonian Institution.

Birney, B. A. 1996. Criteria for successful museum and zoo visits: Children offer guidance. *Curator* 31: 292–316.

Bitgood, S., Ellingsen, E., and Patterson, D. 1990. Toward an objective description of the visitor immersion exhibit. *Visitor Behavior* 5: 11–14.

Borun, M., Chambers, M. B., Dritsas, J., and Johnson, J. I. 1997. Enhancing learning through exhibits. *Curator,* 40(4): 279–93.

Burton, B. E. 1990. The *effects of age, sex, and enclosure characteristics on young people's perceptions of zoo gorillas.* Honors thesis, Department of Zoology, University of Melbourne, Australia.

Carter, T. 1998. Teaching zoo biology and management: A tale of two zoos. *Proceedings of the Annual American Zoo and Aquarium Association Conference, Tulsa, Oklahoma, September 1998*: 142–48.

Coe, J. C. 1995. The evolution of zoo animal exhibits. In C. Wemmer (ed.), *The ark evolving: Zoos and aquariums in transition* (pp. 95–128). Washington, D.C.: Smithsonian Institution Press.

Chicago Zoological Society and Lincoln Park Zoological Society. 1993. *Conservation related perceptions, attitudes, and behavior of adult visitors and non-visitors to Brookfield Zoo and Lincoln Park Zoo.* Unpublished manuscript.

Conway, W. 1969. Zoos: Their changing roles. *Science* 163: 48–51.

———. 1999. Congo: A zoo experiment in participatory conservation. *Proceedings of the Annual Meeting of the American Zoo and Aquarium Association, Minneapolis, Minnesota, September 1999:* 101–4.

Davison, V., McMahon, L., Skinner, T. L., Horton, C. M., and Parks, B. J. 1993. Animals as actors: Take 2. *Annual Proceedings of the American Zoo and Aquarium Association Conference:* 151–56.

Derwin, C. W., and Piper, J. B. 1988. The African rock kopje exhibit: Evaluation and interpretive elements. *Environment and Behavior* 20(4): 435–51.

Falk, J. H., Martin, W. W., and Balling, J. D. 1978. The novel field-trip phenomenon: Adjustment to novel settings interferes with task learning. *Journal of Research in Science Teaching* 15: 127–34.

Finlay, T., James, L. R., and Maple, T. L. 1988. People's perception of animals: The influence of the zoo environment. *Environment and Behavior* 20(4): 508–28.

Ford, J., and Burton, B. E. 1991. Environmental enrichment in zoos: Melbourne Zoo's naturalistic approach. *Thylacinus* 16(1): 12–17.

Glatston, A. R., and Hosey, G. R. 1997. Research in zoos: From behaviour to sex ratio manipulation. *Applied Animal Behaviour Science,* 51: 191–94.

Gold, K. 1997. The conservation role of primate exhibits in the zoo. In J. Wallis (ed.), *Primate conservation: The role of zoological parks* (pp. 43–62). Norman, OK: American Society of Primatologists.

———. 1998. People and animals; experiences in a Dutch rainforest—Apenheul Primate Park. In *Proceedings of the European conference: Humankind in a balanced environment* (pp. 3–8). Vierhouten, The Netherlands: Stichting J.J. Dondorp Fonds.

———. in press. Experiential learning in a Dutch rainforest—Apenheul Primate Park. In *1998 International Zoo Educators conference proceedings, Taipei, Taiwan.*

Gold, K., and Benveniste, P. 1995. Visitor behavior and attitudes towards great apes at Lincoln Park Zoo. *Annual Proceedings of the American Zoo and Aquarium Association Conference:* 152–82.

Greenburg, G. 1987. Zoos as teaching aids for the comparative psychology course. *Applied Animal Behaviour Science* 18: 83–89.

Gutierrez de White, T., and Jacobson, S. K. 1994. Evaluating conservation education programs at a South American zoo. *Journal of Environmental Education* 25: 18–22.

Heinrich, C. J., and Birney, B. A. 1991. Effects of live animal demonstrations on zoo visitors' retention of information. *Anthrozoos* V: 113–21.

Hosey, G. R. 1997. Behavioral research in zoos: Academic perspective. *Applied Animal Behavior Science* 51: 119–207.

International Union of Directors of Zoological Gardens (IUDZG) and International Union for Conservation of Nature and Natural Reserves/Species Survival Commission (IUCN/SSC). 1993. *The world zoo conservation strategy: The role of zoos and aquaria of the world in global conservation.* Gland, Switzerland: World Zoo Organization and the Captive Breeding Specialist Group of IUCN/SSC.

Kellert, S. R. 1980. *Activities of the American public relating to animals.* Washington, D.C.: U.S. Fish and Wildlife Service.

———. 1984. Urban American perceptions of animals and the natural environment. *Urban Ecology* 8: 209–28.

———. 1999. *American perceptions of marine mammals and their management.* Washington, D.C.: Humane Society of the United States.

Kellert, S. R., and Dunlap, J. 1989. *Information learning at the zoo: A study of attitude and knowledge impacts.* Philadelphia: Zoological Society of Philadelphia.

Kleiman, D. G. 1996. Special research strategies for zoos and aquariums and design of research programs. In G. M. Burghart, J. T. Bielitzki, J. R. Boyce, and D. O. Schaeffer (eds.), *The well-being of animals in zoo and aquarium sponsored research* (pp. 15–22). Greenbelt, MD: Scientists Center for Animal Welfare.

Koran, J. J., Jr., and Baker, S. D. 1979. Evaluating the effectiveness of field experiences. In M. B. Rowe (ed.), *What research says to the science teacher* (pp. 26–31). Washington, D.C.: National Science Teachers Association #2.

Kreger, M. D., and Mench, J. A. 1995a. Visitor-animal interactions at the zoo. *Anthrozoos* VIII: 143–58.

———. 1995b. Visitor-animal interactions at the zoo: Animal welfare. *Annual Proceedings of the American Zoo and Aquarium Association Conference:* 310–15.

Lukas, K. E., Marr, M. J., and Maple, T. L. 1998. Teaching operant conditioning at the zoo. *Teaching of Psychology* 25: 112–16.

MacMillen, O. 1994. Zoomobile effectiveness: Sixth graders learning vertebrate classification. *1994 AZA Annual Conference Proceedings:* 181–83.

Maple, T. L., and Hoff, M. P. 1980. *Gorilla behavior.* New York: Van Nostrand Reinhold.

Maple, T. L., McManamon, R., and Stevens, E. F. 1995. Defining the good zoo. In B. G. Norton, M. Hutchins, E. F. Stevens, and T. L. Maple (eds.), *Ethics on the ark* (pp. 219–34). Washington, D.C.: Smithsonian Institution Press.

Maslow, A. 1968. *Toward a psychology of being* (2nd ed.) New York: Van Nostrand Reinhold.

Mench, J., and Kreger, J. 1995. Animal welfare and public perceptions associated with keeping wild mammals in captivity. *1995 AZA Annual Conference Proceedings:* 376–83.

Morgan, J. M., and Gramann, J. H. 1989. Predicting effectiveness of wildlife education programs: A study of students' attitudes and knowledge towards snakes. *Wildlife Society Bulletin* 17: 501–509.

Morris, D. 1968. Must we have zoos? *Life,* Nov. 8, p. 78.

Norton, B. G., Hutchins, M., Stevens, B. F., and Maple, T. L. 1995. *Ethics on the ark: Zoos, animal welfare, and wildlife conservation.* Washington, D.C.: Smithsonian Institution Press.

Ogden, J., Capanzano, C., and Maple, T. L. 1994. Immersion exhibits: How are they proving as educational exhibits? *1994 AZA Annual Conference Proceedings:* 224–28.

Parsons, C. 1999. Do self-guided school groups learn anything? *Proceedings of the Annual Meeting of the American Zoo and Aquarium Association, Minneapolis, Minnesota, September 1999:* 181–91.

People, Places and Design Research. 1992. *The educational value of a family visit to Marine World Africa USA.* Northhampton, MA: Author.

Polakowski, K. J. 1987. *Zoo design: The reality of wild illusions.* Ann Arbor: University of Michigan School of Natural Resources.

Price, E. C., Ashmore, L. A., and McGivern, A. 1994. Reactions of zoo visitors to free-ranging monkeys. *Zoo Biology* 13: 355–73.

Rabb, G. B., and Saunders, C. D. 1999. *God, unicorns, and toilets: Mission-inspired evaluation. AZA Conference Proceedings:* 354–59.

Reade, L. S., and Waran, N. K. 1996. The modern zoo: How do people perceive zoo animals? *Applied Animal Behaviour Science* 47: 109–18.

Regan, T. 1995. Are zoos morally defensible? In B. G. Norton, M. Hutchins, E. F. Stevens, and T. L. Maple (eds.), *Ethics on the ark* (pp. 38–51). Washington, D.C.: Smithsonian Institution Press.

Rhoads, D. L., and Goldsworthy, R. J. 1979. The effects of zoo environments on public attitudes toward endangered wildlife. *International Journal of Environmental Studies* 13: 283–87.

Rosenfeld, A. 1980. *Informal learning in zoos: Naturalistic studies on family groups.* Doctoral dissertation, University of California, Berkeley.

Sherwood, K. P., Jr., Rallis, S. F., and Stone, J. 1989. Effects of live animals versus preserved specimens on student learning. *Zoo Biology* 8: 99–104.

Shettel-Neuber, J. 1988. Second and third-generation zoo exhibits. *Environment and Behavior* 20(4): 452–73.

Sommer, R. 1974. *Tight spaces.* Englewood Cliffs, NJ: Prentice-Hall.

———. 1999. *Lions, tigers, and bears, oh my! From hard architecture to havens for environment–behavior research: The changing role of the zoo.* Paper presented at the Environmental Design and Research Association annual conference, Orlando, Florida, June 1999.

Stoinski, T. S., Lukas, K. E., and Maple, T. L. (1998a). A survey of research in North American zoos and aquariums. *Zoo Biology* 17: 167–80.

———. 1998b. Teaching animal behavior in Africa: Georgia Institute of Technology and Zoo Atlanta's "Field Study in Animal Behavior." *1996 AZA Annual Conference Proceedings:* 224–28.

Stronck, D. R. 1983. The comparative effects of different museum tours on children's attitudes and learning. *Journal of Research in Science Teaching* 20: 283–90.

Swanagan, J. 1993. *An assessment of factors influencing zoo visitors' conservation attitudes and behavior.* Master's thesis, Georgia Institute of Technology, Atlanta.

Think Tank Training Manual. 1995. Unpublished document. Washington, D.C.: National Zoo.

Tunnicliffe, S. D. 1994. Education: A partnership between zoos and visitors—Is it equal? *1994 AZA Annual Conference Proceedings:* 187–93.

———. 1995a. What do zoos and museums have to offer young children for learning about animals? *Journal of Education in Museums* 16: 16–19.

———. 1995b. Zoo talk: The content of conversations of family visitor groups whilst looking at live animals. *1995 AZA Annual Conference Proceedings:* 645–47.

———. 1996. Conversations within primary school parties visiting animal specimens in a museum and zoo. *Journal of Biological Education* 30: 130–40.

Wineman, J., Piper, C., and Maple, T. L. 1996. Zoos in transition: Enriching conservation education for a new generation. *Curator* 39: 94–107.

Wineman, J., and Stoinski, T. S. 1999. *Zoos as living laboratories.* Paper presented at the Environmental Design and Research Association annual meeting, Orlando, Florida, June 1999.

Wolf, R. L., and Tymitz, B. L. 1979. *Do giraffes ever sit?: A study of visitor perceptions at the National Zoological Park.* Unpublished manuscript, Washington, D.C., Smithsonian Institution.

Yerke, R., and Burns, A. 1991. Measuring the impact of animal shows on visitor attitudes. *AAZPA Annual Conference Proceedings, San Diego, California:* 532–39.

———. 1993. Evaluation of the educational effectiveness of an animal show outreach program for schools. *AAZPA 1993 Annual Conference Proceedings:* 366–71.

GEZA TELEKI

6
SANCTUARIES FOR APE REFUGEES

Field reports from Africa and Asia show that great apes are swiftly losing habitats and numbers. These primates likely will be exterminated by the close of the twenty-first century if destruction and persecution by their human cousins are not curbed (Teleki 1989). The root cause is a skyrocketing human population, which jumped from 2.5 billion in 1950 to 5.9 billion in 1998 and is conservatively estimated by the United Nations to reach 7.7 to 11.27 billion by 2050 (PAI 1998). Much of that growth has been and will be in the poor tropical nations also inhabited by apes, where a flood of aid programs fuels the reproductive boom by lowering mortality in all age classes (PAI 1998) without a matching increase in food production (Timberlake 1985). Decreasing infant mortality without decreasing birth rates does not enhance living standards or ensure population survival in the long-term. Grinding poverty and social upheaval add to the problem (Rosenblum and Williamson 1987).

Farming, hunting, trapping, logging, mining, and construction, along with expanding trade supplying pet and entertainment markets as well as zoos and medical laboratories, have cumulative affects on free ape populations and their homelands (Benirschke 1986; Chivers and Lane-Petter 1978; Eudey 1987; Heltne and Marquardt 1989; Lee, Thornback, and Bennett 1988; Mack and Mittermeier 1984; Oates 1985; Wolfheim 1983). Tribal hunting of apes for subsistence and commerce is more lethal now than ever because of easy availability of modern weapons and ammunition as well as growing access to wilderness areas on roads built by logging companies and aid projects (Ammann 1993, 1994b; Ammann and Pearce 1995; Harcourt and Stewart 1980; McRae and Ammann 1997; Sabater Pi 1979; Sabater Pi and Groves 1972; Teleki 1989; chapter 3, this volume). The pace of destruction races ahead despite an array of conservation countermeasures (Oates, 1999; Tuxill 1997). As the pressures escalate, the numbers of refugee primates also increases in all range states (Alp et al. 1993; Ammann 1993, 1994a, 1994b).

The events overtaking free apes are linked to human attitudes that are too often negative or negligent in range states. Vivid evidence of a disregard for pain and death among primates is easily visible in marketplaces, village squares, city streets, hotel lobbies, private garages, and even in roadside bars where apes are chained to concrete slabs, caged for years in bare crates, pestered by youths, dressed in costumes to entertain audiences, and ridiculed by wandering tourists (Alp et al. 1993; Ammann 1993, 1994a, 1994b). The pervasive socioeconomic problems of sub-Saharan Africa and Southeast Asia may partly account for such mistreatment. But persecution of apes also occurs in wealthy nations, in public areas such as shabby roadside menageries or floodlit casino stages and in private places such as film studios or medical laboratories (AWI 1985; Nichols and Goodall 1999; Peterson and Goodall 1993; Ryder 1983; Spiegel 1996). Abuse has been fractionally reduced by recent improvements at some zoos and laboratories, mainly in Europe and North America, where public outcry is influential (Segal 1989).

Environmental and wildlife education is often cited as a key corrective measure. But general education in Africa is slipping backward, swamped by a reproductive rate that practically guarantees a constant majority of uneducated citizens. Conservation education is prone to the same fate. However, sanctuary projects allow local residents to relate positively with individual members of other species, and that experience is transmitted into neighboring communities by employees carrying home the message of respect. The close contact that fosters such respect cannot occur in wilderness areas. Therefore long-term international support of sanctuaries can have some educational value, especially if projects are sited near habitats where conservation of wild populations is underway.

ISSUES

The information and issues reviewed in this chapter stem from 30 years of studying chimpanzees while also trying to help these apes and save their habitats. Although I regard myself as an advocate of sanctuaries, there are aspects of ape rescue efforts that merit a critical look. Is it excusable for members of the zoological park and animal welfare communities to fail to agree on how to help despite pledges to do so, and thus remain inactive where their expertise is most applicable? What is the logic of being uncooperative with neighboring projects in Africa? Sanctuary personnel are Good Samaritans in the general effort to save apes, but they also reflect the flaws that plague other players in a competitive conservation field. The public image of sanctuary work, and also the welfare of the refugees in them, are prone to suffer as a result.

Whether apes needing rescue be labeled as "orphaned," "unwanted," "confiscated," "retired," or as some other flotsam whose fate is abstractly discussed among primatologists (Harcourt 1987), they are refugees from human persecution whose lives are too commonly undervalued, both ethically and biologically (Cavalieri and Singer 1993). I use the term *refugee* because it connotes acute plight caused by some destructive force and also because the traumas experienced by ape refugees are equivalent physically and psychologically to the traumas encountered by displaced human refugees. It is that plight that draws the sympathy leading to creation of ever more sanctuaries (Alp et al. 1993; Ammann 1993; Benirschke 1986; Galdikas 1995; IPPL 1984, 1990; Lemmon 1991; Teleki 1989). Dismissed by many scientists as an ineffective response to species decline, sanctuaries nevertheless will proliferate and the tragedy will attract more media attention.

What are primate sanctuaries and what are their functions? What role can sanctuaries play in the wider endeavor of saving endangered primate species? What management policies—for example, regarding reproduction among the rescued—might best govern sanctuary operations? Most important, is a confined primate a real primate or a mere caricature molded by human contact? Is any sanctuary, however fine it may appear, more than just another form of incarceration? Such questions need to be addressed and resolved if sanctuary projects are to better achieve their professed objectives of helping individuals and saving species.

PROGRESS

The notion of ape sanctuaries first surfaced in the work of Barbara Harrisson (1971), who launched a program to rehabilitate orphan orangutans at the Sepilok Forest Reserve of Malaysia during the early 1960s. Since then primate sanctuaries have exploded in number, particularly during the 1990s, with more than 70 now operating in America, Europe, Africa, and Asia. There are more than 50 focusing on or at least accepting great apes, and others emerge each year. Most of those in sub-Saharan Africa house chimpanzees and most in Southeast Asia house orangutans. But orangutan and chimpanzee sanctuaries differ in basic ways and need to be separately reviewed (see chapter 2, this volume, for a discussion of orangutan sanctuaries). For example, orangutan sanctuaries focus on rehabilitation before release in wilderness areas (Galidikas 1995), whereas the chimpanzee sanctuaries are repositories from which release to self-sustaining freedom has been infrequent (Teleki 1989). Gorilla sanctuaries currently exist only in England, Cameroon, and Congo (Brazzaville). Summaries of some sanctuary projects and discussions of rehabilitation methods appear elsewhere (Ammann 1993, 1994a, 1994b; Borner and Gittens 1978; Hannah and McGrew 1991; IPPL 1990; Lemmon 1987; Rijksen and Rijksen 1979; Teleki 1989). My aim is to take a brief tour through past and present projects that are rescuing or releasing chimpanzees in America (three sites), in England (one site), and in Africa (21 sites, 17 nations).

Outside the range states in Africa, the best known sanctuaries housing chimpanzees are Primarily Primates near San Antonio, Texas, headed by Wally Swett, and Monkey World in Dorset, England, headed by James Cronin. Both accept ex-pets. The former also has many ex-"actors" rescued from entertainers and the latter also houses many ex-props rescued from beach photographers in Spain. Large cages with indoor rooms and fenced outdoor compounds are used for containment. Only Monkey World is commercialized for visitors. Two newer and much smaller sanctuaries in America are the Center for Orangutan and Chimpanzee Conservation in Miami, Florida, and Simian Lodge in Afton, Tennessee. Rehabilitation of apes at these sites focuses on restoring physical and mental health after abusive treatment as well as resocializing to group life after isolated confinement.

Early projects in America that qualified as rehabilitation trials for laboratory apes involved a 1972 to 1973 two-phase release of eight chimpanzees from the Yerkes Regional Primate Research Center onto Bear Island in coastal Georgia (Wilson and Elicker 1976) and a 1975 release of eight LEMSIP (Laboratory of Experimental Medicine and Surgery in Primates) chimpanzees onto a small artificial island owned by Lion Country Safari at West Palm Beach, Florida (Koebner 1982; Pfeiffer and Koebner 1978). The former was closed after some chimpanzees died; the latter remains a tourist attraction. Little was learned about independent survival in these settings, but the projects did demonstrate that even medical research subjects with a frail hold on sanity retain a remarkable potential for recovery given greater freedom and richer social life.

Of the 21 projects in Africa, 8 were abandoned for internal reasons or closed by sociopolitical upheavals. Counted another way, 11 of 15 sanctuaries with fixed facilities still operate, and 2 of 6 release projects (with no fixed facilities) still exist. Of the 11 fixed-facility sanctuaries, 3 use islands, 6 have fenced or walled compounds, and 2 have only rooms or cages. The number of rescued chimpanzees per site ranges from 4 to 70+. Group housing, except for young newcomers or special cases, is standard at all facilities, as is some form of rehabilitation, including resocialization. Projects are located wherever someone happened to place them. Some range states host none and others host several. Release procedures and recommendations are covered in some detail by Brewer (1978, 1980), Hannah and McGrew (1991), and Paredes (1998).

In the Gambia, the Baboon Island Chimpanzee Rehabilitation Project was started in 1968 by Eddie and Stella Brewer with infants caged at Abuko Animal Orphanage in Banjul (Brewer 1978, 1980). Headed since 1977 by Janis Carter, the project includes four groups totaling 50 chimpanzees living on three of five islands in the River Gambia National Park. Fewer than half of these chimpanzees were actually released to the islands, with the rest born on site (Carter 1978, 1981a, 1981b, 1988; also personal communication). The groups are fully integrated and reject human contact, so new refugees are seldom added. Only two recent arrivals have not yet been integrated. This sanctuary is the oldest and one of the largest projects on the continent. It is a haven mostly to youngsters confiscated from local and international trade, but refugees of other provenance have been accepted. Demographic data covering ten years are available onsite. A wildlife education program has long been an important adjunct to the project. This project broke new ground in rehabilitation techniques, and Carter has advised other projects on methods of rehabilitation and release. A systematic published account of the rehabilitation and release strategies used at this site is needed.

In Senegal, the Mt. Asserik Rehabilitation Center is an offshoot of the Baboon Island Project. Five young Mt. Asserik chimpanzees were released to Niokolo Koba National Park in two phases by Stella Brewer between 1972 and 1974. The release was aborted because of attacks on rehabilitants by free chimpanzees (Brewer 1978). Though inconclusive, this case stimulated doubts about attempting open-habitat releases where potential attacks by free apes posed a serious survival risk for rehabilitants (Teleki 1989). It also raised questions about the risks of introducing novel diseases into natural habitats (Harcourt 1987).

In Guinea, a small chimpanzee orphanage was started in 1988 by Charlotte Dorkenoo in her Conakry home. By 1992 there were 30 refugees on site. Protégeons les Chimpanzes de Guinée, set up by Bernard Sertillanges and Angelika Becker, served for a time as the project's umbrella organization. The chimpanzees were to be released to three islands in the Konkoure River, but the project collapsed when Dorkenoo departed from Guinea. The government inherited the chimpanzees, temporarily assisted by Estelle Raballand.

Also in Guinea, a new Chimpanzee Rehabilitation Project was started in 1996 by Janis Carter to provide a home for the 24 remaining individuals (Carter, personal communication). The project, consisting of a complex of outdoor cages at the Parc du Haut Niger near Faranah, funded by the European Union, is a first step toward attempting another release to a wilderness area not yet selected. Rural conservation education is also a major component. After the ending of European Union funds in 1999, Raballand was recruited by the government to restart the program. Basic upkeep is covered by the Humane Society of the United States, and the search for a release site is again underway.

In Sierra Leone, the Tacugama Chimpanzee Sanctuary, started in 1993 by Rosalind Alp in the Western Area Forest Reserve near Freetown, houses 21 chimpanzees, mostly confiscated ex-pets, in a compound with several cages (Alp, personal communication). This project was launched after a 1989 to 1992 urban survey revealed 55 chimpanzees kept as pets in Freetown alone (Alp et al. 1993). The project is severely threatened by sociopolitical upheaval and is in dire financial straits (Alp et al. 1993; also personal communication).

In Liberia, more than 57 ex-laboratory chimpanzees from the New York Blood Center's VILAB II station at Monrovia were released to three coastal islands near the mouth of the Little Bassa River between 1978 and 1986 (Hannah 1986; Prince et al. 1990). Not a true sanctuary project in that acquisition of apes for medical research created its own surplus, this retirement program is comparable to island releases attempted by laboratories in the United States. The mangrove habitat necessitated constant provisioning of the apes. By 1995 many had died from starvation during prolonged nationwide warfare and the rest were recaptured and returned to the station. A plan to export to Asia for further medical use the more than 65 individuals remaining at VILAB II station fortunately has not materialized.

Also in Liberia, construction of an independent Sapo Wildlife Orphanage under government management, scheduled to house chimpanzees and other species before possible release in Sapo National Park, was completed just before being destroyed by armed rebels in the early 1990s (Peal, personal communication). Efforts are underway by the Society for the Conservation of Nature in Liberia to reopen the orphanage.

In Côte d'Ivoire, 20 chimpanzees were released in 1983 by VILAB II to an island in the Bandama River near Azagny National Park; 11 died within a year, probably as a result of a severe diarrheal outbreak, and the remaining nine were moved to a nearby peninsula (Prince et al. 1986). The project ended when transfer to the park was prohibited by wildlife officials fearing disease transmission to resident chimpanzees and potential attacks on tourists (Prince et al. 1990). The fate of the chimpanzees is unrecorded, but none returned to Liberia. As many as five may still be alive (Raballand, personal communication).

In Ghana, Meredith Rucks launched a sanctuary project at Bia National Park in 1972 while stationed there with the Peace Corps. The goal was to rehabilitate six young chimpanzees for eventual release (Rucks 1976). The project stalled when Rucks departed in 1976, and the fate of the apes is unknown.

In Nigeria, the Drill Ranch project, started by Peter Jenkins and Liza Gadsby in 1986 at Calabar and expanded to the Afi Mountain near Cross River National Park (CRNP) in 1995, houses 16 chimpanzees, 8 of whom were released to a natural forest enclosure at CRNP in 1997 (DRP 1996; DRBC 1998). This project is now a permanent orphanage with no plan to release apes in the nearby park. The project advocates birth control for chimpanzees. There is a strong conservation education component.

In Cameroon, Liza Gadsby and Peter Jenkins opened the Limbe Wildlife Center on the grounds of the Limbe town (near Douala) botanical gardens in 1994 (IPPL 1997). There are 20 chimpanzees (as well as gorillas and other primates) in cages and indoor–outdoor enclosures. The chimpanzees are of all ages, some confiscated, some abandoned pets, and some orphans brought in by forestry workers. A fenced compound is being built, and a plan exists to release the apes eventually to an island in the Sanaga River. A wildlife education center was recently completed.

In Gabon, Claude Marcel Hladik (1974) conducted a trial release of nine young ex-laboratory chimpanzees to Impassa Island, on the Ivindo River near Makokou, in 1968 to observe feeding and other adaptations (Hannah and McGrew 1991; Hladik and Viroben 1974). After one unexplained death, the others survived until 1978, when two escaped and six were removed because of a major drop in water level in the river.

In Congo (Brazzaville), Aliette Jamart began rescuing young chimpanzees in the 1960s, with her home in Pointe-Noire serving as an orphanage. Some of the rescued apes were ex-pets but most were youngsters orphaned by bushmeat hunting, which may be killing as many as 3,000 chimpanzees per year in this nation alone (Ammann 1993, 1994a, 1995b; chapter 3, this volume). For some years the project had no international recognition and was financed solely by its founder. Promotion and support by Habitat Écologique et Liberté des Primates (HELP) enabled Jamart to relocate 44 chimpanzees to three islands in the Conkouati Lagoon at the Reserve de Faune de Conkouati between 1991 and 1995 (Penn 1996). An assessment of the reserve's resources and free chimpanzee residents was undertaken to determine the feasibility of releasing rehabilitants into this wilderness area (Tutin 1996). After a program of rehabilitation and medical screening on the lagoon islands, seven chimpanzees were released to the reserve late in 1996, with some injuries (but no deaths) resulting from contact with free chimpanzees (Paredes 1998). The next year seven more rehabilitants were released, and others will follow as long as the program is successful. Provisioning continues on the three islands but not on the mainland. If this project succeeds it will offer solid guidelines for similar endeavors elsewhere.

Also in Congo (Brazzaville), but unconnected to Conkouati, the oil company Conoco built a large chimpanzee sanctuary for the Jane Goodall Institute near Pointe-Noire in 1991 (JGI 1993). With administrative buildings, a complex of interconnected cages, and an electric fence compound enclosing a forest patch on open savanna, this project was designed to accommodate 20 to 25 chimpanzees. However, the facility already in early 2001 had 63 apes of diverse provenance, and new enclosures and compounds are being added. Rehabilitation and resocialization programs are operating and rural conservation education is planned.

In Burundi, the Jane Goodall Institute launched a Chimpanzee Conservation Project in 1990, first headed by Geoffrey Creswell, then by Susanne and Dean Anderson. An orphanage with cages inside an urban compound was built in Bujumbura and soon housed more than 12 confiscated and abandoned ex-pet chimpanzees. Surveys were conducted and plans drafted for a fenced release area on Lake Tanganyika. Warfare and extreme tribal violence in Burundi forced closure of the sanctuary by 1995, and the refugees were flown to Sweetwaters Chimpanzee Sanctuary in Kenya.

In Kenya, Karl Ammann began housing orphan chimpanzees in his home in 1990. By 1993 this became the new Sweetwaters Chimpanzee Sanctuary, with 27 refugees living in two groups at a site near Nanyuki (SCS 1996). The facility, located in the Sweetwaters Game Reserve, which itself is on Ol Pejeta Ranch, includes administrative buildings, indoor housing, and three fenced enclosures of varying sizes (Noon 1997). The aim is to provide a safe haven and lifetime care for 35 to 50 abused, homeless, or orphaned, or confiscated individuals. The project is solidly financed, in part by tourist fees, with a keen awareness of the long-term fiscal responsibilities of caring for rescued apes. A steering committee of professionals oversees operations. There is a strong wildlife education component. Of all African sanctuaries, this is most similar in structure and management to Monkey World in England, and like Monkey World, has the most potential for achieving financial self-sufficiency through charging tourist fees.

In Uganda, the Uganda Wildlife Education Center (UWEC; formerly a wildlife orphanage) at Entebbe has long been the repository for refugee chimpanzees, some abandoned by pet owners but most confiscated from the international wildlife trade (Teleki 1989). Of the 25 individuals housed there, 11 youngsters were released in 1995 to the small Isinga Island in Lake Edward in Queen Elizabeth National Park, under sponsorship of the Jane Goodall Institute (Manning 1996). Isinga Island is sparsely forested, so constant provisioning was necessary, with costs offset by tourist fees. The nine surviving chimpanzees, as well as several who had remained at UWEC, were relocated in 1998 to a new island sanctuary on Lake Victoria, which could be a haven for about 30 individuals (Edroma, Rosen, and Miller 1997).

In Tanzania, the Frankfurt Zoological Society released 17 wild-born but ex-zoo chimpanzees on Rubondo Island of Lake Victoria between 1966 and 1969

(Borner 1985). Because this was the earliest major release project in Africa, inexperience led to several chimpanzee deaths. Rehabilitation before release was minimal, and postrelease monitoring was also limited until 1978. But because the area is large and the habitat suitable for independent survival, a nucleus group established itself in what became the Rubondo National Park. By 1985 those survivors had formed a viable community of at least 20 chimpanzees, which exists today.

In Zambia, which is not inhabited by free chimpanzees, the Chimfunshi Wildlife Orphanage (CWO), started and headed by cattle ranch owners David and Sheila Siddle, is the biggest sanctuary in the world, housing 70 chimpanzees (14 born on-site), many in groups living within one walled and two fenced enclosures (Chambatu 1989; FOC 1998; Herman 1989; Noon 1996; Teleki 1987). A 2,500-acre enclosure, bounded by fence and river, is being constructed. It will accommodate several social groups in a habitat suitable for independent survival. This enclosure is part of a 10,000-acre tract purchased for the apes and deeded to the CWO Trust (FOC 1998). Despite limited outside financial support, no other project in Africa equals this in scope and vision. Completion of the largest enclosure and its buildings will require an additional U.S. $150,000 for construction, including more than 10 miles of electrified game fence (FOC 1998). Many novel methods of rehabilitation emerged from this project (Noon 1996). Young chimpanzees confiscated from wildlife traders entering Zambia from Congo (Kinshasa) formed the nucleus of this sanctuary, but other refugees, including some from overseas and elsewhere in Africa, were added over the years. New additions occur on a regular basis. By providing a haven for confiscated apes, the project encouraged law enforcement so effectively that trade in chimpanzees ceased in Zambia. A major rural education program, benefiting schools and other organizations as well as visitors, has been in place for years.

In South Africa, a small orphanage was launched in Johannesburg in 1997 by Peter Gray on behalf of the Jane Goodall Institute. The orphanage, known as the David Greybeard Project, now houses four young chimpanzees rescued in Angola.

PROBLEMS

Much has been said and written about "ape sanctuaries" without much precision in use of the term. As indicated by the names of projects listed earlier, "sanctuary," "orphanage," and "rehabilitation center" are used synonymously, creating confusion about scope, methods, and release goals. Further, the retirement programs planned for "surplus" laboratory chimpanzees in the United States are also referred to as sanctuaries. The term is even used in new U.S. legislation on retiring federally owned laboratory subjects, the Chimpanzee Health Improvement

Maintenance and Protection Act (H.R. 106-3514). Professional consensus on terminology is needed that clearly distinguishes facilities serving various purposes. The IUCN/SSC's Reintroduction Specialist Group could serve as a forum.

Existing sanctuaries are havens for chimpanzees of widely different provenance, including unwanted adult pets, aging rejects from the entertainment industry, orphans confiscated from commercial wildlife traffic, and ex-inmates of medical laboratories. None specializes in a specific class of refugee. Age and sex rarely determine candidacy, although many entrants happen to be infants. Often only nominal concern is shown for mental norms and potential to resocialize. Entry is mostly on a first-arrive-first-serve basis. Reproduction is seldom prohibited, even though new refugees always exceed vacant spaces. A year or two is often all it takes to fill a new sanctuary to capacity. Triage—rejection or extermination of hopeless and high-risk cases—is not practiced, perhaps to avoid adverse publicity or emotional stress for personnel. Long-term demographic records are seldom available, so recovery rates are not easily compared. Guidelines for human–ape interactions, if any exist, are idiosyncratic and seem governed as much by the nurturing needs of caregivers as by the mental and social needs of the rescued apes.

Facilities range from rooms in private homes to makeshift coops in backyard gardens; from haphazard collections of tiny and rusting cages to large and shiny steel-mesh and concrete-floor enclosures erected at great expense; from spacious fenced or walled compounds to islands of assorted sizes in rivers or lakes. Whether projects are truly pleasant havens for healthy refugees or just storage depots is mostly a matter of finances. With funding both meager and uncertain, long-term planning for the disposition and welfare of rescued apes, who may live more than 50 years, is seldom achievable.

Sanctuary projects routinely emphasize their role as rehabilitation centers offering refugees the means to recover physical and mental health, social adeptness, and survival skills. "Rehabilitation" thus refers to a wide array of supposedly therapeutic measures. But each project sets its own goals and standards in isolation from other projects. Because professional study of the effectiveness of various methods is absent, success is mostly a matter of individual opinion without baselines for comparison.

Most projects do offer some physical and mental health benefits, but few provide complete freedom in a natural habitat, whether bounded or open, where all basic needs of the species are met over the long term. Land that is unoccupied by humans or free apes and can also support self-sustained survival of released apes is extremely scarce in Africa. Islands have so far been the best available seminatural havens for this species. The risks of unleashing contagious diseases in wilderness areas and of fostering dangerous encounters with free apes and rural humans have so far constrained wider releases of rehabilitants into parks and other protected areas. But new and better approaches to release in natural habitats may be pioneered by projects at Conkouati in Congo (Paredes 1998; Tutin 1996) and at Chimfunshi in Zambia (FOC 1998; Noon 1996).

Most literature on ape rehabilitation comes from medical laboratories, where interest focuses on keeping apes content while they undergo assorted experiments (AWI 1985; Segal 1989). Accounts of methods of rehabilitation and release in Africa are mostly anecdotal (Borner 1985; Brewer 1978, 1980; Carter 1981a, 1981b; Prince et al. 1990). Systematic descriptions of field procedures (Noon 1996; Paredes 1998) and informed discussions of management approaches (Hannah and McGrew 1991) are rare. No thorough comparative study of the success of different methods of rehabilitation exists to date.

Whenever a new sanctuary is launched the slow process of trial-and-error learning of what to do and what not to do is repeated. Severe shortages of funds and staff often hinder development of practical construction guidelines, uniform operation standards, and sound management policies. When a survey to examine various sanctuary types in Africa was planned in 1996, funds could be raised to visit only three sites in eastern Africa (Noon 1997). A partial survey of sanctuaries in Western and Central Africa in 1998 by Raballand also had to be personally financed. Another proposed survey of all African sites by Caroline Bocian in 1999 to 2000 received no sponsorship.

Public debate about the merits of sanctuaries flows mostly in the absence of solid information. Because they talk about but do not work in sanctuaries, scientists and conservationists fail to offer much practical guidance on operating them and rarely make arguments in their favor. Thus the institutions with which these professionals are connected neglect to provide essential support for sanctuaries. Although sanctuaries are a hot topic in these circles, useful in attracting public sympathy and in raising funds, their popularity seldom translates into active international support. Nearly every project is perpetually short of funds. Much time and effort that could better be spent on helping refugees is diverted to worrying about funds for daily rations for the rescued apes and for the negligible salaries of local sanctuary workers.

Sanctuaries in primate range states are today supported by only a few small organizations such as the International Primate Protection League, The Jane Goodall Institute, and the Orangutan Foundation International. With some recent exceptions (Friends of Animals, The Humane Society of the United States), mainstream wildlife conservation and animal protection organizations shy away from long-term support because they claim that rescue projects contribute little to species survival. But perhaps they simply regard sanctuaries as too costly and too risky, especially in nations where infrastructure investments may be lost because of socioeconomic upheavals. However, conservation investments are also lost in wealthy nations, and many established national parks in range states do not actually protect endangered primates. The organizations committed to promoting the conservation and welfare of other species are not fully pursuing their own charters if they choose to ignore sanctuaries for endangered apes.

Sanctuaries in Africa cost little to establish and operate when funds are judiciously managed and expended. Several function today with budgets below U.S. $10,000 per year. The most expensive facilities, some topping $100,000 in annual operating costs, are mostly in French-speaking nations. Whatever the price tag, sanctuaries are the sole salvation for apes whose confiscation from trafficking and illegal ownership is vital to international law enforcement. Where no sanctuaries exist, no confiscation of apes occur. In my view, that role alone justifies their existence and merits financial support. But most sanctuaries also attract visitors, including school groups, and thus promote sensitivity to wildlife problems. Moreover, anyone who has seen the widespread suffering of refugee apes and who has witnessed the wonders performed by rescue workers in helping these hapless and homeless individuals recover their dignity and sanity ranks sanctuaries as important in many less doctrinaire ways. After all, chimpanzees did not elect to be refugees, and therefore humans should not choose to neglect them.

SUGGESTIONS

Having aired these concerns, I offer some modest suggestions for improving sanctuary projects with specific actions.

More contact, cooperation, and exchange of knowledge among sanctuaries needs to be fostered. The extensive knowledge and experience of sanctuary workers remains largely isolated in separate minds. Funds should be located to convene annual workshops where relationships can be forged and ideas discussed. The Jane Goodall Institute is well-positioned to coordinate such an organizational development effort, perhaps with support from the American Zoo and Aquarium Association.

A regional African plan for locating future sanctuaries should be developed, based on data on population decline and magnitude of pressures in each range state. An advisory group of professionals from IUCN/SSC's Primate Specialist Group might help draft an action plan with strategic geographical priorities. Funding for a survey of current sanctuary projects would first be necessary.

Institutions might support sanctuaries not only by granting small sums to cover specific short-term needs but also by setting aside endowments over which they retain control while channeling the interest income to sanctuaries. Projects need stable support to guarantee the long-term security and allow the sensible long-range planning required for apes with life spans of more than 50 years.

A comparative review of containment and ancillary structures at existing sanctuaries is needed to standardize low-cost plans for constructing easily maintained facilities and containment techniques such as electrical fencing. An international group of experts could be assembled, with minor funding for travel to a few workshops where sample plans can be drafted for sanctuaries of varying size and purpose.

A systematic comparative study of rehabilitation methods and release procedures is needed to gain maximum value from the accumulated practical experience of rescue workers. The aim would be to produce a training manual for sanctuary staff around the world. Care-giving work is extremely taxing and supersedes other tasks such as research, so sanctuaries cannot be expected to fill this gap alone. Academic institutions would best be suited for locating expertise and funds to undertake the research.

Long-term demographic data on the status and survivorship of rescued apes at all sites must be assembled. Such data are currently fragmented and usually available only on-site. Academic institutions with the expertise to interpret such data are best suited to arranging such a review.

Releases of rescued chimpanzees to islands, to large enclosures, and especially into reserves and national parks need to be monitored in the future to generate a list of tested and workable approaches. A sensible policy on where and where not to release chimpanzees must be formulated. The guidelines of the IUCN/SCC's Reintroduction Specialist Group, which were first applied to the chimpanzee release at Conkouati (Paredes 1998), need to be refined for specific application with great ape refugees. Better communication between sanctuaries and professional organizations, including funding institutions, would assist sanctuary supervisors in locating outside advice and assistance. The Center for Captive Chimpanzee Care, with experience in such matters, could act as coordinator.

Chimpanzees who are to be released to open areas lacking artificial or natural barriers, such as reserves and parks, need to be professionally examined to determine their health status to avoid infection of local residents, both human and ape. Medical protocols need to be standardized. The medical procedures initiated at Conkouati could be a model.

Formulating a general policy on birth control in sanctuaries, especially those in range states, would be valuable. Professional associations (such as the International Primatological Society, the American Society of Primatologists, IUCN/SSP), perhaps with some academic input, can draft position statements for circulation to sanctuaries, offering concrete advice on designing and applying a policy. The even thornier issue of triage also needs to be addressed, again with practical suggestions about how to handle hopeless cases. Sanctuary supervisors are seldom adequately informed about technical aspects of such procedures and would benefit from reviews of the available options and their probable consequences.

A practical way of training sanctuary workers is needed, focusing on how best to rehabilitate rescued apes such that they eventually gain independence from human assistance. Institutions employing staff with experience in managing confined apes could establish an exchange program for sanctuary caregivers, perhaps via an adopt-a-sanctuary relationship. Accredited zoos are most likely to have staff on hand with the required expertise and are well-positioned to raise the funds to carry out such a program.

An issue that is unresolved for some projects is the legal right of ownership and the legal authority to dispose of rescued apes. In Africa, the assumption is often made that host governments retain control. Although this may be true in principle, many African nations have no legislation governing ownership of nonhuman species. Alternatives that serve the best interests of the apes need to be explored, with help from such organizations as the Animal Legal Defense Fund.

ACKNOWLEDGMENTS

Thanks go to the many sanctuaries that supplied information, and especially to Shirley McGreal, Carole Noon, and Estelle Raballand for extensive background information. This report is dedicated to the many unnamed persons who contribute portions of their lives to saving ape refugees.

REFERENCES

Alp, R., Amarasekaran, B., Brima, J., and Tucker, W. 1993. *Chimpanzee rehabilitation in Sierra Leone, West Africa: Survey results and preliminary proposals.* Freetown, Sierra Leone: Ministry of Agriculture, Forestry and Fisheries.

Ammann, K. 1993. Orphans of the forest. *SWARA* (Nov.–Dec.): 16–19.

———. 1994a. The bushmeat babies. *BBC Wildlife* (Oct.): 16–24.

———. 1994b. Orphans of the forest. *SWARA* (Jan.–Feb.): 13–14.

Ammann, K., and Pearce, J. 1995. *Slaughter of the apes: How the tropical timber industry is devouring Africa's great apes.* London: World Society for the Protection of Animals.

Animal Welfare Institute (AWI). 1985. *Beyond the laboratory door.* Washington, D.C.: Author.

Benirschke, K. (ed.). 1986. *Primates: The road to self-sustaining populations.* New York: Springer-Verlag.

Borner, M. 1985. The rehabilitated chimpanzees of Rubondo Island. *Oryx* 19: 151–54.

Borner, M., and Gittins, P. 1978. Roundtable discussion on rehabilitation. In D. J. Chivers and W. Lane-Petter (eds.), *Recent advances in primatology, Volume 2, Conservation* (pp. 101–105). London: Academic Press.

Brewer, S. 1978. *The forest dwellers.* London: Collins.

———. 1980. Chimpanzee rehabilitation: Why and how? *IPPL Newsletter* 7(2): 2–4.

Carter, J. 1978. *Chimpanzee rehabilitation.* Banjul, the Gambia: Wildlife Conservation Department. Mimeographed report.

———. 1981a. Free again. *IPPL Newsletter* 8(1): 2–4.

———. 1981b. A journey to freedom. *Smithsonian* (April): 90–101.

———. 1988. Freed from keepers and cages, chimps come of age on Baboon Island. *Smithsonian* (June): 36–49.

Cavalieri, P., and Singer, P. (eds.). 1993. *The Great Ape Project: Equality beyond humanity.* New York: St. Martin's Press.

Chambatu, P. 1989. The friends. *IPPL Newsletter* 16(3): 14–15.

Chivers, D. J., and Lane-Petter, W. (eds.). 1978. *Recent advances in primatology, Volume 2, Conservation.* London: Academic Press.

Drill Ranch Pandrillus (DRP). 1996. The Drill Ranch chimpanzees. *Drill Ranch Newsletter* 5(3): 2; 6(1): 1–4.

Drill Rehab and Breeding Center (DRBC). 1998. *Drill Rehabilitation and Breeding Center information sheet.* Calabar, Nigeria: Author.

Edroma, E., Rosen, N., and Miller, P. (eds.). 1997. Conserving the chimpanzees of Uganda: Population and habitat viability assessment for *Pan troglodytes schweinfurthii.* Apple Valley, MN: International Union for the Conservation of Nature and Natural Resources/SCC Conservation Breeding Specialist Group.

Eudey, A. A. 1987. *Action plan for Asian primate conservation: 1987–91.* Washington, D.C.: International Union for the Conservation of Nature and Natural Resources/SSC Primate Specialist Group.

Friends of Chimfunshi (FOC). 1998. Steady headway in development of trust land. *Friends of Chimfunshi (SA) Newsletter* 6(2): 2.

Galdikas, B. M. F. 1995. *Reflections of Eden.* Boston: Little, Brown.

Hannah, A. 1986. Observations on a group of captive chimpanzees released into a natural environment. *Primate Eye* 29: 16–20.

Hannah, A., and McGrew, W. C. 1991. Rehabilitation of captive chimpanzees. In H. O. Box (ed.), *Primate responses to environmental change* (pp. 167–85). New York: Chapman and Hall.

Harcourt, A. H. 1987. Options for unwanted or confiscated primates. *Primate Conservation* 8: 111–13.

Harcourt, A. H., and Stewart, K. J. 1980. Gorilla eaters of Gabon. *Oryx* 15(3): 248–51.

Harrisson, B. 1971. *Conservation of nonhuman primates in 1970.* Basel, Switzerland: S. Karger.

Heltne, P. G., and Marquardt, L. A. (eds.). 1989. *Understanding chimpanzees.* Cambridge, MA: Harvard University Press.

Herman, J. 1989. A visit to Chimfunshi Wildlife Orphanage. *IPPL Newsletter* 16(1): 19–21.

Hladik, C. M. 1974. La vie d'un groupe de chimpanzes dans la fôret du Gabon. *Science et Nature* 121: 5–14.

Hladik, C. M., and Viroben, G. 1974. L'alimentation proteique du chimpanzee dans son environment forestier naturelle. *C. R. Academie Science Paris* 279: 1475–78.

International Primate Protection League (IPPL). 1984. The tragic plight of unwanted chimpanzees. *International Primate Protection League Newsletter* 11(2): 2–4.

———. 1990. Primate rehabilitation and release projects. *International Primate Protection League Newsletter* 17(2): 16–18.

———. 1997. Several reports on Limbe Wildlife Center. *International Primate Protection League Newsletter* 24: 1–2.

Jane Goodall Institute (JGI). 1993. *Chimps: A trip to freedom.* Videotape, 30 min. Washington, D.C.: Author.

Koebner, L. 1982. Surrogate human. *Science 82* 3(6): 32–39.

Lee, P. C., Thornback, J., and Bennett, E. L. 1988. *Threatened primates of Africa: The IUCN red data* book. Gland, Switzerland: International Union for the Conservation of Nature and Natural Resources.

Lemmon, T. 1987. The long way back to nature. *BBC Wildlife* 5(4): 172–75.

———. 1991. Chimpanzee under siege. *IPPL Newsletter* 18(2): 19–20.

Mack, D., and Mittermeier, R. A. (eds.). 1984. *The international primate trade.* Washington, D.C.: Trade Record Analyses of Flora and Fauna in Commerce (TRAFFIC).

Manning, C. 1996. The Lake Edward Chimpanzee Sanctuary. *IPPL Newsletter* 23(2): 25–28.

McRae, M., and Ammann, K. 1997. Road kill in Cameroon. *Natural History* 106(1): 36–47, 74–75.

Nichols, M., and Goodall, J. 1999. *Brutal kinship.* New York: Aperture Foundation.

Noon, C. 1996. *The resocialization of chimpanzees* (Pan troglodytes). Doctoral dissertation, University of Florida, Gainesville.

———. 1997. *A report on the use of electric fencing as a primary enclosure barrier for chimpanzees,* Pan troglodytes. Boynton Beach, FL: Center for Captive Chimpanzee Care. Mimeograph report.

Oates, J. F. 1985. *Action plan for African primate conservation: 1986–90.* Washington, D.C.: International Union for the Conservation of Nature and Natural Resources/SSC Primate Specialist Group.

Oates, J. F. 1999. *Myth and reality in the rain forest: How conservation strategies are failing in West Africa.* Berkeley: University of California Press.

Paredes, J. 1998. *The adaptation process of seven released chimpanzees* (Pan troglodytes) *in the Reserve of Conkouati-Congo: Annual report Nov 96–Nov 97*. Pointe-Noire, Congo: Habitat Écologique et Liberté des Primates.

Penn, L. 1996. The chimpanzees of Conkouati. *IPPL Newsletter* 23(1): 3–4.

Peterson, D., and Goodall, J. 1993. *Visions of Caliban: On chimpanzees and people.* Boston: Houghton-Mifflin.

Pfeiffer, A. J., and Koebner, L. J. 1978. The resocialization of single-caged chimpanzees and the establishment of an island colony. *Journal of Medical Primatology* 7: 70–81.

Population Action International (PAI). 1998. What birth dearth? Why world population is still growing. *Population Action International Fact Sheet* 7: 1–4.

Prince, A. M., Brotman, B., Garnham, B., and Hannah, A. C. 1990. Enrichment, rehabilitation and release of chimpanzees used in medical research: Procedures used at VILAB II, the New York Blood Center's laboratory in Liberia, West Africa. *Lab Animal* 19(5): 29–36.

Prince, A. M., Brotman, B., Hannah, A., Donnelly, M., Hentschel, K., and Roth, H. 1986. *Rehabilitation and release to the wild of chimpanzees used in medical research.* New York: New York Blood Center. Mimeographed report.

Rijksen, H., and Rijksen, A. 1979. Rehabilitation: A new approach is needed. *Tigerpaper* 6(1): 16–18.

Rosenblum, M., and Williamson, D. 1987. *Squandering Eden.* New York: Harcourt Brace Jovanovich.

Rucks, M. G. 1976. *Notes on the problems of primate conservation in Bia National Park.* Accra, Ghana: Department of Game and Wildlife. Mimeograph report.

Ryder, R. D. 1983. *Victims of science: The use of animals in research.* Sussex, UK: Centaur Press.

Sabater Pi, J. 1979. Chimpanzees and human predation in Rio Muni. *IPPL Newsletter* 6(2): 8.

Sabater Pi, J., and Groves, C. 1972. The importance of the higher primates in the diet of the Fang of Rio Muni. *Man* 7: 239–43.

Segal, E. F. (ed.). 1989. *Housing, care and psychological well-being of captive and laboratory primates.* Park Ridge, NJ: Noyes.

Spiegel, M. 1996. *The dreaded comparison: Human and animal slavery.* New York: Mirror Books.

Sweetwaters Chimpanzee Sanctuary (SCS). 1996. *Sweetwaters Chimpanzee Sanctuary: Colony size and structure and integration policies.* Nairobi: Kenya Wildlife Service.

Teleki, G. 1987. A visit with Sheila and David Siddle and their sixteen chimpanzees: *IPPL Newsletter* 14(3): 3–7.

———. 1989. Population status of wild chimpanzees (*Pan troglodytes*) and threats to survival. In P. G. Heltne and L. A. Marquardt (eds.), *Understanding chimpanzees* (pp. 312–53). Cambridge, MA: Harvard University Press.

Timberlake, L. 1985. *Africa in crisis: The causes, the cures of environmental bankruptcy.* London: Earthscan.

Tutin, C. E. G. 1996. *Expertise sur les possibilities de reintroduction des chimpanzes en foret naturelle dans le reserve de Conkuoati, Congo: Rapport Final, Mars 1996.* Gland, Switzerland: International Union for the Conservation of Nature and Natural Resources.

Tuxill, J. 1997. The global decline of primates. *World Watch* 10(5): 16–21.

Wilson, M. L., and Elicker, J. G. 1976. Establishment, maintenance, and behavior of free-ranging chimpanzees on Ossabaw Island, Georgia, U.S.A. *Primates* 17(4): 451–73.

Wolfheim, J. H. 1983. *Primates of the world: Distribution, abundance, and conservation.* Seattle: University of Washington Press.

THOMAS L. WOLFE

7
THE RETIREMENT OF RESEARCH APES

With the signing of the Convention on International Trade in Endangered Species of Wild Fauna and Flora (CITES) in 1975, directors and scientists from the seven chimpanzee, *Pan troglodytes*, breeding and research centers in the United States—the Coulston Foundation, the Laboratory for Experimental Medicine and Surgery in Primates, New Iberia Research Center, Primate Foundation of Arizona, Southwest Foundation for Biomedical Research, University of Texas M. D. Anderson Cancer Science Park, and Yerkes Regional Primate Research Center—had to manage the captive population to ensure a stable supply for research without importation from the wild. An ad hoc task force formed by the Interagency Primate Steering Committee (later known as the Interagency Research Animal Committee, IRAC) developed husbandry guidelines that, in 1985, led to a proposed National Chimpanzee Breeding and Research Program. In 1986, the National Institutes of Health (NIH) selected five of the colonies (Coulston, New Iberia, Primate Foundation, M. D. Anderson, and Yerkes) for funding. Awards were also made for genetic monitoring, research in behavior, reproduction, and cryopreservation, and to the International Species Information System (ISIS) for developing a demographic and genetic database combining all five colonies and other colonies on a voluntary basis. From ISIS recommendations, and a selection by each colony director of proven and likely breeders, approximately 315 animals of all ages were selected for the NIH program.

In 1996, after about 400 live births, the NIH assessed the program and found an apparent surplus of chimpanzees in the colonies. This led the NIH to ask the National Research Council (NRC) of the National Academy of Sciences whether chimpanzees were still important for biomedical research. If they were determined to be so, how many and how should they be provided? If animals were found that were no longer needed for research or breeding, the NRC was to provide cost-effective recommendations for their future. Euthanasia and retirement to sanctuaries were two possibilities. The NRC formed a panel of experts in chimpanzee biology, behavior, veterinary medicine, colony management, virology, genetics, demography, animal welfare, and ethics. The panel produced a report, *Chimpanzees in Research: Strategies for Their Ethical Care, Management, and Use* (NRC 1997), that concluded that chimpanzees had made important biomedical contributions in the past and were likely to do so in the future, especially for understanding and preventing diseases caused by respiratory syntitial viruses, hepatitis viruses, and the human immunodeficiency virus (HIV). That chimpanzees had not responded as anticipated to HIV challenge and the success of the breeding program were cited as principal reasons for the oversupply of chimpanzees. Yet the panel felt that chimpanzees might still play a critical role in the development of HIV vaccines, as well as in understanding new, highly infectious zoonotic agents.

In the mid-1980s, concurrent with the establishment of the chimpanzee research and breeding program, the NIH established an Animal Models of AIDS Committee (AMC). This was a time when chimpanzees were thought to hold the key to understanding the pathogenicity of HIV, and numerous investigators were seeking to use chimpanzees for this critical research. The purpose of the AMC was to review all HIV research sponsored by the Public Health Service (PHS) that used chimpanzees, to avoid unnecessary duplication and to ensure that the animals were being used in the highest quality research with good probability of achieving significant findings. This original mandate was later expanded to include all chimpanzee use. The NRC committee encouraged the AMC to seek to expand studies of basic chimpanzee physiology and behavior for the betterment of both chimpanzee health and well-being and the scientific understanding essential for the use of chimpanzees in studying future human diseases.

The NRC report recommended that overall management of the approximately 1,500 research chimpanzees in the United States, or at least the 1,000 used by the NIH and other federal agencies, be centralized in the office of the director at the NIH. The purpose would be to monitor research use, to assign animals to approved research protocols, to limit breeding to that which was needed to support the colonies and research, to monitor standards of housing and care, and to identify animals no longer needed for breeding or research. It was recommended that a Chimpanzee Management Program (ChiMP) Office be established at the NIH for these purposes. The office would have oversight by a council of nongovernment experts with a wide range of expertise in captive chimpanzee management, veterinary medicine, reproductive biology, demography, population genetics, biomedical research, animal welfare, and ethics.

The NRC report recommended against euthanasia for population control. However, many colony managers and scientists argued to the committee that without suitable funding to support existing colonies, the health and quality of life of the entire colony might be threatened, requiring euthanasia as a humane option of last resort.

The NRC report suggested three options for chimpanzees no longer needed for breeding or research: Send them to existing private sanctuaries, let them remain in government-sponsored chimpanzee facilities remodeled for cost-effective long-term care, or transfer them to publicly managed sanctuary facilities—which did not yet exist. Each option has strengths and weaknesses, and although the report endorsed the concept of sanctuaries and encouraged the NIH and Congress to explore those possibilities, it did not make a recommendation for which option to follow.

There are two key issues in regard to retirement of research chimpanzees. The first is how and where to house approximately 300 animals with a confirmed potential for transmitting zoonotic diseases (NRC 1997). It can be reasonably argued that some cost savings can be achieved by housing them in proximity to existing chimpanzee research facilities and sharing the costly infrastructure. Otherwise, sanctuaries housing diseased animals must either independently provide for human safety or risk not complying with the occupational health and safety, veterinary care, facilities, and animal maintenance requirements of the U.S. Department of Agriculture (USDA) and the Occupational Safety and Health Administration (OSHA).

The second is whether any chimpanzees will ever be declared unnecessary for breeding or research. Although unable to provide an estimate, the NRC committee believed that there are more animals in the captive population than needed for research or breeding, and thus some might qualify for retirement.

Following careful review of the published report, the NIH endorsed the study and charged the National Center for Research Resources (NCRR) with its implementation. In accordance with the report, the NCRR is seeking to redefine the breeding and research program. As recommended in the NRC chimpanzee report, NCRR/NIH is proceeding to triage the U.S. biomedical population into four categories: breeding, research, reserve for research and breeding, and potential risk. Notably absent from this list are animals no longer needed for research or breeding, which logically might be transferred to sanctuaries for lifetime care. It is important to emphasize that the NIH has consistently stated that NIH funds will not be used to support animals that are not available for research or breeding. Under that condition, it therefore seems reasonable to assume that animals managed by the ChiMp office will not include animals identified for sanctuaries. There is, however, one possible exception: the animals in the NIH's chimpanzee research and breeding program. NIH officials have indicated they feel a strong moral commitment for the welfare of these animals and have voiced a willingness to consider transfer of ownership to the government. From this pool, it is likely that some would be identified as not needed for research or breeding and potentially subject to transfer to some type of long-term care facility.

RETIREMENT

Subsequent to publication of the NRC chimpanzee report in 1997, a consortium of protectionist organizations and individuals from the research community sought congressional support for a bill to create a sanctuary system for the permanent retirement of research chimpanzees not needed for research or breeding. Their work was rewarded in the 106th Congress when H.R. 106-3514 was passed, "to amend the Public Health Service Act to provide for a system of sanctuaries for chimpanzees that have been designated as being no longer needed in research conducted or supported by the Public Health Service, and for other purposes" (2000). President Bill Clinton signed the Chimpanzee Health Improvement, Maintenance, and Protection Act on December 20, 2000, with the admonition that, "I sign this measure with reservations concerning flaws in the bill that the next Administration and the Congress should correct to ensure the viability and effectiveness of the proposed sanctuary system." His reservations are shared by the NIH and by many scientists. At issue is timely access to sanctuary animals by biomedical scientists. NIH feels that taxpayers should not fund chimpanzees that are not available to contribute to critical human health issues. If, for example, a sanctuary animal that was previously in studies of HIV/AIDS shows clinical signs of HIV/AIDS, scientists would be eager to study that animal with the hope that it might provide further insight into the disease and thereby improve the health of humans around the world. In President Clinton's (2000) words,

> The Act puts severe constraints on use of a chimpanzee for further research, once it has been declared "surplus" and accepted into the sanctuary system. Before it could thereafter be used, other than for noninvasive behavioral research, the Secretary [of the Department of Health and Human Services] must determine that extremely stringent criteria are met concerning the indispensability of that particular chimpanzee and the key nature of the research. In addition, the board of directors of the nonprofit entity operating the sanctu-

ary must determine that the research design minimizes physical and mental harm to the chimpanzee—a determination that can be set aside only if the Secretary finds it arbitrary or capricious. Finally, the Secretary's and board's determinations must be published for a public comment period of not less than 60 days.

These are issues that remain unresolved.

Sanctuaries, as defined by H.R. 106-3514 and most protectionists, are long-term retirement facilities managed by nongovernment entities. They are considered to be "one-way" facilities; the animals would never go back to biomedical research without stringent conditions being met. A number of sanctuaries already exist but none currently has facilities for the large number of research chimpanzees potentially available. Although some of these facilities have accepted research animals, they were founded primarily for housing ex-pet and performing animals. Existing sanctuaries struggle for funding. Some find it difficult to turn away any needy new animal, but additional animals require space and caregivers' time, and this is often provided at the expense of reducing the welfare of the other animals in the colony. The proper balance between the welfare of individual animals and that of the group is difficult to reach and, when the focus is to help a new animal, there is a risk of reducing enriched space and sanitation for the remainder of the animals to an unacceptable level. To address what is thought to be a surplus of research chimpanzees, more than one new sanctuary has been proposed that will provide an appropriate balance of space, sanitation, health care, and management such as that found in state-of-the-art zoos and research facilities. Unfortunately, no such facility is yet developed to the point of accepting animals, but at least one has acquired land and is developing construction plans. It is hoped that such a facility, with lifetime endowment for all animals, will become available in time to provide some research animals with the rich social environment they are used to and deserve for the remainder of their lives.

Facilities managed by government entities, such as the chimpanzee breeding and research centers, could provide another type of long-term care. Animals in these facilities that are determined not to be needed for research or breeding today might be placed in adjacent or nearby facilities and taken out of the research pool. However, the animals might be returned to research if they are needed for medical research at some time in the future. In this way, these animals might be thought of as "two-way" or in temporary retirement. Animals that are a potential health risk to humans or other animals are among those likely to be in this category.

The economies are obvious, and the development of any new retirement facility should consider this possibility.

All types of long-term care facilities have strengths and weaknesses. Sanctuaries have the opportunity to remove chimpanzees from biomedical research. The NRC's chimpanzee committee believes that there are more than needed in the U.S. research population. However, the NIH is opposed to supporting chimpanzees with taxpayer dollars. It is often stated by biomedical researchers that because chimpanzees have been used to solve diseases of the public, the animals should now be the responsibility of public groups who would like to see them in sanctuaries. Although there might be some validity to that opinion, complex issues are seldom so easily resolved.

A key concept is that the majority of chimpanzees in research and breeding facilities that are government-supported are potentially suitable for placement in a sanctuary are privately owned, and proper management is a concern of the current owners, who must "donate" the animals to the sanctuaries. They feel strongly that the animals are in rich, social settings now and cared for by experienced technicians, behaviorists, and veterinarians. The NRC report sets high standards for the current owners, calling for enriched, social housing. Chimpanzees, including those in research, should not be singly housed more than six months at a time without specific justification and approval by an oversight body, such as an institutional animal care and use committee. Every effort should be made to pair- or group-house all animals. Veterinary care, including the ability to separate single animals from their social group for observation, treatment, or integration into a more compatible social group, is emphasized. Because the chimpanzee research and breeding centers receive funds from the government and are licensed as research centers, they are inspected by the USDA and must comply with the Public Health Service (PHS) Policy on Humane Care and Use of Laboratory Animals (PHS 1996) and the Guide for the Care and Use of Laboratory Animals (hereinafter Guide; NRC 1996). Also, most are accredited by the Association for Assessment and Accreditation of Laboratory Animal Care International (AALAC). Their personnel are well-trained and the centers have strong programs in the many areas called for in the Guide, such as occupational health and safety, disaster planning, sanitation, veterinary medicine, animal well-being, administrative commitment, and facilities maintenance. Of critical importance is that modern chimpanzee facilities have the knowledge, experience, space, and facilities to form new social groups and make adjustments to that group as needed. Well-trained behaviorists and behavioral technicians are employed to study and maintain the social harmony of groups and to ensure the safety and well-being of each individual (NRC 1998). How well sanctuaries stack up against these criteria will be an important consideration. Financial stability is essential in ensuring success, but for many present owners the current welfare of the animals trumps promises of lifetime care in substandard facilities. For animals transferred to sanctuaries, the NRC report provides detailed recommendations for housing and husbandry, recommends adoption of an accreditation program, and suggests that the USDA conduct routine animal welfare inspections. H.R. 106-3514 includes a provision that sanctuaries comply with the Animal Welfare Act.

The NIH has expressed interest in supporting only animals available for research, as mentioned earlier, and thus is not actively seeking a role in placing privately owned animals not needed for research or breeding into sanctuaries. Numbers tell the story. In 2001 the NIH owned one third of the chimpanzees in research and breeding facilities. The remainder are privately owned by the chimpanzee research and breeding centers.

REHABILITATION

Rehabilitation refers to a process of adaptation to a new environment, such as the transfer of former performing animals to zoos or research, for research animals into sanctuaries, or for captive animals into a wild environment. To address the latter first, there is a strong consensus of both captive and wild chimpanzee biologists that research chimpanzees should not be transferred to Africa or even out of the United States, except for bona fide exhibition or research. The danger associated with introducing new diseases into wild populations, either directly to native populations of chimpanzees or via other vectors, is too great. Captive animals have little or no experience with the risks in the wild and will, in all probability, suffer greatly before dying or being killed. Such a venture has a low probability of success, comes with a high price tag, and is an inhumane form of euthanasia.

Performing animals moved to sanctuaries or zoos might well be in need of rehabilitation. But is this notion appropriate when discussing movement of current research chimpanzees to sanctuaries? In years past, housing conditions for research chimpanzees were much different, and infants and juveniles grew up not knowing how to be chimpanzees. Few of these conditions remain today, and chimpanzee research facilities strive to rear their animals socially. Most current facilities are excellent and provide the animals with a wide range of movement and social interaction. Therefore, many past reasons for rehabilitation are no longer valid. Yet the term *rehabilitation* is still used by some to imply that research, and possibly zoological, institutions remain in the dark ages, have not been listening, and just "don't get it." This is simply not true. There are many remarkable zoos and animal parks and exemplary research and breeding facilities in the United States and other countries; it is difficult to see what rehabilitation is needed for the majority of the population. As do the best of the zoo facilities, the best research facilities provide opportunity for the animals to interact socially, express a wide range of species-typical behaviors, and in some cases have a lofty view of the surrounding countryside. When mothers cannot or will not rear their young, or foster mothers are not available, infants sometimes have to be nursery-reared (Fritz and Fritz 1985). When nursery rearing is necessary there are facilities and trained, compassionate care staff available. This is an extremely labor-intensive, time-consuming, and expensive process, but youngsters reared in this manner in recent years have grown up to be normal adults and competent breeders and mothers (Bard 1994; Maki, Fritz, and England 1993). Chimpanzee centers provide some type of transitional caging for nursery-reared youngsters or older animals that have trouble adapting to a social environment. Enriched Prim-A-Dome™ structures provide the flexibility to enable access to outdoor play and social groups for many animals. There are always exceptions and extenuating circumstances, but there are many competent chimpanzee biologists, veterinarians, and behaviorists in this country to handle these situations in a compassionate and humane manner when they occur. (See also Fritz, Wolfle, and Howell 1999, for discussions regarding the care and management of captive chimpanzees.)

CONCLUSION

These comments are meant to provide a somewhat different picture of the life and conditions of research chimpanzees than is sometimes portrayed. Picture animals living in large, multigenerational social groups in enriched environments with climbing structures, food puzzles, nesting materials, swings, foraging opportunities, deep bedding in the night quarters, and on-the-spot veterinary medical care. A facility director, colony manager, veterinarian, animal behaviorist, and a trained and dedicated animal care staff oversee this structure. Furthermore, the days are gone in which animals on study were maintained singly for extended duration in small unenriched isolation cubicles. With very few exceptions, animals in research facilities live a rich life of social interaction in which they have ample facilities to express a wide range of species-typical behaviors. This is not said to justify their use in research but to constructively focus attention on the real issues and away from any notion that the goal is to "release tortured animals from their cruel captors."

Achieving the goals of the NIH, Congress, and the public will require several efforts. The NIH's first responsibility is to improve the health of the citizens of the United States through biomedical research. It has, however, also established highly regarded standards for humane care and use of research animals and has a rich history of dealing with the public on matters of using animals in research. This prestigious institution will take seriously the ethical and moral issues associated with the use of chimpanzees in research and establish dialogue with those who have interest in identifying chimpanzees suitable for placing in some type of long-term care facility. Selecting chimpanzees not needed for research or breeding should not be difficult, and most chimpanzee colony directors have likely identified potential candidates already. With passage of H.R. 106-3514, the government just now seeks to find nonprofit entities to establish, construct, and manage sanctuaries. Funding for construction and maintenance is to consist of matching funds provided by the nonprofit entity and the government. To oversee the ChiMP program, the NCRR includes on its National Advisory Council a wide range of expertise as recommended in the NRC report (NRC 1997).

Protectionist organizations should join with the chimpanzee colony directors and federal agencies in developing a plan for the establishment and management of sanctuaries in accordance with H.R. 106-3514. Groups interested in developing chimpanzee sanctuaries are at a disadvantage by not knowing potential numbers of available chimpanzees, yet at least one has acquired property and developed facility plans and budgets. The NIH has expressed an obligation for chimpanzees it has supported in research and breeding and would benefit by discussing retirement options with protectionists and others. By so doing, the NIH might considerably reduce its long-term obligation for animals no longer needed, and thus gain public support. Neither group can resolve these issues by themselves. The owners of the animals must also come to the table and enter into these discussions.

Colony directors and interested members of the public, including protectionists, should work with the NIH to achieve agreed on plans for retiring research chimpanzees in accordance with H.R. 106-3514. Without this broad cooperation, many animals not included in NIH support grants will become unsupported, risking a decision of euthanasia. The greatest threat to cooperation is in the sincere difference of opinion regarding timely access to sanctuary animals by investigators. As discussed, options appear to be available that would result in a resolution of this seeming stalemate. If ever there was a time, and opportunity, for researchers, zoological institutions, and protectionists to come to the table together on behalf of humankind's closest relative, this is it. We must put individual agendas behind us, stop issuing unilateral manifestoes, and seek ways by which to achieve the goal desired by so many. When rhetoric replaces good judgment, the issues become clouded and the animals become victims rather than beneficiaries. We owe it to the chimpanzees, living and dead, who have contributed so much to enrich our lives through research and education. H.R. 106-3514 provides an opportunity to achieve these goals.

REFERENCES

Bard, K. A. 1994. Very early social learning: The effect of neonatal environment on chimpanzees' social responsiveness. In J. J. Roeder, B. Thierry, J. R. Anderson, and N. Herrenschmidt (eds.), *Current primatology: Vol. II. Social development, learning and behavior* (pp. 339–46). Strasbourg, France: Université Louis Pasteur.

Chimpanzee Health Improvement, Maintenance, and Protection Act. 2000. Pub. L. No. 106-551.

Fritz J., and Fritz, P. 1985. The hand-rearing unit: Management decisions that may affect chimpanzee development. In C. E. Graham and J. A. Brown (eds.), *Clinical management of infant apes* (pp. 1–34). New York: Alan R. Liss.

Fritz, J., Wolfle, T., and Howell, S. 1999. Chimpanzees. In T. Poole (ed.), *UFAW handbook on the care and management of laboratory animals, 7th edition, Volume I, Terrestrial Vertebrates* (pp. 643–58). Hertz, UK: Universities Federation of Animal Welfare.

Maki S., Fritz, J., and England, N. 1993. An assessment of early differential rearing conditions on later behavioral development in captive chimpanzees. *Infant Behavior and Development* 16: 373–81.

NRC. 1996. *Guide for the care and use of laboratory animals. Committee to revise the guide for the care and use of laboratory animals.* Washington, D.C.: Institute of Laboratory Animal Resources, National Academy Press.

———. 1997. *Chimpanzees in research. Strategies for their ethical care, management, and use. Committee on Long-Term Care of Chimpanzees.* Washington, D.C.: Institute for Laboratory Animal Research. National Academy Press.

———. 1998. *The psychological well-being of nonhuman primates. Committee on well-being of nonhuman primates.* Washington, D.C.: Institute of Laboratory Animal Resources, National Academy Press.

Public Health Service (PHS). 1996. *Public health service policy on humane care and use of laboratory animals.* Washington, D.C.: U.S. Department of Health and Human Services. (Pub. L. No. 99-158, Health Research Extension Act, 1985)

Section 3

HISTORY AND EVOLUTION

RAYMOND CORBEY

8
NEGOTIATING THE APE–HUMAN BOUNDARY

Between 1650 and 1780, about 650 mostly young men were convicted of bestiality and executed in Sweden, together with hundreds of cows, mares, and other animals (Liliequist 1990). What today would be regarded as relatively harmless experiments with youthful sexuality in the context of a rural adolescent subculture was in those days perceived as a capital offense: a sin against the natural, sacred order of God's creation, a demonic transgression of the sacrosanct human–animal boundary. When Carolus Linnaeus, the eighteenth-century Swedish naturalist, classified humans together with apes in the same anatomical order, that of the Anthropomorpha or Primates, and included a nonhuman primate in the genus *Homo,* he did something similar. Other naturalists took offense at the unprecedented rapprochement of human and beast in the various editions of *Systema naturae,* even though Linnaeus repeatedly stressed that, notwithstanding all morphological similarities, their invisible, reasonable soul put humans high above animals.

Apes are the most prominent inhabitants of the borderland between human and beast in Western imagination. They are the animals perceived to be closest to humans. In ancient times it was because of their general human-like appearance; since the seventeenth century it was because of specific anatomical similarities to humans; in the eighteenth century, as well as in recent decades, it has been because of their presumed linguistic capacities (Wokler 1995). Since the nineteenth century, it has been because of evolutionary reasons. Recently discovered biochemical similarities, as well as the understanding that nonhuman primates are also toolmakers and capable of deceiving others, have strengthened the perception of affinity (e.g., Corbey 1996).

This chapter explores some episodes of the complex and fascinating struggle concerning the status of human and nonhuman primates that has taken place since Linnaeus, in various sciences as well as in Western cultural imagination in general, thus providing some historical and intellectual background to recent discussions on the moral status of the great apes (Cavelieri and Singer 1994). The history of scientific approaches to primates since the eighteenth century is characterized by an enormous increase in empirical knowledge pertaining to their systematics, evolution, genetics, ecology, behavior, and cognition. But at the same time it is, as we shall see, a permanent struggle with the most significant, most heavily tabooed dividing line within nature; an enduring activity of negotiating—drawing, denying, policing, bridging, displacing—the metaphysical, religious, and moral boundary between humans and their closest relatives in nature. Indeed, an alternation of humanizing and bestializing moves with respect to both apes and humans.

THE APE'S ESSENCE

In the mid-eighteenth century, Linnaeus was the first to rank humans emphatically with other primates in the many editions of his authoritative *Systema naturae.* The reactions provoked by that act reveal the preoccupation, and its basis, of most of his contemporaries with human distinctiveness and dignity. Johan Georg Gmelin from Petersburg, for one, immediately challenged the inclusion of human among the Anthropomorpha, pointing to Genesis 1:26ff: Humans are *imago Dei,* the only living being created in the image of God (Gmelin 1746/1976). Gmelin, Wallerius, Klein, Haller, and Pennant all disagreed with Linnaeus. In his *Histoire Naturelle de l'Homme,* Buffon in Paris pointed to the divine breath that penetrates the human body and, following Descartes, conceived of humans as *H. duplex:* consisting of a body that is similar to that of the brute apes but also consisting of mind, which is a unique feature, connected with the faculty of speech (Leclerc de Buffon 1749). To confuse a human with a beast, one must be as poorly enlightened as a beast, he sneered, probably also against Jean-Jacques Rousseau, among others, who saw anthropoid apes as some sort of natural man, as humans in their "natural state," in principle capable of speech (Rousseau 1755).

Blumenbach in Göttingen, Germany, then split the order of the Primates established by Linnaeus into two, getting rid of the uncomfortable closeness of human and ape. Humans belonged in the separate biological order of two-handed Bimanus, not in that of the four-handed Quadrumana, he claimed (Blumenbach 1776). This had to do with a new approach to classification, taking not only morphology but also functions, corresponding to the Creator's intentions, into account: Our grasping hands and upright gait set humans apart; we were designed to stand and walk upright. The preoccupation with human uniqueness was a guiding principle of Blumenbach's natural history of humans, which, as he wrote in a letter to Haller in 1775, set out "to defend the rights of mankind and to contest the ridiculous association with the true ape" (Dougherty 1984, 64–66).

Despite this criticism, Linnaeus's natural history was, if anything, Christian science, a *scientia divina.* Linnaeus was an orthodox Lutheran and almost certainly agreed when Haller in 1746, in a review of Linnaeus's *Fauna Suecica,* called him a second Adam naming the animals, even though that was meant sarcastically (von Haller 1787/1971). Linnaeus saw nature as God's creation, a strictly ordered, immensely diverse hierarchy or scale of beings, with humans on top, and it was his sacred mission, he believed, to reconstruct and codify that order.

In the Christian cosmology that was constitutive for much of eighteenth-century natural history, humans were seen as *imago Dei,* the only creature created in God's own image: As *H. dominator,* it is immediately added in Genesis 1:28, humans have dominion over the rest of nature, including animals and apes. The then-predominant view of apes was still close to the medieval icon of *simia figura diaboli*—the hideous, frivolous monkey as the image of the devil. The Christian view of the world, taking the ontologies of Plato and Aristotle into account, explains structure and change in nature in terms of eternal, immutable essences, called *souls* in the case of living beings. Only humans have rational souls, and that constitutes their essence (*essentia*) and "essential" difference from apes. The influence of this Christian metaphysics of apes on eighteenth-century natural history converged with that of seventeenth-century rationalist, Cartesian philosophy, which held humans to be the only being possessing reason, thus keeping up the human–animal boundary.

There was yet another source of unfavorable views of animals in general and nonhuman primates in particular, in addition to Judeo-Christian cosmology, Greek ontology, and modern rationalism. Modern European citizens looked on themselves as "civilized" persons and acted as such, behaving, dressing, eating, defecating, making love, and so on, in a proper, "civilized" manner. In this context, animals and their "uncontrolled," "beastly" behavior provided models of how not to behave and were associated by citizens with their own "beastly" bodies and bodily functions, which a "civilized" person was to keep under control. Europeans thus articulated their own identity in terms of the exemplary otherness of uncivilized animals: We are not like them. Monkeys especially, because of their ambiguous, uncanny similarity to humans, provided a prominent model of undesirable otherness in European culture (see Corbey 1994).

These various backgrounds converged, as they still do today, to a strictly drawn theoretical, moral, and practical human–animal boundary. Given these backgrounds, we begin to understand why eighteenth-century Europeans reacted so severely to rural youngsters playing sexual games with cattle or even to natural historians associating humans closely with other primates.

APE ANCESTRY

The replacement of the idea of eternally fixed essences of species by that of transmutation by random variation and selective retention implied that humans descended from apelike ancestors. Indeed, as in the preceding centuries the great apes were gradually discovered and ultimately distinguished from one another in the context of worldwide trade and the expansion of European nation states, so from the mid-nineteenth century onward a long series of early hominids were discovered: Neanderthals in Europe and *H. erectus* in Indonesia, followed by several species of australopithecines and *H. habilis* in Africa, among others. How did these discoveries affect how Europeans saw themselves, apes, and the rest of nature?

Again, human and ape were brought in painfully close association, closer even than in Linnaeus's time, because of a postulated genealogical link. Just as in the case of Linnaeus, Charles Darwin himself was the first to worry about this new challenge to traditional European cosmology, threatening to turn it upside down. A remark jotted in his notebooks in 1838, "The Devil under form of baboon is our Grandfather" (Barrett and Gautrey 1987, 128), reminds of the medieval *simia figura diaboli* icon, and in a letter to Joseph Hooker, Darwin referred to the evolutionist view of nature as that of "a Devil's Chaplain" (Desmond and Moore 1992, 449). His concern for human dignity and fear for his own reputation as a God-fearing and law-abiding citizen made him postpone dealing explicitly with human descent until 1871, when he published *The Descent of Man*.

How intensely Darwin wrestled with the problem of human's place in nature, not just biologically but also in a metaphysical and moral sense, is evident from many ambiguous as well as ambivalent passages in that book. "Man may be excused," he wrote at the end, "from feeling some pride at having risen . . . to the very summit of the organic scale; and the fact of his having thus risen, instead of having been aboriginally placed there, may give him hope for a still higher destiny in the distant future." Despite his "god-like intellect" (the *imago Dei* idea again) and other noble qualities, "man still bears in his bodily frame the indelible stamp of his lowly origin" (Darwin 1877, 619). In a similar vein, Thomas Huxley, one of Darwin's close supporters, stipulated that our "reverence for the nobility of manhood will not be lessened by the knowledge that man is, in substance and structure, one with the brutes" (Huxley 1863, 112). Because of that structural, evolutionary unity with brute animals Huxley reinstated the Linnaean primate order that had been split by Blumenbach.

Like Linnaeus in another context, the evolutionists were confronted with contemporaries who, like themselves, perceived the threat to human dignity posed by the "grim and grotesque procession" of ape ancestors, in the Duke of Argyll's words (Argyll 1868), but dealt differently with that menace in the context of their interpretations of biological data. One of those contemporaries was the anatomist, Richard Owen, well-versed in German holistic and teleological *Naturphilosophie,* for whom not just the spiritual but also the morphological separation of human and ape was a moral and metaphysical imperative during the whole of his career. In search of an anatomical feature that would permit a classification with humans apart, not just in a different order but in a different subclass at the head of creation, Owen (1857) proclaimed the hippocampus minor, a lobe of the brain, to be such a feature, unique to humans; Huxley, however, proved him wrong (Huxley 1861).

Impending apes and apemen were successfully kept at a distance by casting the newly discovered evolutionary process as a progress toward humanness, epitomized by European middle-class civilization, as its apex and natural goal—a widespread conviction toward the end of the nineteenth century. In this temporalized form, the traditional idea of a moral hierarchy of creatures with humans on top lingered. The question posed by Benjamin Disraeli in a speech in Oxford in 1864, "Is man an ape or an angel?", answered by him with, "My Lord, I am on the side of the angels" (quoted in Monypenny and Buckle 1929) now came to be generally decided in favor of apes. That "yes" to apes, however, was a "yes, but . . . ," because apish ancestors were kept at arm's length, together with apes and apish "contemporary ancestors," by situating them only at the beginning of the ascent *toward* humanness.

Apish, monstrous others loom large in many scientific, literary, and political writings from around the turn of the twentieth century. Early hominids, great apes, humans from prestate societies, or indeed one's political opponents were quite consistently depicted as unable to restrain their beastly impulses, as prone to violence, rape, incest, and cannibalism, again epitomizing uncivilized otherness. The Darwinian perception of nature as red in tooth and claw, a struggle for life in a quite literal sense, supported the image of primeval, wild, apish otherness, to be subdued by civilized control.

THE BEAST WITHIN

Not much later, an unrestrained animal other surfaced in one of the most influential views of human behavior, psychology, and culture of the twentieth century: psychoanalysis. As a Lamarckian, and under the influence of Ernst Haeckel's "biogenetic law" or "principle of recapitulation," Sigmund Freud held the harsh experiences of primeval humans to be constitutive of present-day human souls. Among those experiences, according to him, were primal patricides and incestuous sexual desire, giving rise to, among other things, the Oedipus complex, neurosis, religious behavior, and the characteristic behavior of crowds. "Incest, cannibalism and the lust for killing," he wrote in *The Future of an Illusion*, are "born afresh with every child" (Freud 1961; cf. Corbey 1991). Neurosis especially was a so-called atavism, a regression, not just ontogenetically but also phylogenetically, and Freud believed the behavior of neurotics, early humans, contemporary "primitives," and modern children to be comparable with respect to the role of "primitive" impulsiveness and a deficient sense of reality.

Civilization, in this view, is possible only by taming human's dark, ineradicable animal nature, by domesticating the beast within. Human behavior is the outcome of a struggle between civilized control, exercised by society through the child's real or internalized parents, and wild primeval impulses. Freud, an avid reader of the evolutionist anthropology of his day, concluded his 1915 discussion of the phenomenon of war with the statement that "[in] this respect, as in many others, the man of prehistoric times survives unchanged in our unconscious" (Freud 1957, 296). In the psychoanalytic conception of human nature, one of the clearest examples of an articulation of human identity in terms of animal otherness, the human–animal boundary is drawn within ourselves; the encroaching primeval beast within has to be controlled and humanized.

The "beast-in-man," usually depicted as apelike and to some extent an avatar of earlier, Platonic, Pauline, and Protestant dualistic views of human nature, is a forceful, omnipresent metaphor in the twentieth century, depicting human nature before, or deprived of, culture. King Kong in the 1933 film classic is close to what this beastly other was imagined to look like, and similar apelike monsters popped up in fascist anti-Communist propaganda and Communist antifascist propaganda. Many pictorial or literary descriptions of early hominids and great apes during the first half of the twentieth century presented the same image and were inspired by the same master narrative of an ascent toward humanness.

FROM IGNOBLE TO NOBLE?

Since the 1960s, the rapid expansion of research on many aspects of nonhuman primates, not least their social life and cognition, did much to change the predominantly negative image of the generalized apish other in Western imagination, a change that parallels that from a Hobbesian to a Rousseauesque perception of the "natural state" of humankind two centuries earlier. The well-known photograph where young Jane Goodall and a chimpanzee reach their pointing fingers toward each other, repeated by Dian Fossey and by actress Sigourney Weaver playing her in a successful film on Fossey's research among mountain gorillas (*Gorillas in the Mist*), sums it up neatly: This was a highly symbolic gesture, explicitly aimed at bridging the human–animal boundary. These apes were cast as happy rather than brutish beasts, living in relative harmony with their kin in an East African Eden instead of having to survive in the harsh jungle. The regular occurrence of violence among chimpanzees, discovered more recently, was a crack in that particular mirror for Western humans.

The well-known work of primatologist Frans de Waal revealed the complexity of social and political behavior among nonhuman primates and its close similarity to that of humans with respect to aggression, reciprocal altruism, and mechanisms of conflict control (de Waal 1982). It helped to combat the widespread inclination of seeing our bad habits as exclusively animal and our good ones as exclusively human. De Waal increasingly stressed that apes are basically *Good Natured* (de Waal 1996), as the title of his 1996 book has it, both on a technical, scientific, and philosophical level. His interpretations of *Peace-Making among Primates* (de Waal 1989) in terms of concepts normally used for humans, such as politics, friendship, empathy, and forgiveness, have not gone uncriticized, although at least some of the reproach of anthropomorphism would seem to be another expression of the anthropocentric concern with human uniqueness that has accompanied the study of nonhuman primates since Linnaeus.

Harvard primatologist Richard Wrangham's Hobbesian approach to primates is diametrically opposed to de Waal's stress on peacemaking. His 1996 book on *Demonic Males,* written for a large audience in cooperation with Dale Peterson, emphasizes male violence and aggression as pivotal survival strategies: Males are selected by females for exploitive and aggressive behaviors, leading to competitive success. "We are cursed," Wrangham and Peterson wrote, "with a demonic male temperament and a Machiavellian capacity to express it," a "5-million years stain of our ape past" (Wrangham and Petersen 1996, 258). It should be added, however, that both Wrangham and Peterson's *Demonic Males* and de Waal's *Good Natured* have been received with mixed feelings by primatologists and anthropologists.

Although since the 1960s there has been a shift toward generally more positive views of nonhuman primates, small-scale non-Western peoples, and early hominids, the total picture is more complex. A Hobbesian perception of human and primate nature persisted in certain tendencies in the work of ethologists and sociobiologists, and, in a sense, in Richard Dawkins's "selfish genes" metaphor, explaining altruism as a form of selfishness on a more basic level. Feminist paleoanthropologists and primatologists, on the other hand, counterbalanced the "man the aggressive hunter" approach with a new, "woman the peaceful gatherer" research paradigm in the 1970s. This added to a tendency appearing at about the same time to depict early hominids as happy families in peaceful East African landscapes, replacing the monstrous primeval apemen of earlier generations, armed with clubs and struggling violently for survival.

Remarkably, violence and peacefulness as interpretive viewpoints, in scientific research as well as on the level of popular imagination, have an analogous role to play in ethnography. Traditionally, violence has been one of the main ascribed characteristics of non-Western humans, often perceived as apish "lower races" or "contemporary ancestors," and nonhuman primates alike. Both categories were seen as primitive, brute, and unrestrained and associated with the savage beginnings of humankind's progress to civilization. The shifting balance of negative and positive views of primates in recent decades is quite parallel to characterizations, during that same period, of certain peoples as explicitly aggressive and fierce or, alternatively, unambiguously gentle and peaceful.

Combining insights from sociobiology with cultural materialist ones, the anthropologist Napoleon Chagnon, for example, in his research on the Amazonia Yanomami, stresses the inclusive fitness of male warriors in the complex interrelationship between individuals, groups, and their natural environment: The more women they have access to, the better the proliferation of their genes (Chagnon 1997). This picture of the Yanomami as vengeful aggressors, beating up women and warring constantly, has been criticized as at least one-sided, and the Chewong and Semai Senoi from Malaysia, as well as the Sakkudei from Indonesia, have been thrown in the balance as decisively peaceful peoples (e.g., Howell and Willis 1989).

Comparable shifts took place in views of the Kalahari Desert !Kung. Although they used to be regarded as one of the most primitive and lamentable races of humankind, slotted between Caucasian humans and the "lowly" apes, they came to be hailed as gentle and harmless "noble savages," who, unlike Western humans, lived in close harmony with one another and with nature. Here too a crack appeared in the mirror, just like in the case of the chimpanzees, when their high homicide rates were pointed out by biologically orientated authors. Something similar happened to Margaret Mead's 1928 idyllic portrayal of *Coming of Age in Samoa:* Her underestimation of the role of jealousy, abuse, rape, and violence in the life of adolescents in Samoa was criticized as a culturalist bias by, again, a biologically orientated anthropologist (Freeman 1983).

ENCROACHING APES

Although speech has traditionally been seen as the outward appearance of mind, a faculty unique to humans, research on the linguistic abilities of bonobos and other apes has undermined the idea of *H. loquens,* humans as the only animal capable of language, closely linked with two other pervasive human self-definitions, that of *H. symbolicus* and that of *H. sapiens.* Self-recognition of their reflections by chimpanzees and dolphins also suggested similarity to the mind of humans, as did studies on tactical deception by nonhuman primates (e.g., Parker, Mitchell, and Bocca 1994).

The notion of a chasm between symbolic human culture and the rest of the living world is another instance of the human–animal boundary. In the eyes of many, if not most, anthropologists, that notion lies at the foundation of cultural anthropology as a discipline, serving to legitimize its autonomy with regard to the biological sciences, including physical anthropology. Learned, arbitrary, and variable symbolic meaning imposed on the environment was, and still is, held to set humans apart, and therefore requires a special methodology and indeed special *human* sciences. "Believing . . . that man is an animal suspended in webs of significance he has himself spun," wrote an influential cultural anthropologist, Clifford Geertz, giving voice to this conviction, "I take culture to be those webs, and the analysis of it to be therefore not an experimental science in search of law but an interpretive one in search of meaning" (Geertz 1973, 5).

Sociobiology and behavioral ecology, on the other hand, start from an assumption that is diametrically opposed to the *H. symbolicus* view of anthropologists: the uniformity of all behaving organisms, including humans. The attempts of these disciplines to deal with complex human behavior in the same way as with that of other primates and other animals met with furious resistance from cultural anthropologists. In their introduction to the volume they edited on *Understanding Behavior: What Primate Studies Tell Us about Human Behavior* (1991), James Loy and Calvin Peters complained about the fundamental reluctance of the human sciences to give serious attention to behavioral data from animals, or to take an evolutionary approach, a reluctance that is a direct corollary of their disciplinary identity. Everybody who has worked in an anthropology department will in some way have experienced the divergence between (physical) anthropologists who do and (cultural) anthropologists who do not think primate studies can tell us much about human behavior and culture.

Homo faber, "human as toolmaker," was a widespread human self-definition. When in the early 1960s archaeologist Louis Leakey heard that his pupil Jane Goodall had discovered tool use among chimpanzees in Tanzania, he reportedly remarked that now the notion of tool had to be redefined, or that of human, or else chimpanzees had to be accepted as human (Cole 1975). In an influential 1969 article titled, "Culture: A *Human* Domain," anthropologist Ralph Holloway claimed the imposition of arbitrary form on the environment to be "specific and unique to human behavior" and to be identifiable by the appearance of stone tools in the archaeological record (Holloway 1969). Paleoanthropologist Ian Tattersall complained as recently as 1994 about the "power of toolmaking to mesmerize paleoanthropologists into classifying the makers of any stone tools, however crude, in the genus *Homo*" (Tattersall 1994). However, much has been discovered on tool use and cultural traditions in nonhuman species since the 1960s, and *Pan* the toolmaker now stands side by side with *Homo* toolmakers.

REBUTTED AGAIN

For some, such discoveries pertaining to the linguistic, cognitive, and technological abilities of animals, especially nonhuman primates, have further problematized the human–animal boundary as it was traditionally drawn, whereas for others it has necessitated fortifying that boundary by redefining it. Chimpanzees may use symbols, the latter group say, but we should not just look for symbols but for syntactically ordered symbols; and even if they have syntax, do they use it to be reflexive about language, like we do? They may use tools, they say, but we should not just look for tools but for tools made *with* tools; they may have cultural traditions in the sense of intergenerationally transmitted, learned solutions to problems, but should we not look for the transmission of *symbolic* meaning?

Matt Cartmill, a physical anthropologist himself, has sharply criticized the persistent anthropological focus on human uniqueness as the phenomenon to be explained. Such supposedly unique human features as large brains, language, conceptual thinking, and upright bipedalism, he wrote,

> are uniquely human by definition rather than as a matter of empirical fact. Much scientific effort and ingenuity has gone into redefining such characteristics whenever discoveries about other animals have posed a threat to human uniqueness. (Cartmill 1990, 173)

Cartmill elaborated on a string of examples from his own field, physical anthropology. The uniquely large human brain, by anthropocentric definition, for example, came under attack several times. The much larger brains of elephants and whales were dealt with by shifting the criterion to brain size relative to body size. Unfortunately, according to that criterion, humans are surpassed by squirrels and other small animals, so a new one was proposed that was corrected for allometry. That move brought porpoises uncomfortably close, a threat to human uniqueness that again was neutralized by taking metabolic rates into account (Cartmill 1990).

The fallacy is assuming that the human form of a particular characteristic is its defining feature and the privileged standard against which to judge all species. The same fallacy is shown by Barbara King to be present in much work on the evolution of intergenerational information transmission and language, where, as she shows, language precursors are typically sought in a human-oriented way (King 1994, especially chap. 6). Such strategies, once again, keep apes at a comfortable distance, even in recent scientific research.

APES AND METAPHYSICS

There is more than one sense in which one can speak of the metaphysics of apes. Primarily they are important characters, exemplary others, in the grand metaphysical narratives and articulations of human identity in Western tradition. Over the past few decades, philosophy of science has become very aware of the fact that metaphysical or ontological assumptions are to be found in the very core of theoretical approaches in natural sciences as well as human sciences. Such usually implicit but very basic conceptual presuppositions are germane not only to disciplinary identity but also to concrete interpretations of data. When interpreting data in primatology, human origin studies, or research on the evolution of language and cognition, for example, much depends on how we conceive of *language, species,* or *intention.*

In that second sense, "metaphysics of apes" refers to ontological assumptions that structure our scientific approaches to human and nonhuman primates, their evolution, behavior, communication, and minds (see Corbey 1998). Such assumptions are predicated on the metaphysics of apes in the first sense: the way in which traditional Western religion and philosophy, Aristotelian, Cartesian, Kantian, phenomenological, hermeneutic, and so on, conceived of nonhuman primates. A third sense of the phrase, finally, that is relevant in this context is "metaphysics" in the sense of the views apes themselves have of the world, of morals, of us, of themselves.

Although ontological assumptions have a role to play in all scientific approaches, be it Linnean systematics, linguistics, or behavioral ecology, we should, in my opinion, be careful not to judge the history of our scientific dealings with nonhuman primates too quickly and simplistically in terms of a progress from, as one well-known anthropologist once put it, a data-poor and nonsense-rich era to a more sophisticated, data-rich era. The foregoing anthology from that history shows no such unequivocal progress, but, if anything, the persistence of metaphysics in the core of scientific theories. There are now, as there always have been, enormous differences "in the preconceptions, assumptions, and biases that different workers [bring] to bear on the resolution of problems that, on the surface at least, were thought to be held in common," as the editors of a volume on such preconceptions in the field of human origins studies say (Clark and Willermet 1997, 1; cf. Corbey and Roebroeks 2001). Such differences can at times be so vast as to preclude any common basis for discussion.

Analogously, there is no easy way of disqualifying scientific approaches that keep apes at a distance as somehow less scientific than those that stress continuities. Much of the scientific resistance against the idea of linguistic competence in apes, for example, is very sophisticated methodologically. Defenders of the idea of ape language, on the other hand, adhere to approaches that are as refined but take a different, often more hermeneutic or interpretive, view of what language is.

MOVING BEYOND

We have seen how the sacrosanct traditional markers of humanness, as an ontological and as a moral category, were compromised and redefined time and again, not only by rural Swedish youngsters but also, showing that science is also culture and metaphysics, by natural historians, by human scientists, and by linguists. Human identity was at stake and was defended against apes and apishness. Categorizing the living world was a means of keeping other animals at a distance and functioned as an ideological justification of human behaviors toward them (Corbey 1997). Such categorizations were, and still are, embedded in grand metaphysical narratives that install moral hierarchies or the still forceful remnants of such narratives: that of God and His creation, or, alternatively, that of the ascent of humans, or at least certain humans, beyond the animality of their apish ancestors toward "civilization" and true humanness.

The very idea of *essential* difference, although abandoned by twentieth-century biology, still pervades much of Western thought, laws (see chapter 15, this volume) and practices, inextricably connected with the traditional metaphysical view that species have eternal, immutable, discrete essences that form a moral hierarchy. Next to a growing awareness of our intimate connections with other apes, the preoccupation with an unambiguous, pure human identity persists, as does the rebuttal of whatever may contaminate that purity. Essentialism and anthropocentrism as avatars of the Western metaphysical tradition are reinforced by our commonsense habit of perceiving the world in terms of a moral order of clear-cut natural kinds, the integrity of which is not to be tinkered with by, for example, gene transfer or organ transplantation between animals and humans, cloning organisms, or indeed classifying nonhuman and human primates too closely together.

It may be time to move from drawing to bridging the boundary between humans and the rest of nature more definitely, by thinking against the grain of most of Western tradition and cultural attitudes, not necessarily as a definitive persuasion but as a move our era calls for, a morally relevant gesture of solidarity toward fellow beings who may thus play a new role: that of missing links between humans and the rest of nature. A sharp insight into the traditional, anthropocentric metaphysics of apes that still inspires much of human dealings with them, in society and in science, is indispensable in that context.

REFERENCES

Argyll, the Duke of. 1868. *The reign of law.* London: Strahan.

Barrett, P. H., and Gautrey, P. J. (eds.). 1987. *Charles Darwin's notebooks, 1836–1844.* Cambridge: British Museum—Natural History with Cambridge University Press.

Blumenbach, J. F. 1776. *De generis humani varietate nativa.* Göttingen, Germany: Vidua Abr. Vandenhoeck.

Cartmill, M. 1990. Human uniqueness and theoretical content in paleoanthropology. *International Journal of Primatology* 11: 173–92.

Cavalieri, P., and Singer, P. (eds.). 1994. *The Great Ape Project: Equality beyond humanity.* New York: St. Martin's Press.

Chagnon, N. 1997. *Yanomame* (5th ed.). Fort Worth, TX: Harcourt Brace.

Clark, G. A., and Willermet, C. M. (eds.). 1997. *Conceptual issues in modern human origins research.* New York: Aldine de Gruyter.

Cole, S. 1975. *Leakey's luck: The life of Louis Seymour Bazett Leakey, 1903–1972.* London: Collins.

Corbey, R. 1991. Freud's phylogenetic narrative. In R. Corbey and J. Leerssen (eds.), *Alterity, identity, image: Selves and others in society and scholarship* (pp. 37–56). Amsterdam: Rodopi.

———. 1994. Ambiguous apes. In P. Cavalieri and P. Singer (eds.), *The great ape project: Equality beyond humanity* (pp. 126–36). New York: St. Martin's Press.

———. 1996. Roots, backgrounds and contexts of primatology. A bibliographic essay. *Primate Report* 45: 29–44.

———. 1997. Inventaire et surveillance: L'appropriation de la nature à travers l'histoire naturelle. In C. Blanckaert (ed.), *Le Muséum au premier siècle de son existence* (pp. 541–57). Paris: Archives du Muséum National d'Histoire Naturelle.

———. 1998. De l'histoire naturelle à l'histoire humaine: Comment conceptualiser les origines de la culture? In A. Ducros, J. Ducros, and F. Joulian (eds.), *La nature est-elle naturelle? Histoire, épistémologie et applications récentes du concept de culture* (pp. 223–38). Paris: Éditions Errance.

Corbey, R., and Roebroeks, W. (eds.). 2001. *Studying human origins: Disciplinary history and epistemology.* Amsterdam: Amsterdam University Press.

Darwin, C. 1877. *The descent of man, and selection in relation to sex.* London: Murray.

Desmond, A., and Moore, J. 1992. *Darwin.* London: Penguin Books.

de Waal, F. 1982. *Chimpanzee politics: Power and sex among apes.* London: Jonathan Cape.

———. 1989. *Peace-making among primates.* Cambridge, MA: Harvard University Press.

———. 1996. *Good natured: The origins of right and wrong in humans and other animals.* Cambridge, MA: Harvard University Press.

Dougherty, F. (ed.). 1984. *Commercium epistolicum J.F. Blumenbachii.* Göttingen, Germany: Niedersächsische Staats und Universitätsbibliothek.

Freeman, D. 1983. *Margaret Mead and Samoa: The making and unmaking of an anthropological myth.* Cambridge, MA: Harvard University Press.

Freud, S. 1957. Thoughts for the times on war and death. In *Standard edition of the complete psychological writings of Sigmund Freud* (Vol. XIV, pp. 273–302). London: Hogarth Press and the Institute of Psycho-Analysis.

———. 1961. The future of an illusion. In *Standard edition of the complete psychological writings of Sigmund Freud* (Vol. XXI, pp. 5–56). London: Hogarth Press and the Institute of Psycho-Analysis.

Geertz, C. 1973. *The interpretation of cultures.* New York: Basic Books.

Gmelin, J. G. 1976. Letter to Linnaeus of 19 December 1746. In *Letters on natural history to Carolus Linnaeus* (Vol. V, pp. 41–42). London: Linnaean Society. (Original work published 1746)

Holloway, R. 1969. Culture: A *human* domain. *Current Anthropology* 10: 395–412.

Howell, S., and Willis, R. (eds.). 1989. *Societies at peace: Anthropological perspectives.* London: Routledge.

Huxley, T. 1861. On the zoological relations of man with the lower animals. *Natural History Review* 1: 67–84

———. 1863. *Evidence as to man's place in nature.* London: Williams and Norgate.

King, B. 1994. *The information continuum: Evolution of social information transfer in monkeys, apes and hominids.* Santa Fe, NM: SAR Press.

Leclerc de Buffon, G.-L. 1749. *Histoire Naturelle, Générelle et Particulière* (Vol. II, pp. 429–603 and Vol. III, pp. 305–539). Paris: Imprimierie Royale.

Liliequist, J. 1990. Peasants against nature: Crossing the boundaries between man and animal in seventeenth- and eighteenth-century Sweden. *Focaal: Tijdschrift voor Antropologie* 13: 28–54.

Loy, J., and Peters, C. 1991. *Understanding behavior: What primate studies tell us about human behavior.* New York: Oxford University Press.

Mead, M. 1928. *Coming of age in Somoa: A study of sex in primitive society.* New York: Editions for the Armed Services.

Monypenny, W. F., and Buckle, G. E. 1929. *The life of Benjamin Disraeli* (Vol. II). London: John Murray.

Owen, R. 1857. On the characters, principles of division, and the primary groups of the class mammalia. *Proceedings of the Linnaean Society* 2: 1–37.

Parker, S. T., Mitchell, R. W., and Bocca, M. L. (eds.). 1994. *Self-awareness in animals and humans.* Cambridge: Cambridge University Press.

Rousseau, J. J. 1755. *Discours sur l'origine et les fondements de l'inégalité parmi les hommes.* Amsterdam: M. M. Rey.

Tattersall, I. 1994. What do we mean by human—And why does it matter? *Evolutionary Anthropology* 3: 114–17.

von Haller, A. 1971. *Tagebuch seiner Beobachtungen über Schriftsteller und über sich selbst.* Frankfurt am Main: Athenaeum Reprints. (Original work published 1787)

Wokler, R. 1995. Enlightening apes: Eighteenth-century speculation and current experiments on linguistic competence. In R. Corbey and B. Theunissen (eds.), *Ape, man, apeman: Changing views since 1600* (pp. 87–100). Leiden, Germany: Department of Archaeology of Leiden University.

Wrangham, R., and Peterson, D. 1996. *Demonic males: Apes and the origin of human violence.* Boston: Houghton-Mifflin.

RUSSELL H. TUTTLE

9
PHYLOGENIES, FOSSILS, AND FEELINGS

Apes and people are hominoid, anthropoid, primate, mammalian, chordate animals. That is, systematically we are common members of the Linnaean kingdom Animalia, phylum Chordata, class Mammalia, order Primates, suborder Anthropoidea, and superfamily Hominoidea. Accordingly, phrases such as "primate and human evolution" and "humans and animals" are redundant because humans are primates and animals. If one wishes to highlight *Homo sapiens* among the Animalia in titles and discussions, zoologically correct vernacular phrasings for the two examples are, "human and nonhuman primate evolution" and "humans and other animals."

Systematic subdivision of the Hominoidea into families, subfamilies, tribes, genera, and other Linnaean categories has been and will continue to be fraught with problems and controversy. Schemes that are classically based on comparative morphology of extant forms and a few fossil bits face formidable challenges from cladistically minded molecular biologists and a new generation of paleoprimatologists who are backed by phalanxes of new and fortified fossil species. Because of mosaic evolution—bodily organs changing at different rates in the various lineages—morphologists may be forced to choose heads or tails and further to rationalize their phylogenic models against evidence from DNA sequences and grosser chromosomal features or other molecular data.

It is probably tactically unwise to refer to humans as "other apes," based on removal of African apes from the Pongidae and placing them in the Hominidae. In such a scheme the correct vernacular reference for chimpanzees, bonobos, and gorillas is "other people," which would not set well among many persons. Further, it is both taxonomically and literally incorrect to designate chimpanzees and bonobos *Homo troglodytes* and *Homo paniscus,* respectively, so that one could refer to people as a "third chimpanzee" (Diamond 1992). Indeed, if this change were sanctioned by the *International Code of Zoological Nomenclature,* chimpanzees and bonobos would be other people, which would be difficult for many to accept.

Both paleoanthropology and molecular evolutionary biology are booming. Major discoveries of fossil humanoids and precedent primates are broadcast annually. Moreover, developmental biology promises to provide vital links among genetics, environmental factors, morphology, and behavior, which will allow more secure weighting of traits in phylogenic modeling.

Paleoanthropology is a difficult historical science in which experimentation and finely controlled comparisons are rarely possible. Nonetheless, one commonly encounters the term *reconstruction* in evolutionary literature, particularly with regard to phylogenic schemes. In fact, an implication of reconstruction is misleading. Instead, we attempt to devise *models* because one rarely has enough bits for a reconstruction. Experts can reassemble fossil skulls and other bones and artifacts if enough pieces are recovered. But even in fossil restorations, a good deal of modeling may be required in lieu of missing pieces.

In the speculative realms of paleobehavioral and paleoecological modeling, we actually propose *scenarios* instead of *models* in the sense that physical scientists use the terms. Indeed, paleoanthropologists are almost always reduced to writing scientifically informed stories because behavior and habitats are transitory and leave only tantalizing traces. The key phrase is *scientifically informed.* Persuasive paleoanthropological scenarios differ from science fiction in that the former are bounded by facts. The extant apes and other animals, including humans, provide us with a wealth of suggestive anatomical, behavioral, and ecological data that complement patchy information from archaeology and paleontology.

HOMINOID PHYLOGENIES AND THE MIOCENE FOSSIL RECORD

Although modern paleoanthropological research has provided abundant data, there are notable gaps in knowledge of the human career. This is exponentially the case also for the careers of extant apes. During the past century a trove of fossil human and human-like creatures has been collected in Africa, Eurasia, and Australia. They are used by paleoanthropologists to model when and how we acquired our peculiar morphological and behavioral characteristics. Unfortunately, novel fossil discoveries often generate new speculation and controversy instead of solving existing puzzles.

It is encouraging that among scientifically informed persons both within and outside the academy there is a steady erosion of arrogance about our place in nature. For example, schemes such as that of G. Elliot Smith (1927), which boldly highlighted the Hominidae and *Homo* as the central crown branch in primate evolution, have been replaced by cladograms in which humans, represented by mere dots at the far edge of the figure, are shown equally with African apes, orangutans, gibbons, and extinct morphotypes (Tuttle 1986, 25–26).

The vital span in the Late Miocene, 10 to 5 million years ago (MA), during which molecular biologists, paleoanthropologists, and comparative morphologists generally agree that the common ancestors of modern African apes and humans lived, appears to be nearly void of compelling stem species. Candidates from earlier Miocene times have also failed to fulfill this role. There is a remote chance that we are overlooking the stem species because of temporal, zoogeographical, or morphological bias.

During the Miocene Epoch, Africa and Eurasia hosted prolific radiations of anthropoid primates with dentitions and, to a lesser extent, skulls that harbinger those of extant hominoids. In 1988, Fleagle conservatively favored the view that the Early and Middle Miocene catarrhine radiation in Eastern Africa may contain links to the remote ancestors of modern gibbons, great apes, and humans. Recently, however, venturesome paleoprimatologists have warmed to certain Eurasian Late Miocene species as ancestral to the lineages of extant great apes and humans, and they have sidelined ephemerally popular forms, including *Proconsul, Dendropithecus, Micropithecus, Kenyapithecus,* and *Afropithecus* (Andrews et al. 1996; Harrison 1993).

Currently, the most attractive candidate for Late Miocene ancestry of the human lineage via *Australopithecus* is *Graecopithecus freybergi* (*Ouranopithecus macedoniensis*). Whereas de Bonis (1987a, 1987b, 1987c) and coworkers (de Bonis and Koufos 1993, 1997; de Bonis and Melentis 1984; de Bonis et al. 1990) considered *Graecopithecus,* under the alias *Ouranopithecus,* to be ancestral only to the lineage containing *Australopithecus,* including *Paranthropus,* and *Homo,* Andrews et al. (1996) accepted the possibility that *Graecopithecus freybergi* is close to the ancestry of *Pan, Gorilla,* and *Australopithecus* sensu lato. De Bonis (1987a) preferred *Dryopithecus* as ancestral to *Pan* and *Gorilla.*

Based on an extensive cladistic analysis of fossil and extant species, Begun and Kordos (1997) and Begun, Ward, and Rose (1997) also concluded that *Dryopithecus* and *Ouranopithecus* are the closest Miocene species to *Pan, Gorilla, Australopithecus,* and *Homo.* Moreover, the most parsimonious cladogram based on their list of 239 skeletal traits, plus whether life history is rapid or prolonged, places *Pan* and *Australopithecus* closer to one another than either is to *Gorilla.* Their morphological finding mirrors the results of some molecular evolutionary biologists (Ruvolo et al. 1994; Sibley and Ahlquist 1987), which suggests that *Pan troglodytes,*

P. paniscus, and *H. sapiens* are closer to one another than any of them is to *Gorilla gorilla,* and particularly *Pongo pygmaeus.* It is noteworthy that the most parsimonious cladogram need not represent the actual phylogeny of the great apes and humans, because "there are many alternative hypotheses that are minimally less parsimonious . . . fewer than 1% more steps" (Begun et al. 1997, 404).

For more than a decade it appeared that consensus might be reached that the Miocene ancestor of *P. pygmaeus* could be found among *Sivapithecus* spp., which lived in South Asia 12.7 to 6.8 MA (Ward 1997) and which were thought to occur also in Turkey and Europe. Recently recovered cranial remains of the Turkish species *Sivapithecus meteai* indicate not only that it is an unlikely ancestor of *Pongo* but also that it should retain its maiden name, *Ankarapithecus meteai* (Alpagut et al. 1996; Andrews et al. 1996). Similarly, other Turkish and European species of *Sivapithecus* have been renamed *Griphopithecus alpani* and *G. darwini,* respectively (Andrews et al. 1996). Finally, because of discordance between cranial and postcranial evidence, there is lingering doubt that South Asian *Sivapithecus* are viable ancestors to *Pongo,* which leaves them, like *Pan* and *Gorilla,* an insecurely attached terminal branch in hominoid dendrograms (Begun et al. 1997; Pilbeam 1997; Pilbeam et al. 1990; Ward 1997). Budding nearby is *Lufengpithecus lufengensis* (Wu 1987), an 8-MA Chinese species, which might be grafted in place of *Sivapithecus* as a Late Miocene ancestor to *Pongo* (Schwartz 1997), but data and analyses are not complete enough to support this placement (Begun et al. 1997).

RADIANT PLIO-PLEISTOCENE HOMINIDS AND HUMAN EVOLUTION

Because molecular biological studies suggest that humans and the African apes are more closely related to one another genetically than any of them is to orangutans, gibbons, or monkeys, some scientists are inclined to place the African apes with humans in the Hominidae. Contrarily, other researchers, who stress the bodily and behavioral differences between people and apes, continue to reserve the Hominidae for humans and fossil species that have humanoid physical characteristics (Tuttle 1986). I follow the latter scheme.

The chief adaptive complex shared by the Hominidae is obligately terrestrial, fully upright bipedal posture and gait. Some anthropologists further consider all fossil members of the Hominidae to be human, and others restrict humankind to hominids that are and were dependent on culture, as represented archaeologically by stone tool technologies and later by control of fire, fabricated shelters, intentional burial of the dead, bodily adornment, rock art, and figurines. Accordingly, mere evidence of bipedalism is not sufficient for a fossil hominid to be human.

The chief criteria for humankind are behavioral traits that first appear with and were later elaborated by hominids with expanded brains and reduced faces and dentitions, whereby we identify the genus *Homo* anatomically. In brief, because humanity began with *Homo,* all humans are hominids, but not all hominids were humans (Tuttle 1999).

From the 3.5-MA humanoid footprint trails at Laetoli, Tanzania, we know that at least one species of Pliocene hominid was obligately bipedal. But the extent to which other parts of their anatomy were humanoid cannot be inferred from prints alone. Further, we do not know how much earlier obligate terrestrial bipedalism evolved (Tuttle 1987).

By 4.4 MA hominids with many apelike dental and upper-limb features inhabited Africa, as evidenced by *Ardipithecus ramidus* at Aramis, Ethiopia. Although a crushed partial skeleton was recovered, it has not been reassembled and studied to the extent that its locomotor behavior is determined. *Ardipithecus* lived in a closed woodland. If it was not an obligately terrestrial biped, it could be grouped with apes instead of the hominids (Tuttle 1999).

Two major adaptive radiations of Hominidae followed *Ardipithecus:* one of *Australopithecus,* between 4.1 and 1 MA, and a second of *Homo* between 2.5 MA and 100 KA. African forests were reduced by global climatic changes, which allowed open woodlands and grasslands to become more extensive. They provided novel habitats for new species and greater numbers of antelopes, pigs, and other animals for adventurous hominids to scavenge and perhaps to kill. Tubers, seeds, and grasses also may have become more common in the diets of hominids that foraged in the open areas (Tuttle 1999).

Large cats, dogs, and hyenas also flourished in the new environments. They not only provided meat for alert scavengers but also posed a threat to hominids with whom they competed and on whom they probably preyed (Tuttle 1999).

The initial radiation of *Australopithecus* probably occurred in forests or in mosaic localities of riverine forest with adjacent woodlands and more open areas. *Australopithecus* includes at least eight species, which are diagnosed chiefly by differences in their teeth, jaws, and other cranial parts when available: *Australopithecus anamensis* in Kenya; *Australopithicus afarensis* in Ethiopia and perhaps Tanzania; *Australopithicus bahrelghazali* in Chad; *Australopithicus africanus* in South Africa; *Australopithicus garhi* in Ethiopia; *Australopithicus (Paranthropus) aethiopicus* in Ethiopia and Kenya; *Australopithicus (Paranthropus) boisei* in Kenya and Tanzania; and *Australopithicus (Paranthropus) robustus* in South Africa (Asfaw et al. 1999; Gore 1997b; Johanson and Edgar 1996; Tuttle 1999).

Australopithecus is characterized by ape-sized brains (400–500 cc, 24.4–30.5 cu in.; Conroy et al. 1998) and large molar teeth with thick enamel. The early species and *A. africanus* have protruding jaws, but the faces of *Paranthropus robustus* and *P. boisei* are more vertical. The latter two species are also characterized by relatively small incisors and canine teeth, particularly when viewed against their enormous cheek teeth and molarized premolar teeth. *A. afarensis* and species of *Paranthropus* also commonly have sizable bony crests atop and on the back of the skull, to which powerful chewing and neck muscles attached (Tuttle 1999).

For half a century the species of *Paranthropus* were known as robust australopithecines, and *A. africanus* were called gracile australopithecines. When sufficient noncranial skeletal parts were found, it became clear that the names were unwarranted because their bodily size ranges overlapped notably. Females are smaller than the males in species of *Australopithecus* and *Paranthropus,* for which there are quantities of specimens (McHenry 1992).

The robustly skulled species of *Paranthropus* may have eaten tougher foods than those of the more gracile-skulled *Australopithecus.* Indeed, Robinson (1963) suggested that *Paranthropus* were vegetarians, and *A. africanus* had more meat in their diet. Tooth-wear patterns in *A. afarensis* indicate that they may have stripped vegetation by manually pulling it across their front teeth (Tuttle 1999).

Locomotor skeletal remains of *Australopithecus* show apish features that are related to arboreal activity, including curved finger and toe bones, an elongated pisiform bone in the wrist, and iliac blades of the hip bones facing posteriorly (McHenry 1991; Tuttle 1981). Accordingly, it is reasonable to conclude that they continued to climb trees to forage, rest, sleep, and escape from terrestrial predators, rivals, and pests (Tuttle 1981, 1994b).

The phylogenic relationships among the eight species of *Australopithecus* and *Paranthropus* and the direct ancestors of *Homo,* which presumably arose from one of them, are all unknown. *Paranthropus aethiopicus* may have evolved from *A. afarensis,* which then gave rise to *P. boisei* and *P. robustus.* Alternatively, *Australopithecus africanus* may be ancestral to *P. robustus* in Southern Africa and only *P. boisei* descended from *P. aethiopicus* in Eastern Africa (Tuttle 1999). Species of *Paranthropus* are unlikely basal ancestors to *Homo,* and it is arguable whether *Homo* arose directly from *A. afarensis* or from a smaller toothed variant of *A. africanus* (Tuttle 1988).

There is no consensus on which specimens belong in each species, but the radiation of *Homo* is thought to have produced the following: *H. habilis* at Olduvai Gorge, Tanzania, and perhaps several localities in Kenya and South Africa; *H. rudolfensis* in northern Kenya and perhaps Malawi; *H. ergaster* in northern Kenya and the Republic of Georgia (Gabunia et al. 2000); *H. erectus* in Africa, Asia, and perhaps Europe (Ascenzi et al. 1996); *H. antecessor* in Spain and perhaps Eritrea (Abbate et al. 1998); *H. heidelbergensis* in Europe and possibly Ethiopia and South

Africa; *H. neanderthalensis* in Europe and western Asia; and *H. sapiens* globally (Gore 1997a, 1997b; Johanson and Edgar 1996; Tuttle 1999).

Although there is no agreement about which of the three early species of *Homo, H. rudolfensis, H. habilis,* or *H. ergaster,* gave rise to the four younger species, the candidacy of African *H. ergaster* is gaining popularity among paleoanthropologists, who accept it as a species distinct from *H. erectus. H. heidelbergensis* may have arisen from *H. ergaster, H. erectus,* or *H. antecessor,* and any or none of them could have been ancestral to the two latest species of *Homo: H. neanderthalensis* and *H. sapiens.* Neanderthalian populations, particularly as represented by specimens from western Europe, are generally thought not to be ancestral to modern humans (Tuttle 1999).

Species of *Homo* have smaller molars and premolar teeth (36.6–146.4 cu in.) and larger brains (>600–2400 cc) than those of *Australopithecus.* Apish features in the locomotor skeletons of *H. habilis* suggest regular arboreal activity, though they were obligately bipedal when moving on the ground. Foot bones from *H. erectus, H. ergaster, H. antecessor,* and *H. heidelbergensis* are few, but other skeletal parts indicate that, like *H. neanderthalensis,* they were basically adapted to a human pattern of bipedal locomotion (Tuttle, 1988, 1994a, 1999).

Many scenarios for the development of bipedalism have been suggested (R. Tuttle, Webb, and N. Tuttle 1991), and the selective factors that are stressed in each are not mutually exclusive, particularly if one introduces them at different times in the hominid career. An earlier history of bipedal movement and foraging on branches and climbing vertically on tree trunks and vines would prepare the earliest hominids for bipedal terrestrial foraging on shrubs and low tree branches and short-distance travel in open woodland habitats (Tuttle 1994b, 1999).

Experiments show that gibbons, which naturally run bipedally on large branches, expend more energy in bipedal branch-walking and climbing vertical supports than when running bipedally on the ground. Bipedal foraging, crouching, and squatting would select for robust heels and perhaps short toes and the unique pedal arch that serves us so well in running and trekking (Tuttle 1994b, 1999; Tuttle, Hallgrimsson, and Stein 1998).

Bipedalism is advantageous over quadrupedism in open tropical areas because less bodily surface is exposed to direct sunlight, and the delicate brain is kept cooler away from the ground, where there is more circulation of air. Reduction of bodily hair and the proliferation of sweat glands in areas that benefit most from moving air also developed during eras of routine vigorous activity in hot open areas. These physiological refinements probably occurred sometime after 2 MA, during the period of brain enlargement, elaboration of technology, and the deployment of *Homo* out of Africa (Tuttle 1994b).

We do not know when our ancestors essentially quit the trees. Until they had innovated secure technological defenses, including control of fire, formidable weapons, and secure ground shelters, they probably used tree platforms for nightly repose and may have conducted other activities, such as food collection and processing, grooming, play, sex, and defensive maneuvers on them. The earliest record of stone artifacts is 2.5 MA in Ethiopia. Although these crude flakes and chipped nodules could assist food processing, they (and tree branches) would not be effective defensive weapons against large individual predators and groups of smaller social carnivores, unless launched from arboreal vantage points (Tuttle 1992, 1994b, 1999).

Heavier stone tools and more diversified tool kits are associated with *H. erectus* and especially with *H. heidelbergensis, H. neanderthalensis,* and *H. sapiens* (Gamble 1994; Gore 1997b; Johanson and Edgar 1996).

Although well-built hearths are rare until 50 KA, concentrations of charcoal, burnt bones, seeds, and artifacts occur in China and France as early as 460 KA. They arguably indicate (James 1989) that *H. erectus* or *H. heidelbergensis* or members of both species used fire. If claims for 1-MA hominid control of fire in South Africa are confirmed, *Paranthropos robustus* or *H. habilis* could be added as fire keepers (Gamble 1994; Johanson and Edgar 1996; Tuttle 1999).

Archaeological traces of human-made shelters appear rarely in Middle Paleolithic Europe from 60 KA, and become common in the Upper Paleolithic 40 to 10 KA, particularly in regions with notable seasons of inclement weather. Although some Neanderthalians buried their dead, there is little evidence of mortuary ceremony in their graves (Gamble 1994; Johanson and Edgar 1996; Tuttle 1999).

During the Upper Paleolithic, many other features and artifacts that characterize the modern human condition flourished: stylized burial of the dead; sewn, decorated clothing and other bodily adornment; elegant realistic and abstract paintings, engravings, and figurines; and a greater variety of tools in materials other than stone (Gamble 1994; Johanson and Edgar 1996; Tuttle 1999).

The first appearances and development of symbolically based speech and spirituality are highly elusive to paleoanthropologists because they leave no morphological or unarguable archaeological trace before the innovation of writing and ritual paraphernalia. Some paleoanthropologists believe that cerebral regions associated with human speech can be detected on the inner surface of fossil hominid braincases and that flexion of the base of the skull indicates a roomy humanoid vocal tract. Moreover, some experts interpret Upper Paleolithic cultural remains to indicate religious beliefs among their makers. Actually, too little is known about the neurological substrates of speech, symbolic capabilities, and other markers of the several human intelligences to presume to read them from inside battered fossil braincases, models of ancient hominid vocal tracts, or Pleistocene remnants of technology and art (Tuttle 1999).

PROXIMITY AND PARITY

The remarkable genetic closeness of apes to people sparks questions about how we are to interact with them. During the past century, people have observed and interacted with apes in a wide variety of contexts, ranging from natural to cruelly bizarre. In revulsion to the latter and in response to the former, many scientists and nonscientists are compelled to lobby vigorously for better treatment of apes and other nonhuman animals. Chimpanzees, gorillas, orangutans, and bonobos are highly persuasive poster mammals in the general animal rights movement. Accordingly, Cavalieri and Singer (1993; also Cavalieri 1996) codified the case for bonobos, chimpanzees, gorillas, and orangutans in a "declaration on great apes," which is supported by an impressive roster of scientists and humanists (Tuttle 1994a; see also chapter 16, this volume).

The often-cited 98.4 percent overall similarity between chimpanzee and human DNAs is an important datum in the call for equal rights of apes with people. Subscribers to the Great Ape Project also cite the performance of apes in artifactual language studies to support their claims of equal rights of apes to life, liberty, and nonpersecution (Tuttle 1994a).

It is probably imprudent for advocates of the Great Ape Project to premise their case on current genetic information. Although chimpanzees and humans may express approximately 98.4 percent overall similarity in their nuclear DNAs, this still leaves many individual genes that differ. Critically, we do not know the functional meaning of the genetic data or how it is expressed developmentally via interactions with complex environmental factors. Were genetic studies to reveal discrete, profound differences between humans and the African apes and if such genetic factors were to be weighed heavily in decisions about the relative status of apes versus people, the apes might be in jeopardy (Tuttle 1994a).

Neither fossils nor genes are a critical factor. We are faced with a knotty ethical and moral dilemma, and it is philosophers and ethicists who must lead us to greater understanding of the nature of the problem and to practical, humane solutions. Because I have little expertise in the methods used by these colleagues I leave the task to them. Before other evolutionary biologists join the fray, they should first read detailed accounts by historians of science on the pitfalls of past reliance on presumed biological facts to inform ethical and moral issues.

It is difficult to predict the impact of the Great Ape Project on public policy. The best hope of its subscribers is to expand lobbies that protect the lives and dignity of these magnificent creatures and to extend the mandate to embrace all animals that evidence humanoid sensibilities and personalities. Like people, they can return great joy to those who love them and attempt to understand them on their own terms. The bottom branch is that policymakers must bend to a largely emotional appeal that deeply prods our moral, spiritual being, which is the essence of humanity. I am skeptical that the Great Ape Project holds the answer (Tuttle 1994a, 1998).

Biologically informed evolutionary anthropologists must continue to patrol the boundary of *H. sapiens,* and all participants in the great ape debate should avoid analogies to injustices that privileged persons visit on slaves, intellectually challenged persons, women, children, and gender extenders, who are fighting so hard for specific parity and social justice. Such comparisons neither promote the equal dignity of all humans nor lead to a full appreciation of the adaptive complexes and novel capabilities of apes (Tuttle 1994a). Apes surely merit a full agenda of ape rights based on their keen cognitive capacities, sensitivity, sociability, and potential longevity. They have paid precious dues for the advancement of medical and basic scientific knowledge. The world would be much poorer without them, both because they are fascinating fellow beings and because we have barely begun to plumb how they can inform questions about the human career and condition.

ACKNOWLEDGMENTS

I thank our dear family friend and treasured colleague Ben Beck for organizing and inviting me to this timely discussion. I am grateful also to the other authors of this volume for their stimulating chapters. Special thanks go also to Raymond Corbey for a thoughtful reading and comments on this manuscript. Our hosts at the Disney Institute during the conference that led to this volume were gracious and most helpful in distracting me from the Mickey Mouse overkill in our cushy accommodations.

REFERENCES

Abbate, E., Albianelli, A., Azzaroli, A., Benvenuti, M., Tesfamariam, B., Bruni, P., Cipriani, N., Clarke, R. J., Ficcarelli, G., Macchiarelli, R., Napoleone, G., Papini, M., Rook, L., Sagri, M., Tecle, T. M., Torre, D., and Villa, I. 1998. A one-million-year-old *Homo* cranium from the Danakil (Afar) depression in Eritrea. *Nature* 393: 458–60.

Alpagut, B., Andrews, P., Fortelius, M., Kappelman, J., Temizsoy, I., Celebi, H., and Lindsay, W. 1996. A new specimen of *Ankarapithecus meteai* from the Sinap Formation of central Anatolia. *Nature* 382: 349–51.

Andrews, P., Harrison, T., Delson, E., Bernor, R. L., and Martin, L. 1996. Distribution and biochronology of European and Southwest Asian Miocene catarrhines. In R. L.

Bernor, V. Fahlbusch, and H. W. Mittmann (eds.), *The evolution of Western Eurasian Neogene mammal faunas* (pp. 168–207). New York: Columbia University Press.

Ascenzi, A., Biddittu, I., Cassoli, P. F., Segre, A. G., and Segre-Naldini, E. 1996. A calvarian of late *Homo erectus* from Ceprano, Italy. *Journal of Human Evolution* 31: 409–23.

Asfaw, B., White, T., Lovejoy, O., Latimer, B., Simpson, S., and Suwa, G. 1999. *Australopithecus garhi:* A new species of early hominid from Ethiopia. *Science* 284: 629–35.

Begun, D. R., and Kordos, L. 1997. Phyletic affinities and functional convergence in *Dryopithecus* and other Miocene and living hominoids. In D. R. Begun, C. V. Ward, and M. D. Rose (eds.), *Function, phylogeny, and fossils* (pp. 291–316). New York: Plenum Press.

Begun, D. R., Ward, C. V., and Rose, M. D. 1997. Events in hominoid evolution. In D. R. Begun, C. V. Ward, and M. D. Rose (eds.), *Function, phylogeny, and fossils* (pp. 389–415). New York: Plenum Press.

de Bonis, L. 1987a. L'origine des hominidés. *L'Anthropologie (Paris)* 91: 433–54.

———. 1987b. Les primates de l'ancien monde du Paléocène au Miocène. In M. Marois (ed.), *L'évolution dans sa réalité et ses diverses modalités.* Paris: Foundation Singer-Polignac Masson.

———. 1987c. Les racines des Hominidae: Les recherches in Grèce. *La Vie des Sciences, Comptes rendus, série générale* 4: 287–304.

de Bonis, L., Bouvrain, G., Geraads, D., and Koufos, G. 1990. New hominid skull material from the late Miocene of Macedonia in northern Greece. *Nature* 345: 712–14.

de Bonis, L., and Koufos, G. 1993. The face and the mandible of *Ouranopithecus macedoniensis:* Description of new specimens and comparisons. *Journal of Human Evolution* 24: 469–91.

———. 1997. The phylogenic and functional implications of *Ouranopithecus macedoniensis.* In D. R. Begun, C. V. Ward, and M. D. Rose (eds.), *Function, phylogeny, and fossils* (pp. 317–26). New York: Plenum Press.

de Bonis, L., and Melentis, J. 1984. La position phylétique d'*Ouranopithecus. Courier Forschungsinstitut Senckenberg* 69: 13–23.

Cavalieri, P. 1996. The Great Ape Project. *Etica & Animali* 8: 3.

Cavalieri, P., and Singer, P. 1993. *The Great Ape Project.* London: Fourth Estate.

Conroy, G. C., Weber, G. W., Seidler, H., Tobias, P. V., Kane, A., and Brunsden, B. 1998. Endocranial capacity in an early hominid cranium from Sterkfontein, South Africa. *Science* 280: 1730–31.

Diamond, J. 1992. *The third chimpanzee.* New York: HarperCollins.

Fleagle, J. T. 1988. *Primate adaptation & evolution.* San Diego, CA: Academic Press.

Gabunia, L., Vekua, A., Lordkipanidze, D., Swisher, C. C. III, Ferring, R., Jutus, A., Nioradze, M., Tvalchrelidze, M., Antòn, S. C., Bosinski, G., Jöris, O., deLumley, M. A., Majsuradze, G., and Mouskhelishvili, A. 2000. Earliest Pleistocene hominid cranial remains from Dmanski, Republic of Georgia: Taxonomy, geological setting, and age. *Science* 288: 1019–25.

Gamble, C. 1994. *Timewalkers.* Cambridge, MA: Harvard University Press.

Gore, R. 1997a. Expanding worlds. *National Geographic Magazine* 191(5): 84–109.

———. 1997b. The first steps. *National Geographic Magazine* 191(2): 72–99.

Harrison, T. 1993. Cladistic concepts and the species problem in hominoid evolution. In W. H. Kimbel and L. B. Martin (eds.), *Species, species concepts, and primate evolution* (pp. 345–71). New York: Plenum Press.

James, S. R. 1989. Hominid use of fire in the lower and middle Pleistocene. *Current Anthropology* 30: 1–26.

Johanson, D., and Edgar, B. 1996. *From Lucy to language.* New York: Simon and Schuster.

McHenry, H. M. 1991. First steps? Analysis of the postcranium of early hominids. In Y. Coppens and B. Senut (eds.), *Origine(s) de la Bipédie chez les Hominidés* (pp. 133–41). Paris: Éditions du Centre National de la Recherche Scientifique.

———. 1992. Body size and proportions in early hominids. *American Journal of Physical Anthropology* 87: 407–31.

Pilbeam, D. 1997. Research on Miocene hominoids and hominid origins. The last three decades. In D. R. Begun, C. V. Ward, and M. D. Rose (eds.), *Function, phylogeny, and fossils* (pp. 13–28). New York: Plenum Press.

Pilbeam, D., Rose, M. D., Berry, J. C., and Shah, S. M. I. 1990. New *Sivapithecus* humeri from Pakistan and the relationship of *Sivapithecus* and *Pongo. Nature* 348: 237–39.

Robinson, J. T. 1963. Adaptive radiation in the australopithecines and the origin of man. In F. C. Howell and F. Bourlière (eds.), *African ecology and human evolution* (pp. 385–416). New York: Wenner-Gren Foundation for Anthropological Research.

Ruvolo M., Pan, D., Zehr, S., Goldberg, T., Disotell, T. R., and von Dornum, M. 1994. Gene trees and hominoid phylogeny. *Proceedings of the National Academy of Science. USA* 91: 8900–8904.

Schwartz, J. H. 1997. *Lufengpithecus* and hominoid phylogeny. Problems delineating and evaluating phylogenically relevant characters. In D. R. Begun, C. V. Ward, and M. D. Rose (eds.), *Function, phylogeny, and fossils* (pp. 363–88). New York: Plenum Press.

Sibley, C. H., and Ahlquist, J. E. 1987. DNA hybridization evidence of hominoid phylogeny: Results from an expanded data set. *Journal of Molecular Evolution* 26: 99–121.

Smith, G. E. 1927. *The evolution of man.* London: Oxford University Press.

Tuttle, R. H. 1981. Evolution of hominid bipedalism and prehensile capabilities. *Philosophical Transactions of the Royal Society, London* B 291: 89–94.

——— 1986. *Apes of the world.* Park Ridge, NJ: Noyes.

———. 1987. Kinesiological inferences and evolutionary implications from Laetoli bipedal trails G-1, G-2/3, and A. In M. D. Leakey and J. M. Harris (eds.), *Laetoli: A Pliocene site in northern Tanzania* (pp. 503–23). Oxford: Clarenden Press.

———. 1988. What's new in African paleoanthropology? *Annual Review of Anthropology 1988* 17: 391–426.

———. 1992. Hands from newt to Napier. In S. Matano, R. H. Tuttle, H. Ishida, and M. Goodman (eds.), *Topics in primatology, Vol. 3, Evolutionary biology, reproductive endocrinology and virology* (pp. 3–20). Tokyo: University of Tokyo Press.

———. 1994a. A trans-specific agenda. *Science* 264: 602–603.

———. 1994b. Up from electromyography: Primate energetics and the evolution of human bipedalism. In R. S. Corruccini and R. L. Ciochon (eds.), *Integrative paths to the past: paleoanthropological advances in honor of F. Clark Howell* (pp. 269–84). Englewood Cliffs, NJ: Prentice-Hall.

———. 1998. Global primatology in a new millennium. *International Journal of Primatology* 19: 1–12.

———. 1999. Hominids. In *Encyclopedia of paleontology.* Chicago: Fitzroy Dearborn.

Tuttle, R. H., Hallgrímsson, B., and Stein, T. 1998. Heel, squat, stand, stride: Function and evolution of hominoid feet. In A. Rosenberger, J. Fleagle, H. McHenry, and E. Strasser (eds.), *Primate locomotion* (pp. 435–48). New York: Plenum Press.

Tuttle, R. H., Webb, D. M., and Tuttle, N. I. 1991. Laetoli footprint trails and the evolution of hominid bipedalism. In Y. Coppens and B. Senut (eds.), *Origine(s) de la Bipédie chez les Hominidés* (pp. 203–18). Paris: Éditions du Centre National de la Rechevche Scientifique.

Ward, S. 1997. The taxonomy and phylogenetic relationships of *Sivapithecus* revisited. In D. R. Begun, C. V. Ward, and M. D. Rose (eds.), *Function, phylogeny, and fossils* (pp. 269–90). New York: Plenum Press.

Wu, R. 1987. A revision of the classification of the Lufeng great apes. *Acta Anthropologica Sinica* 6: 263–71.

ROGER FOUTS

10
DARWINIAN REFLECTIONS ON OUR FELLOW APES

Our perspective of the world determines how we behave. If we thought that the world was flat, we would not try to sail around it. I propose that we are in the middle of a major change in perspective with regard to our species' place in nature and our relationship to our fellow organic beings. A great deal of evidence has been accumulating in the 150 years since Darwin's introduction of his theory of evolution that stimulates and supports this change. The question we face is whether we have the integrity and courage as honorable scientists to embrace the empirical facts and accept the implications of this great change. It is this courage and integrity that Linnaeus manifested when he decided to practice science rather than theology and group humans with other Anthropomorpha (later changed to Primates) based on the best scientific evidence then available. Thomas Pennant's surprisingly frank criticism of Linnaeus echoes criticisms that we face today: "My vanity will not suffer me to rank mankind with apes, monkeys, and bats" (quoted in Spencer 1995, 16).

We have certainly made great scientific advancements since Linnaeus (1707–1778) was born nearly 300 years ago. We have the electron microscope and we are well-involved in the molecular world of DNA. Molecular biologists clearly recognize our close relationship to our fellow apes (for overviews see Diamond 1992; R. Fouts and Mills 1997). Based on this and other empirical evidence, the most recent edition of the Smithsonian's definitive classification, *Mammal Species of the World* (Wilson and Reeder 1993), has demonstrated the courage of Linnaeus and moved our fellow great apes into the family Hominidae, previously reserved for humans alone. This means that the previous exclusive use of that term for ourselves and our fully bipedal ancestors has gone the way of the Bimanus and

Quadrumanus classifications of Blumenbach (Spencer 1995). Some of us are slowly beginning to embrace the empirical evidence and reject our old perspective of a vertical chain of being in which our species is separated from and far above our fellow apes and the rest of the organic beings. I do not think that it would be too unreasonable to conjecture that in 25 years Wilson and Reeder will be classifying *Pan* as *Homo.*

Our worldview affects our language. For example, if we see our own species as separate from nature or somehow superior to our fellow organic beings then we will seek ways to emphasize the differences, imagined or real. For example, the very title of the workshop on which this volume is based implicitly ignores the new classification of our fellow apes in favor of the old. Wilson and Reeder's classification has dispensed with the great ape family (Pongidae) as a grouping based on erroneous premolecular science. As a result, "great apes and humans" is now as meaningless as "rational soul and brute soul beings" or "rational humans and automata."

"CHAIN OF BEING," "MAN IS THE MEASURE OF ALL THINGS," AND OTHER ANACHRONISMS

The ancient Greek platonic view and the more recent Cartesian view holds that "man" is superior to all other beings, including women. The Greek view was the more traditional chain-of-being approach, which is like a ladder on which the inferior creatures are placed below the superior male human. However, if the defective beings struggled enough to leave their bodies behind and live in their minds they might evolve or be lucky enough to eventually incarnate into a man. Descartes's view was slightly different in that a definite gap or difference in kind existed between man and the defective automata below him. It was still a chain of being that ordered our fellow animals in a descending fashion on a scale of imperfection, but these imperfect automata were quite distinct and different in kind from man because they lacked reason. The main difference was that Descartes gave no allowance for any of the unthinking, unfeeling automata ever to bridge that gap. Both views dissociated the rational mind from the body and held that only man had rationality. The views differ with regard to the body. For Plato and Aristotle the body was ruled by the emotional brute soul, whereas Descartes relegated the body and non–European-man animals to the unthinking, unfeeling status of machines.

I wish to address two of these important elements: One is the "ladder" notion that brings with it the implicit "man is the measure of all things" evaluation, and the second is the dissociation of mind from body. The first notion of a unilinear ladder-like evolution is still prevalent. This is because our conception of our fellow organic beings generally does not come from asking them what they are about. This is in contrast to Darwin, who admired our fellow organic beings for their own adaptations, without making value judgments or odious comparisons. More typically we note something that we as humans do, and then without bothering to ask our fellow organic beings we assume that they lack it or are only capable of it in an inferior and rudimentary manner. We as dominators did not limit this evaluation method to other species. For example, Gill (1997) noted that early European immigrants to the United States viewed the complex and indigenous cultures of the Native Americans as "primitive." Likewise the European immigrant viewed Native American languages as simple noun-based linguistic codes that relied heavily on sign language and their adult cognitive capacities as comparable to those of white children. How many times have we heard the claim that a chimpanzee, gorilla, bonobo, or orangutan has the intelligence of a young human child? This type of comparison implies that our fellow apes are mentally neotenic versions of ourselves, nothing more than "developmentally delayed" humans with no hope of ever crossing the brink of "human superiority." The arrogant mistake is that those holding these positions are confusing "differences" with "defects." Chimpanzees are not defective humans. Likewise, they are not evolving into humans. They are perfectly fine chimpanzees, and they have millions of years of survival to prove it. The same is true for their fellow hominids. As I attempted to follow chimpanzees in their natural habitats, one of the things that I discovered about our species was our worthiness of the title "defective chimpanzee." Can you imagine a chimpanzee with the mentality of a 2- or 3-year-old human surviving in the jungle? I hope that one can see that such comparisons are meaningless and only serve to move the illusory Cartesian line between ourselves and our fellow hominids. These comparisons simply move the chimpanzees from nonrational brute to mentally retarded child; the unbridgeable gap created by such comparisons is still there. Only for Cartesians do such "lines" and "brinks" exist. For the Darwinian it is a horizontal continuity. Darwin makes this point with the statement,

> we must also admit that there is a much wider interval in mental power between one of the lowest fishes, as a lamprey or lancelet, and one of the higher apes, than between an ape and man; yet this interval is filled up by numberless gradations. (Darwin 1877/1989, 70)

Granted, members of our species can survive in the jungle. But do we fare as well as the chimpanzees? I noted that in the forest the chimpanzees could lose even the most experienced field assistants. Would our inferior performance in the jungle justify chimpanzees' exploitation of us? Are we incapable of making comparisons between different species or even different individuals without attaching value judgments?

DARWINIAN ALTERNATIVES TO ARROGANCE AND HUMANISM

Darwin's great contribution was replacing arrogant humanism and the delusional vertical ladder of being (with or without gaps) with the empirically sound horizontal Darwinian bush. On the ends of the buds of this bush there are extant humans, chimpanzees, gorillas, orangutans, dogs, cats, fish, cockroaches, pigeons, pear trees, and all manner of organic beings. Darwin displays a profound adoring respect for all the organic beings on this bush:

> How have all those exquisite adaptations of one part of the organization to another part, and to the conditions of life, and of one organic being to another being, been perfected? . . . In short, we see beautiful adaptations everywhere and in every part of the organic world. (Darwin 1859/1991, 46)

With his awe and reverence of nature, Darwin understood our place as one of humility rather than arrogance: "Natural Selection . . . is a power incessantly ready for action, and is as immeasurably superior to man's feeble efforts, as the works of Nature are to those of Art" (Darwin 1859/1991, 47). Likewise, he did not create gaps and separations to exclude us from nature, but saw us as of one blood with our fellow animals. His concept of "species" illustrates this well:

> I look at the term species as one arbitrarily given, for the sake of convenience, to a set of individuals closely resembling each other, and that it does not essentially differ from the term variety, which is given to less distinct and more fluctuating forms. The term variety, again, in comparison with mere individual differences, is also applied arbitrarily, for convenience' sake. (Darwin 1859/1991, 40)

It would seem that for Darwin the distinction between chimpanzee and human would be one of convenience, an arbitrary one. Darwin's distinction has been proven by the fuzzy genetic differences between ourselves and our fellow hominids. The difference between me and a gorilla, an orangutan, a bonobo, or a chimpanzee is one of degree. The gaps exist only in the minds of Cartesians, not in nature. Yet on the basis of these arbitrary distinctions and on Euro-American value judgments, we justify depriving other organic beings of the right to be themselves for the sake of our education and entertainment or our conjectured or real medical needs. We commit the error of mechnomorphism by assuming that our fellow organic beings are more similar to a bicycle or a toaster than they are to us. Who draws these arbitrary lines and on what genuinely objective basis, if they should be drawn at all? Where have we drawn these lines in the past in the name of medical advancement or economic gain? Would you take the life of a black slave to save the life of your white wife or let her die without a transplant? Certainly, if you were a Caucasian and if your wife was as well, the genetic difference between you and the black slave would be greater than it was between you and your wife. Does this degree of genetic difference justify taking the slave's life? According to what Chancellor Harper of the University of South Carolina said in 1838, the answer would be yes: "If there are sordid, servile, and laborious offices to be performed, is it not better that there should be sordid, servile, and laborious beings to perform them?" (quoted in Spiegel 1996, 39). In my own lifetime the question of whether or not to exploit supposed "inferior" beings for the benefit of supposed "superior" beings was answered in the Tuskegee Syphilis Study, in which scientists studied the results of untreated syphilis in an involuntary experiment done on black cotton-field workers for 40 years until it was exposed by the press (Spiegel 1996). We should never forget the answer and we should never stop being ashamed of that answer.

Darwin saw nature for what it was. He accepted nature's harshness when he said,

> we do not see or we forget, that the birds which are idly singing round us mostly live on insects or seeds, and are thus constantly destroying life; or we forget how largely these songsters, and their eggs, and their nestlings, are destroyed by birds and beasts of prey. (Darwin 1859/1991, 47)

This view of nature has been used to justify our exploitation of our fellow hominids. Darwin also recognized another way of existing, or coexisting when he clarified what he meant by the "struggle for existence." He did not see this solely as an aggressive act between organic beings but as a metaphor that included "dependence of one being on another, and including (which is more important) not only the life of the individual, but success in leaving progeny" (Darwin 1859/1991, 47). One might argue that in addition to the cooperation and peaceful coexistence in nature there is also competition, and therefore it is natural for us to exploit our fellow hominids. More to the point, what is the difference between what we do and what the songbird does? My answer comes from the poet and essayist Wendell Berry (1990), who examined the goal of a harmony between nature and human economy that will preserve both. He pointed out that this is a traditional goal, but the world is now divided between those who adhere to this ancient purpose and those who intentionally do not. He maintained that this division is far more portentous for the future than any of the national or political or economic divisions we see threatening the world today:

> The remarkable thing about this division is its relative newness. The idea that we should obey nature's laws and live harmoniously with her as good husbanders and stewards of her gifts is old. And I believe that until fairly recently our destructions of nature were more or less unwitting—the by-products, so to speak, of our ignorance or weakness or depravity. It is our present principled and elaborately rationalized rape and plunder of the natural world that is a new thing under the sun. (Berry 1990, 108)

CHIMPANZEE SIGN LANGUAGE

Ape language research has played a significant role in helping the scientific community embrace this Darwinian perspective of continuity. Project Washoe, begun by Allen and Beatrix Gardner at the University of Nevada, Reno, in 1966, was the first and longest ongoing successful attempt to teach a human language to a nonhuman being. Previous attempts had failed because inappropriate expectations were placed on the apes, such as attempting to teach them spoken English (Furness 1916; Hayes and Hayes 1951; Witmer 1909). For example, Hayes and Hayes (1951) attempted to teach spoken English to the chimpanzee Viki. After seven years of training, Viki had a vocabulary of four largely voiceless words (Hayes and Hayes 1951; Hayes and Nissen 1971). Despite the fact that their vocal ability is limited, chimpanzees can freely move their hands, indicating that a gestural language would be better suited to the chimpanzees' abilities. The Gardners confirmed this as Washoe acquired a vocabulary of 132 gestural signs in 51 months (R. Gardner and B. Gardner 1978). The Gardners' research, along with several later projects, have resulted in challenges to the evolutionarily naive and anthropocentric claims of human uniqueness. Following the Gardners' research, other investigators have looked at language acquisition in other great ape species such as gorillas (Patterson and Cohn 1990), an orangutan (Miles 1980, 1990), and bonobos (Savage-Rumbaugh and Lewin 1994). Some investigators did not use the Gardners' cross-fostering approach and instead used training methods to teach specific language-like abilities in different modes, such as plastic tokens or lexigram boards (Premack 1971; Rumbaugh 1977; Savage-Rumbaugh 1986).

Cross-Fostering Research

Washoe was cross-fostered by the Gardners and their students, who used only American Sign Language (ASL) in her presence (B. Gardner and R. Gardner 1971,

1974; R. Gardner and B. Gardner 1969, 1989). *Cross-fostering* is members of one species raising a member of another. In Washoe's case, this was in the context of a home and family rather than in an operant conditioning laboratory. She learned the signs of ASL during the first years of the project in a natural social environment comparable to that of a human deaf child.

> She asked for goods and services, and she also asked questions about the world of objects and events around her. When Washoe had about eight signs in her expressive vocabulary, she began to combine them into meaningful phrases YOU ME HIDE, and YOU ME GO OUT HURRY were common. She called her doll, BABY MINE; the sound of a barking dog, LISTEN DOG; the refrigerator, OPEN EAT DRINK; and her potty-chair, DIRTY GOOD. Along with her skill with cups and spoons, and pencils and crayons, her signing developed stage for stage much like the speaking and signing of human children. (R. Gardner and B. Gardner 1989, 6)

The Gardners continued their work with other cross-fostered chimpanzees. Moja, Pili, Tatu, and Dar were each raised in the same type of environment as Washoe but with more deaf human signers and each other to sign with (R. Gardner and B. Gardner 1989). Taking the two projects together, the Gardners found that the chimpanzees showed the same basic stage-by-stage acquisition of language as hearing and deaf children (B. Gardner and R. Gardner 1994; R. Gardner and B. Gardner 1994). The Gardners also used double-blind testing conditions to demonstrate that the chimpanzees communicated information to humans with ASL, that there was high interobserver reliability in identifying the signs made by the chimpanzees under these testing conditions, and that the five chimpanzees used their signs in natural language categories (e.g., the sign TREE for any tree, BABY for any baby or doll; B. Gardner and R. Gardner 1985).

Post-Reno Research

In 1970, Washoe moved to the Institute for Primate Studies (IPS) at the University of Oklahoma, where there were other captive chimpanzees. Confirming earlier findings, these chimpanzees, Booee, Bruno, Cindy, and Thelma, also acquired ASL signs with the help of human caregivers (R. Fouts 1973). Other home-reared chimpanzees associated with the IPS, such as Lucy and Ally, were also taught ASL signs. Ally was able to comprehend and produce novel prepositional phrases, demonstrating his ability to use appropriate sign order (R. Fouts, Shapiro, and O'Neil 1978). Lucy was able to categorize and conceptualize fruits as different from vegetables and to describe novel objects by combining signs already existing in her vocabulary—for example, referring to a watermelon as CANDY DRINK and DRINK

FRUIT and to an old bitter radish CRY HURT FOOD after she took a bite of it and spit it out (R. Fouts 1975). Like Lucy, Washoe also combined signs to produce novel phrases to describe objects for which she did not have signs in her vocabulary, such as swans and Brazil nuts, which she referred to as WATER BIRD and ROCK BERRY (R. Fouts and Rigby 1977). Ally also could be taught ASL signs by using vocal English words as exemplars and then in a blind condition transfer these signs to their physical referents (R. Fouts, Chown, and Goodin 1976). Later Shaw (1989) reconfirmed the ability of chimpanzees to comprehend spoken English when she tested Tatu's ability to translate vocal English words into their ASL signs under a blind testing condition.

Cross-fostering and the use of a natural language allowed sign language to become a normal part of the chimpanzees' lives, not just a means for requesting food (R. Gardner and B. Gardner 1989). Today, Washoe, Moja, Tatu, Dar, and one new addition, Loulis, continue to use sign language in a variety of contexts and situations at the Chimpanzee and Human Communication Institute (CHCI). Within this natural social context scientists have been able to demonstrate the cultural transmission of signs between chimpanzees. Ten-month-old Loulis was adopted by Washoe and acquired signing and other skills from her and the other signing chimpanzees in his community. To ensure that Loulis would acquire his signs only from the chimpanzees, the humans in his presence restricted their signing to seven specific signs or used vocal English to communicate with him and the other chimpanzees. Washoe was observed to teach Loulis through modeling, molding, and signing on his body. Most of his signs appeared to be delayed imitations of signs he had seen Washoe and the other chimpanzees use in similar contexts. The staff and I at CHCI were able to examine the development of social behavior, communication, and other skills in Loulis without disrupting these skills. In this way, we obtained a comprehensive record of the cultural transmission of signing (R. Fouts and D. Fouts 1989; R. Fouts, D. Fouts, and Van Cantfort 1989; R. Fouts, Hirsch, and D. Fouts 1982).

Remote videorecording was used to verify Loulis's use of signs. This demonstrated that Loulis, as well as the other four chimpanzees, would use their signs to communicate with each other when no humans were present. The chimpanzees mainly signed about social issues such as play, reassurance, and grooming (R. Fouts and D. Fouts 1989; R. Fouts, D. Fouts, and Schoenfeld 1984).

Remote videorecording was also used to examine how the chimpanzees sign to themselves, which is called private signing. This method was used to record 368 instances of the chimpanzees using their signs in a nonsocial fashion to sign to themselves. These instances were classified into nine different functional categories, as has been done with humans' private speech (Furrow 1984). As in humans, a few of the categories accounted for the majority of the chimpanzees' private signing. The referential category, an utterance that refers to an object or event that is present (for example Tatu signing THAT FLOWER to a magazine picture of a flower) accounted for 59 percent of the instances. The informative category, an utterance that refers to an object or event that is not present (for example Washoe signing DEBBI to herself when Debbi was not present) accounted for 12 percent (Bodamer et al. 1994). The remaining functional categories in which signing by the chimpanzees was observed were instrumental (8 percent), describing their own activity (8 percent), imaginary (5 percent), expressive (3 percent), interactional (2 percent), regulatory (2 percent), and attentional (1 percent; Bodamer et al. 1994).

Imaginary play in the five chimpanzees was also studied with remote videorecording. Six instances were observed in 15 hours of recording. There were two types: *animation,* in which an inanimate object is treated as if it is animate; and *substitution,* in which an object is given a new identity. For example, when Moja took a purse, placed it over her foot, and then referred to it as THAT SHOE (Jensvold and Fouts 1993).

Remote videorecording was also used to examine the five chimpanzees' nighttime behavior. The chimpanzees were more active at this time than previously assumed. There were even a few instances of signing in their sleep (Williams 1995).

Cianelli and Fouts (1998) found that the chimpanzees often used emphatically signed ASL signs during high-arousal interactions such as fights and active play. For example, after separating Dar and Loulis during a fight and with all the chimpanzees still screaming, Washoe signed COME HUG to Loulis who responded with the sign NO. These results indicate that the chimpanzees' signing is very robust.

Other studies have involved more traditional methods but are subject-paced—in other words, the chimpanzees are not required or forced to participate but do so voluntarily. Davis (1995) found that when a human distorted a sign while signing to a chimpanzee and the distortion was low, the chimpanzee would restore the sign to its nondistorted form. All signs have a place where they are made, a specific hand configuration and a specific movement associated with them. For example, the sign SPOON has a hand configuration of the index and middle fingers extended from a fist, these two fingers touch the middle of the open palm of the opposite hand, which is the "place" the sign is made, and then the two fingers are moved toward the mouth. In this study, the place was distorted so that in low distortion for the sign SPOON the two fingers might touch the wrist instead of the palm. Low distortion was defined as moving the place where the sign is made from 1.02 to 4.06 in. (2.54 to 10.16 cm) from the standard place where it is usually made correctly. If distortion was high (9.14 to 12.19 in. or 22.86 to 30.48 cm from the standard place), the chimpanzees typically did not respond. This flexibility in ability to comprehend and adjust to distorted speech is similar to that of humans.

Beaucher (1995) examined the semantic range of how a signing chimpanzee would categorize novel food stimuli that were manipulated on three dimensions. First the chimpanzees were presented with Oreo® cookies and Saltine® crackers, to verify that the chimpanzees would reliably refer to these two food items with the signs for COOKIE and CRACKER. She then devised novel food items that varied on the dimensions of sweetened to nonsweetened, salty to not salty, and circle to square. She found that Washoe significantly used the sweet and nonsalty dimension when comparing novel food items. These results indicated that Washoe was using a prototype model, which requires that the research participant abstract attributes or dimensions of a category when applying them to novel objects.

Krause and Fouts (1997) examined referential pointing, eye gaze, and pointing accuracy in a study with Moja and Tatu. The first experiment examined how Moja and Tatu attracted the attention of a human and then directed the human to an out-of-reach object. The chimpanzees would first get the visual attention of the human (e.g., make a noise to get the human to turn toward them and look at them) before they would point to the out-of-reach object. Once the human was looking at the chimpanzee, Moja or Tatu would alternate their eye gaze direction between the human and the out-of-reach object. The chimpanzees' use of attention-getting devices, mutual gaze, and concomitant gaze alternation between the out-of-reach object and the human was taken as evidence that the chimpanzees' pointing was indeed referential and communicative. A second experiment tested the chimpanzees' accuracy of pointing toward objects when the objects were placed in close proximity to each other. Humans were able to accurately respond to the chimpanzees' pointing under this condition. It was interesting to note that in the second experiment, which required a greater degree of precise pointing, both chimpanzees showed a left-hand bias not found in the first experiment.

R. Fouts and colleagues (1997) have begun a long-term research project examining the use of gestural dialects and idiolects in wild and captive chimpanzees. This study examined Washoe, Moja, Tatu, Dar, and Loulis's interactions for non-ASL sign gestures (nonverbal gestures) that were used during communicative interactions. They used eight dialectical gestures and idiolectical gestures. For example a subtle "chin tip" (a quick raising of the head with a directional component) was observed in chase games, where the direction of the tip (left or right) would indicate direction of the chase and the active "chin tipper" would be the chaser.

Wild chimpanzees are known to use different gestural dialects in different communities. McGrew and Tutin (1978) reported observing a "grooming-hand-clasp" in the Kasoje community of chimpanzees in the Mahale Mountains that was different from the grooming gesture used by the Kasakela chimpanzees at Gombe, 50 mi (80 km) to the north. The Kasoje chimpanzees approached each other at the beginning of a grooming session, and each of the participants simultaneously raised an arm over their heads and then grasped each other's wrists. The Kasakela chimpanzees raised one arm straight into the air above their heads to solicit grooming. This is an apparent dialectical difference. McGrew and Tutin observed the Kasoje "grooming-hand-clasp" an average of once every 2.4 hours, higher than the frequency of sexual behavior, agonistic behavior, and food sharing. Thus it plays a very important role in the Kasoje chimpanzees' social behavior. It is remarkable that this behavior has never been observed in the thousands of hours of observation (more than 35 years) of the Kasakela chimpanzees at Gombe. This, along with the discoveries of variation in food preferences, "tool kits," and other differences between communities is seen by some as evidence of culture (McGrew 1992).

Nishida (1987) observed a behavior he called "leaf-clipping," which occurs during courtship in the Kasoje community. Apparently for a male to successfully consort with a female, she must cooperate so they can slip away unnoticed. In leaf-clipping, a male will make eye contact with an estrus female while placing a leaf in his mouth and then tear the leaf and let it drop. The male then quietly leaves and the estrus female follows him. McGrew (1992) stated that the male used the leaf-clipping gesture as a subtle tool to get the attention of females. Nishida noted that this gesture was another cultural dialect difference between the two communities, because it has not been reported at Gombe despite extensive study of sexual behavior (Goodall 1986; McGinnes 1979; Tutin 1979).

At the CHCI, we are beginning a study of different communities of wild chimpanzees in hopes of examining other gestures that might be dialectical. The discovery of gestural dialects among the wild chimpanzees demonstrates that gestural communication and acquisition of different gestures are natural for wild or captive chimpanzees. Darwin (1889/1998) anticipated this as well. Darwin held that although many of our facial expressions are inherited and demonstrate the continuity of species, our gestures are largely acquired and have dialectical variation.

A MODERN DARWINIAN MODEL: FUZZY LOGIC

Fuzzy logic seems like an oxymoron to a Western mind. Logic is supposed to be clear-cut and concise and certainly not fuzzy. Though we attempt to avoid fuzziness in science, it is still a part of our lives. We have washing machines and vacuum cleaners that incorporate fuzzy logic in their programming to constantly adjust to the kind and amount of dirt. Even our cars' computers use fuzzy logic in the ABS braking systems. Japan, not shackled by Western thought, has embraced fuzzy logic. For example, the subway trains in Sandai use it to make starts and stops undetectable (NcNeill and Freiberger 1993).

Modern videocameras have "digital image stabilizers" to compensate for the hand-held camera jiggle. The camera's computer edits out the jiggle by comparing the current frame with the previous ones, and if the whole picture has shifted it adjusts. The computer is using fuzzy logic to distinguish jiggles from object movement. Our brain accomplishes the same end with regard to our vision. Our eyes are continually shifting focus automatically about five times every second. We call these shifts *saccades.* We never notice them, yet the brain, like the camera's computer, is editing out the jiggle (NcNeill and Freiberger 1993).

Why have we rejected such a great model for understanding behavior? It is those ancient Greeks again, Plato and Aristotle, and their intolerance of paradox and vagueness. Consider Woodger's Paradox, which states that an animal can only belong to one taxonomic family. As Woodger, a biologist, stated, "Therefore, at many points in evolution, a child must have belonged to a completely different family from its parents. But this feat is basically impossible genetically" (quoted in McNeill and Freiberger 1993, 27). Rather than asking the question of how do we get different species or families, the better question would be, "Why do we persist in imagining gaps in nature that do not exist?"

We find that in the real world both life and behavior are made up of vague boundaries. When we begin to examine these fuzzy edges, the vagaries turn out to be not just a marginal annoyance but a vast domain. C. S. Peirce (1839–1914) was a philosopher who helped launch experimental psychology and advanced Boolean logic by applying it to electric circuits. McNeill and Freiberger stated,

> Peirce laughed at the "Sheep & Goat separators" who split the world into true and false. Rather he held that all that exists is continuous, and such continuums govern knowledge. . . . size is a continuum . . . time is a continuum. . . . speed and weight form spectrums, as do effort, distance, and intensities of all sorts. Politeness, anger, joy, and other feelings and behaviors come in continuums. Consciousness itself is a continuum, varying . . . from high alertness through coma. . . . (1993, 28)

That covers the sciences! Vagueness is ubiquitous and not a mark of faulty thinking.

Richard Dawkins, in his article titled "Gaps in the Mind" (1993), described a "ring species." He used the herring gull, *Larus argentatus,* and the lesser black-backed gull, *L. fuscus*, as an example. He stated that in Britain the two species are quite different in color and clearly distinct. However, when one traces the herring gull population westward around the North Pole to North America and then through Alaska and across to Siberia and back to Europe, the herring gulls gradually become less and less like British herring gulls and become more like British lesser black-backed gulls. It turns out that the lesser black-backed gulls are the other end of a ring that started out as herring gulls. Now at every stage along the way the gulls are similar enough to their neighbors so that they will interbreed with them, until the continuum reaches Europe again where the herring gull and the lesser black-backed gull never interbreed. Dawkins pointed out that all pairs of related species are potentially ring species, but in the herring and black-backed gull case all the intermediates are still alive.

With regard to chimpanzees and humans Dawkins (1993) presented one of my favorite examples. He asked that you imagine holding your mother's hand. Now imagine that your mother takes her mother's (your grandmother's) hand, and she in turn takes her mother's hand until you are standing at the end of a line of your grandmothers. Allowing 3.03 ft (1 m) per person, 20 years for each generation, and 6 million years of separate evolution, in a surprisingly short distance of less than 186 mi (300 km) your "grandmother" will take the hand of one of Washoe's "grandmothers." This common grandmother would have had at least two daughters, one your grandmother, and her sister, Washoe's grandmother. The two sisters stand facing each other holding their mother's hands. Washoe's grandmother has a daughter who stands facing her cousin, who in turn has a daughter and so on as this now double line of cousins winds its way back until you are standing face to face with Washoe. That is biological reality.

Looking at this reality, I ask where did we go awry in our thinking and ethics? Again, it was the ancient Greeks. Plato gave us the notion of the "ideal," which implicitly carries the notion that if you are not "ideal" then you are "not ideal," just as the notion of "perfect" carries with it the notion of "imperfect." His student Aristotle gave us the Law of the Excluded Middle, or more correctly the Fallacy of the Excluded Middle: "A cannot be both B and non-B," therefore, "A must be either B or non-B." This provided us with the false sense of "certainty" and "absolute prediction." True and false became our absolutes. Even Aristotle saw problems with this logic. In his *Metaphysics* he recognized that "more" and "less" are still present in the nature of things. But it was Plato's "ideals" that caught on and led to "essentialism," one of the main barriers to the theory of evolution. Essentialism held that each species is completely distinct from all other species and is based on an eternal static essence. The variations were nothing more than imperfections of the underlying ideal model. This model placed permanent gaps in the phylogenetic scale. But these gaps were really gaps in the mind and in the scholarship of humans. Today this archaic superstition still survives. Science still implicitly clings to the arrogant notion that we are somehow different and superior to the "have-nots." Whereas Darwin's principle that all biological functions vary in degree rather than kind is certainly accepted with regard to blood and bone, the mind still remains embattled, and at the center of that battle is language. Descartes is not dead and Aristotle's excluded middle is not outmoded. As examples of science embracing the delusional notion that we are superior and different in kind from other animals, consider the following statements, based on no evidence, by Chomsky (1972, 70) and Skinner (1988, 466), respectively: "Possession of human language is associated with a specific type of mental organization, not simply a higher degree of intelligence," and "No other species has developed the verbal environment we call a language. I doubt that the Gardners have ever seen one chimpanzee show another how to sign."

APE LANGUAGE IMPLICATIONS

Ape language research has demonstrated that apes can acquire and communicate with the signs of ASL and other artificial communication systems (chapter 13, this volume). Chimpanzees can pass their signing skills on to the next generation, demonstrating cultural transmission of their acquired language. They use their signs to converse spontaneously with each other when no humans are present, they sign to themselves, and use their signs during imaginary play. Ape language behavior is rich enough to provide texts that could be analyzed for a number of linguistic traits that are shared with human language.

Why has this research been so scientifically controversial when Darwinism maintains that the cognitive difference between apes and humans is one of degree? It is because many scientists still adhere to the Aristotelian superstitions and Cartesian Dark Ages notion that "humans are outside of nature," different in kind from our fellow animals. Although this arrogant position may be popular and handy to justify exploitation and abuse, it is out of touch with biological reality and serves little else than to puff up human pretensions. A second problem is that the questions regarding ape language should not have been framed in the Aristotelian–Cartesian dichotomy of, "Is it language or is it not?" but rather, "To what degree do we share common communicative characteristics with our fellow animals, and what are the adaptive functions that the similarities and differences serve in helping the individual and species survive in the environments for which they are adapted?" The scientific evidence presented in this chapter clearly demonstrates that the difference between chimpanzees and humans is one of degree, just as it is with all of our fellow animals. This evidence is consistent with the Darwinian notion of continuity. The chimpanzee and other fellow apes just happen to be our next of kin in our phylogenetic family.

THE DOUBLE-EDGED SWORD

It is ironic that the chimpanzees' extreme similarity to humans has served to work against their welfare. For example, the biomedical community has used chimpanzees in research on the AIDS virus, organ-transplant research, hepatitis research, and brain injury research. The biomedical community justifies this research because the chimpanzee's physiology and biology are so similar to that of humans. Yet at the same time they ignore the ethical and moral responsibility for the damage they do to the chimpanzee by relying on the Cartesian view that we are different in kind. The biomedical community wants to have it both ways. Members of the biomedical community cannot go on maintaining that they are being responsible members of our moral human community while at the same time treating chimpanzees as if they are nothing more than "hairy test-tubes." To do so is to walk up to Dawkin's line of mothers and daughters holding hands and to arbitrarily break the chain, telling one that she will be a biomedical research subject to help the other live. It does not make any ethical sense, especially considering that the chimpanzee is an endangered species. It does not make rational sense in light of *our* species' overpopulation.

The zoo community faces a similar dilemma. The zoo community wishes to educate the public to learn to respect our fellow apes (chapter 5, this volume), but to do so they sentence the apes and their future offspring to a life in prison. The zoo community supports conservation efforts (chapter 18, this volume), but keeping animals in cages implicitly supports the Euro-American dominator and the chain-of-being models—with humans on top. Some zoos have enthusiastically embraced enrichment programs, others have not. Some zoos are designing exhibits that take the apes' needs into consideration and look like natural areas—but despite their looks, they are still prisons. In general, zoos have been moving in a direction that is positive for their fellow hominids. I see this volume as evidence that some members of the zoo community are serious and sincere about developing a new perspective with regard to their fellow hominids.

A NEW PERSPECTIVE, A NEW WORLD, A PRACTICAL HARMONY

Before change can effectively take place, we must not only recognize the necessity of moving in a new direction but also recognize how the old direction was wrong and turn away from it. Only then can we begin to make headway in realizing this new way of being within our shared world. Undertaking such a change was one of the most difficult things I have ever done. I had to recognize that I was a part of a research project, in the ignorance of the time, that was party to a baby being taken from her mother and the killing of her mother. It was a project, in its ignorance, that condemned a young girl to a life where she could never fully reach the potential for which she was born, and would always be out of place, and would always be considered inferior. It was a project that took a young girl from her culture and family where she could have learned and given so much. It was a project that condemned her to life in prison, even though she had never committed a crime. It is for these reasons that I have publicly stated that I would never again support or be a part of a project that necessitates the taking of an infant chimpanzee from her or his mother or his or her species. It is for these reasons that I have publicly stated that research projects that would do this are morally wrong. Projects that do this today cannot hide behind the ignorance that existed before the 1970s. It is for these reasons that I have publicly stated that it is wrong to breed such persons into captivity to serve human purposes. I had to come to grips with the reality that even though the originators of the project had the ignorance of their time to justify the project, I do not have this convenience. I have to take responsibility for my actions and their actions given today's knowledge of wild chimpanzee culture and chimpanzee mental and emotional life. I have to accept the Darwinian fact that Washoe is a person by any reasonable definition and that the community of chimpanzees from which she was stolen are a people. I have to accept the responsibility for unjustly imprisoning a relative of mine who has done nothing wrong. I have to accept the fact that I cannot undo the damage that has been done to her. I have to accept the fact that I cannot return her to her family, nor bring her mother back to life. Because of these things, I act. Because the five chimpanzees for whom I am responsible are marooned in this prison for life, I insist that their interests and well-being be our first priority. In their home, human arrogance of any sort is forbidden. They only take part in research if they wish; they are not bribed with food or forced with threat or socially harangued into submission. If Loulis spits on you, you have two choices: Ignore it or walk away. It is his home and if he wants to spit on you he can. Enrichment is a full-time effort at CHCI, and it goes on all day with events such as birthday celebrations, social interactions such as chase games, food events such as forages, and objects such as magazines and toys.

We educate at CHCI as well, but it is an active process whereby the visitors first learn how our species is exploiting its relatives and driving them toward extinction. The visitors are taught to take the chimpanzees on their terms by requiring them to approach the viewing area as if they are "uninvited guests" with the proper nonverbal submissive tentative behavior. We educate the visitors about our responsibility to stop the exploitation and to respect our relatives.

As a scientist I act on behalf of the chimpanzees. I am willing to speak out in favor of better prison conditions, against biomedical research, in favor of sanctuaries, against logging and forest destruction, in favor of protecting preserves, against the bushmeat trade, in favor of poaching patrols, against using apes in entertainment, in favor of the Great Ape Project, against captive breeding, and so on.

In what way would the biomedical community change if it embraced the implications of the Darwinian perspective of biological reality? The members of the community would unite to end all research that is in any way invasive, corrosive, manipulative, or harmful to the chimpanzees. They would stop breeding chimpanzees into captivity. They would establish humane sanctuaries to allow the chimpanzees to live out their lives in peace where the chimpanzees' well-being and interests are first and foremost. Within those sanctuaries, they would individualize treatments to socialize and rehabilitate the chimpanzees who have been damaged socially and psychologically. Finally, as a responsible gesture and a form of reparation for past depravities visited on the chimpanzee and perhaps most critical to the chimpanzees' survival as a species, the biomedical community would invest large portions of their profits into establishing habitat protection for chimpanzees in Africa.

In what way would the zoo community change if it embraced the implications of the Darwinian perspective of biological reality? First, the intentions of zoos toward our great ape populations would have to change. They would have to put the well-being of the individual apes and ape species as their first and foremost priority. This would involve working toward the eventual abolition of zoos and a proactive stance in protecting natural habitats. They would abandon breeding and replace it with active support and establishment of in situ preserves and parks. They would turn their exhibits into sanctuaries where the needs of the individual ape would come first and public education and humane nonmanipulative, noninvasive scientific study would be secondary. Their educational programs would include information about the apes as well as our species' role in their exploitation, both in captivity and in the wild. Their educational programs would encourage our species to respect apes and take them on their terms. They could use their positions as honorable professionals to speak out against the use of apes in biomedical research, against the use of apes in entertainment, against the logging companies destroying the rain forest, against the purchasing of hardwoods taken from the rain forest, and any other human endeavors that harmed or exploited our fellow hominids.

Those doing such things might blanch at the risk of being called activists. I have come to realize that one would be less of an activist and more of a healer, in the sense that a clinical psychologist or psychiatrist is a healer. After all, would a psychiatrist be called an activist for treating a delusional patient who thought that he or she was Napoleon? Taking the road I have mapped would be doing the same thing, but at the species level instead of the individual level. It is simply asking our species to embrace the empirical realities of Darwin and continuity, to get a grip on the reality that our species is not outside of nature and that we are not gods. We might lose the illusory heights of being demiurges, but this new perspective would offer us something greater: the full realization of our place in this great orchestra we call nature.

REFERENCES

Beaucher, J. A. 1995. *Categorization models used by chimpanzees when forming basic level categories.* Unpublished master's thesis, Central Washington University, Ellensburg.

Berry, W. 1990. *What are people for?* San Francisco: North Point Press.

Bodamer, M., Fouts, R. S., Fouts, D. H., and Jensvold, M. L. A. 1994. Functional analysis of chimpanzee (*Pan troglodytes*) private signing. *Human Evolution* 9: 281–96.

Chomsky, N. 1972. *Language and mind.* New York: Harcourt Brace Jovanovich.

Cianelli, S. N., and Fouts, R. S. 1998. Chimpanzee to chimpanzee American Sign Language communication during high arousal interactions. *Human Evolution* 13: 147–59

Darwin, C. 1989. The descent of man, and selection in relation to sex: Part one. In P. H. Barrett and R. B. Freeman (eds.), *The works of Charles Darwin* (Vol. 21, pp. 3–212). New York: New York University Press. (Original work published 1877)

———. 1991. *The origin of species by means of natural selection.* New York: Prometheus Books. (Original work published 1859)

———. 1998. *The expression of the emotions in man and animals.* New York, Oxford: Oxford University Press. (Original work published 1889)

Davis, J. Q. 1995. *The perception of distortions in the signs of American Sign Language by a group of cross-fostered chimpanzees* (Pan troglodytes). Unpublished master's thesis, Central Washington University, Ellensburg.

Dawkins, R. 1993. Gaps in the mind. In P. Cavalieri and P. Singer (eds.), *The Great Ape project. Equality beyond humanity* (pp. 80–87). New York: St. Martin's Press.

Diamond, J. 1992. *The third chimpanzee.* New York: Harper Perennial.

Fouts, R. S. 1973. Acquisition and testing of gestural signs in four young chimpanzees. *Science* 180: 978–80.

———. 1975. Communication with chimpanzees. In I. Eibl-Eibesfeldt and G. Kurth (eds.), *Hominization and behavior* (pp. 137–58). Stuttgart: Gustav Fisher Verlag.

Fouts, R. S., Chown, B., and Goodin, L. 1976. Transfer of signed responses in American Sign Language from vocal English stimuli to physical object stimuli by a chimpanzee (*Pan troglodytes*). *Learning and Motivation* 7: 458–75.

Fouts, R. S., and Fouts, D. H. 1989. Loulis in conversation with the cross-fostered chimpanzees. In R. A. Gardner, B. T. Gardner, and T. Van Cantfort (eds.), *Teaching sign language to chimpanzees* (pp. 293–307). Albany: State University of New York Press.

Fouts, R. S., Fouts, D. H., and Schoenfeld, D. 1984. Sign language conversational interactions between chimpanzees. *Sign Language Studies* 34: 1–12.

Fouts, R. S., Fouts, D. H., and Van Cantfort, T. 1989. The infant Loulis learns signs from cross-fostered chimpanzees. In R. A. Gardner, B. T. Gardner, and T. Van Cantfort (eds.), *Teaching sign language to chimpanzees* (pp. 280–92). Albany: State University of New York Press.

Fouts, R. S., Haislip, M., Iwaszuk, W., Sanz, C., and Fouts, D. H. 1997. *Chimpanzee communicative gestures: Idiolects and dialects.* Paper presented at the Rocky Mountain Psychological Association meetings in Reno, NV, April 18–20.

Fouts, R. S., Hirsch, A. D., and Fouts, D. H. 1982. Cultural transmission of a human language in a chimpanzee mother–infant relationship. In H. E. Fitzgerald, J. A. Mullins, and P. Page (eds.), *Psychobiological perspectives: Child nurturance series* (Vol. 3, pp. 159–93). New York: Plenum Press.

Fouts, R. S., and Mills, S. 1997. *Next of kin.* New York: William Morrow.

Fouts, R. S., and Rigby, R. 1977. Man-chimpanzee communication. In T. Sebeok (ed.), How *animals communicate* (pp. 1034–54). Bloomington: Indiana University Press.

Fouts, R. S., Shapiro, G., and O'Neil, C. 1978. Studies of linguistic behavior in apes and children. In P. Siple (ed.), *Understanding language through sign language research* (pp. 163–85). New York: Academic Press.

Furness, W. H. 1916. Observations on the mentality of chimpanzees and orangutans. *Proceedings from the American Philosophical Society* 55: 281.

Furrow, D. 1984. Social and private speech at two years. *Child Development* 55: 355–62.

Gardner, B. T., and Gardner, R. A. 1971. Two-way communication with an infant chimpanzee. In A. Schrier and F. Stollnitz (eds.), *Behavior of nonhuman primates* (pp. 117–84). New York: Academic Press.

———. 1974. Comparing the early utterances of child and chimpanzee. In A. Pick (ed.), *Minnesota symposium on child psychology* (Vol. 8, pp. 3–23). Minneapolis: University of Minnesota Press.

———. 1985. Signs of intelligence in cross-fostered chimpanzees. *Philosophical Transactions of the Royal Society* B308: 159–76.

———. 1994. Development of phrases in the utterances of children and cross-fostered chimpanzees. In R. A. Gardner, B. T. Gardner, B. Chiarelli, and F. X. Plooij (eds.), *The ethological roots of culture* (pp. 223–55). Boston: Kluwer Academic.

Gardner, R. A., and Gardner, B. T. 1969. Teaching sign language to a chimpanzee. *Science* 165: 664–72.

———. 1978. Comparative psychology and language acquisition. *Annals of the New York Academy of Sciences* 309: 37–76.

———. 1989. A cross-fostering laboratory. In R. A. Gardner, B. T. Gardner, and T. Van Cantfort (eds.), *Teaching sign language to chimpanzees* (pp. 1–28). Albany: State University of New York Press.

———. 1994. Ethological roots of language. In R. A. Gardner, B. T. Gardner, B. Chiarelli, and F. X. Plooij (eds.), *The ethological roots of culture* (pp. 199–222). Boston: Kluwer Academic.

Gill, J. H. 1997. *If a chimpanzee could talk; and other reflections on language acquisition.* Tucson: University of Arizona Press.

Goodall, J. 1986. *The chimpanzees of Gombe: Patterns of behavior.* Cambridge, MA: Harvard University Press.

Hayes, K. J., and Hayes, C. 1951. The intellectual development of a home-raised chimpanzee. *Proceedings of the American Philosophical Society* 95(2): 105–09.

Hayes, K. J., and Nissen, C. H. 1971. Higher mental functions of a home-raised chimpanzee. In A. Schrier and F. Stollnitz (eds.), *Behavior of nonhuman Primates* (pp. 59–115). New York: Academic Press.

Jensvold, M. L. A., and Fouts, R. S. 1993. Imaginary play in chimpanzees (*Pan troglodytes*). *Human Evolution* 8(3): 217–27.

Krause, M. A., and Fouts, R. S. 1997. Chimpanzee (*Pan troglodytes*) pointing: Hand shapes, accuracy, and the role of eye gaze. *Journal of Comparative Psychology* 111(4): 330–36.

McGinnes, P. R. 1979. Sexual behavior of free-living chimpanzees: Consort relationships. In D. A. Hamburg and E. R. McCown (eds.), *The great apes* (pp. 429–39). Menlo Park, CA: Benjamin Cummings Press.

McGrew, W. C. 1992. *Chimpanzee material culture: Implications for human evolution.* Cambridge: Cambridge University Press.

McGrew, W. C., and Tutin, C. E. G. 1978. Evidence for a social custom in wild chimpanzees? *Man* 13: 234–51.

McNeill, D., and Freiberger, P. 1993. *Fuzzy logic.* New York: Simon and Schuster.

Miles, H. L. 1980. Acquisition of gestural signs by an infant orangutan (*Pongo pygmaeus*). *American Journal of Physical Anthropology* 52: 256–57.

———. 1990. The cognitive foundations for reference in a signing orangutan. In S. T. Parker and K. R. Gibson (eds.), *"Language" and intelligence in monkeys and apes: Comparative developmental perspectives* (pp. 511–39). Cambridge: Cambridge University Press.

Nishida, T. 1987. Local traditions and cultural transmission. In B. Smuts, D. Cheney, R. Seyfarth, R. Wrangham, and T. Struhsaker (eds.), *Primate societies* (pp. 462–74). Chicago: University of Chicago Press.

Patterson, F. G., and Cohn, R. H. 1990. Language acquisition by a lowland gorilla: Koko's first ten years of vocabulary development. *Word* 41(2): 97–143.

Premack, D. 1971. Language in chimpanzee? *Science* 172: 808–22.

Rumbaugh, D. 1977. *Language learning by a chimpanzee.* New York: Academic Press.

Savage-Rumbaugh, E. S. 1986. *Ape language: From conditioned response to symbol.* New York: Columbia University Press.

Savage-Rumbaugh, E. S., and Lewin, R. 1994. *Kanzi: The ape at the brink of the human mind.* New York: John Wiley and Sons.

Shaw, H. L. 1989. *Comprehension of the spoken word and ASL translation by chimpanzees.* Unpublished master's thesis, Central Washington University, Ellensburg.

Skinner, B. F. 1988. Signs and countersigns. *Behavioral and Brain Sciences* 11(3): 466.

Spencer, F. 1995. *Pithekos* to *Pithecanthropus:* An abbreviated review of changing scientific views on the relationship of the *Anthropoid* apes to *Homo.* In R. Corbey and B. Theunissen (eds.), *Ape, man apeman: Changing views since 1660* (pp. 13–27). Lieden, Netherlands: R. Corbey, Department of Prehistory, Lieden University.

Spiegel, M. 1996. *The dreaded comparison: Human and animal slavery.* New York: Mirror Books.

Tutin, C. E. G. 1979. Mating patterns and reproductive strategies in a community of wild chimpanzees *(Pan troglodytes schweinfurthii). Behavioral Ecology and Sociobiology* 6: 29–38.

Williams, K. 1995. *Comprehensive nighttime activity budgets of captive chimpanzees* (Pan troglodytes). Unpublished master's thesis, Central Washington University, Ellensburg.

Wilson, D., and Reeder, D. M. (eds.). 1993. *Mammal species of the world* (2nd ed.). Washington, D.C.: Smithsonian Institution Press.

Witmer, L. 1909. A monkey with a mind. *Psychological Clinic* 3: 179–205.

Section 4

ETHICS, MORALITY, AND LAW

SARAH T. BOYSEN AND VALERIE KUHLMEIER

11
CONCEPTUAL CAPACITIES OF CHIMPANZEES

For more than two centuries, humans have been captivated by the specter of ourselves when looking into the eyes of great apes, and we have continuously sought to understand the striking similarities and glaring differences between ourselves and our pongid kin. As we have come to better appreciate the intellectual capacities and emotional sensitivity of great apes, particularly through recent studies of their social complexity, conceptual abilities, and behavioral flexibility, an even stronger case must be made for preserving ape populations in situ and maintaining captive groups in optimal environments. There may be time to preserve these irreplaceable animals, but only if a sense of extreme urgency prevails and the plight of the great apes in the wild and captivity is recognized as a crisis.

CHIMPANZEE COGNITION

Our ape of choice is the chimpanzee, *Pan troglodytes*, with whom we have spent the better part of the past 25 years. Over the years, chimpanzees that we have had the privilege of knowing continue to surprise and enlighten us, and with Gary G. Berntson, we have sought to share what we have learned from our ape partners with both the scientific community and the public whenever possible.

To have the opportunity to further investigate some of the issues and questions that remained or were provoked after working on two different chimpanzee language projects between 1974 and 1980, the Primate Cognition Project was initiated with Berntson in 1983 at The Ohio State University. As the number of chimpanzees, and later capuchin monkeys, pigs, and dogs, we studied increased, the name of the project was changed to the Comparative Cognition Project to recognize species diversity. Our initial two chimpanzees, both 3-year-old males, were loaned and later donated by Dr. Fred King, now retired as director of Yerkes Regional Primate Center, Emory University. Kermit and Darrell, 20 years old as of this writing, continue to actively and enthusiastically participate in our studies, and they also have a strong alliance for maintaining stable social structure among the seven adult chimpanzees in our current group.

All other adult chimpanzees in the Ohio State colony came to us when there was no other optimal long-term care available for them. These included Sarah, who arrived at age 28 from the laboratory of David Premack (e.g., Premack 1971, 1986) at the University of Pennsylvania. Sheba came at age 2 after her entertainment value for zoo public relations declined; Bobby arrived at 22 months from a private owner who was promoting him at state fairs and shopping malls to have his picture taken with children. Another species-isolated chimpanzee, Abby, arrived as an adult at age 20 from a private household. Finally, Digger came to us recently at age 7 from a biomedical laboratory that was divesting its collection of 225 chimpanzees. More recently, we adopted two chimpanzees at ages 12 and 16 months, Keeli and Ivy, whose long-term participation in biomedical research would have almost been certain, because they had been bred in colonies supported in part by the NIH Chimpanzee Breeding Program (see chapter 7, this volume); as well, we added 7-month-old Harper and 11-week-old Emma, for a total of 11 chimpanzees in the current group housed at The Ohio State University Chimpanzee Center.

Our cognitive studies focused on the acquisition of numerical competence in chimpanzees (e.g., Boysen 1993; Boysen and Berntson 1989) and a range of tasks exploring conceptual abilities such as individual recognition (Boysen and Berntson 1986, 1989), cross-modal discriminations, cause-and-effect relations in tool use (Limongelli, Boysen, and Visalberghi 1995), and stone tool-making. More recently, in addition to continued studies of numerical competence, we have been exploring different approaches to representational capacities in the chimpanzee, such as the comprehension-of-scale models compared to their real-world counterparts (Kuhlmeier, Mukobi, and Boysen 1999).

THE IMPORTANCE OF DETERMINING IF CHIMPANZEES CAN ACQUIRE NUMBER CONCEPTS

Though it may never be entirely possible to raise the idea of numerical skills in animals without visions of Clever Hans, the debunked and infamous "counting horse" (Rosenthal 1965), we nevertheless were interested in pursuing studies of the chimpanzees' potential for understanding a nonlinguistic representational symbol system. Following training on one-to-one correspondence, we taught the chimpanzees to associate collections of candies with their corresponding Arabic numeral, and subsequently the reverse association, such that the animals were required to select the matching candy array when a numeral was presented (Boysen

1993; Boysen and Berntson 1989). Thus between the two structured tasks the chimpanzees had productive use of number symbols for labeling arrays and could demonstrate number comprehension if presented with a symbol by then selecting the corresponding quantity of gumdrops. Because the chimpanzees always got to eat the array of candies after each trial, these two protocounting tasks quickly became very popular. Our next effort was to create a more pragmatic way for the chimpanzees to use numbers. Food items were hidden in several locations around the lab, with the idea that the chimpanzee would move around to the various locations and see how many foods were hidden. She was then to report, by selecting the correct number placard from among 0 to 4 (current counting repertoire at the time of testing) the total number of food items she had found. Because the foods were transported, and were also not visible to the chimpanzee when she made her selection from the Arabic numeral options, some learning and innovation would be necessary for her to eventually "sum" the number of items and remember the total long enough to select the matching numeral.

Although we had anticipated that this task was likely within the intellectual grasp of our animals, we certainly did not expect Sheba, a 6-year-old female, to immediately catch on to the various novel task demands. In fact, she performed significantly above chance from the very first trial, the very first session, and thereafter (Boysen and Berntson 1989). This type of rudimentary summation process, regardless of how simplistically one dissects it in terms of encoding and retrieval mechanisms required for its completion, implies that Sheba had some basic understanding of enumeration or counting and some grasp of the ordinal nature of the numbers. However, the most astonishing feature of this experiment was the revelation that Sheba was able to demonstrate such rudimentary addition without specific training, based solely on the two structured counting tasks that she had been taught. Such emergent skills are common for children and represent what minds can do: Integrate, under novel circumstances, previously acquired information and generate new concepts, cognitive processes, and innovative, creative reasoning. Sheba demonstrated that chimpanzee minds are similarly capable of such high-level, emergent cognitive capacities (see also chapter 13, this volume).

QUANTITY JUDGMENTS: LESS CAN MEAN MORE

More recently, we have been exploring a limitation in the chimpanzees' number-related abilities. The chimpanzees' inability to learn a particular task (discussed next) has provided us with more hypotheses and insights into chimpanzee cognitive capacities than some of their earlier triumphs. The quantity judgment task was simple in design: Two dishes that contained different quantities of candy (e.g., one versus four gumdrops) were available for selection by one chimpanzee, as her partner watched. Once a dish was chosen, the experimenter intervened and provided the candy in the chosen dish to the chimpanzee's partner, who readily complied by accepting and eating the goodies. The candy in the remaining dish was given to the chimpanzee who had made the initial choice.

Over trials, the selecting animal should have come to realize that to benefit the most from this game, she should always pick the dish that had the smaller amount of candy. Once she made a choice, the candy in the selected dish would be given to her passive partner, leaving the chimpanzee who chose with whatever remained in the second dish. In the vocabulary of animal conditioning, this was a simple quantity-discrimination task, with a reversed reinforcement contingency. Based on their behavior, both cognitive and otherwise, the chimpanzees in our group regarded it as simply unfair. They absolutely could not catch on to this game, with its backward counterintuitive rules, and thus they persisted in selecting, trial after trial, session after session, the dish containing the greater number of candies. Of course, this meant that their partners reaped the benefits of their choices, and they were stuck with the smaller reward nearly every time. Needless to say, the chimpanzees did catch on to the fact that the other chimpanzee in this game was getting the bigger bounty, and eventually, each time the experimenter divided up the candy, the selecting chimpanzee protested vigorously by vocalizing or by pounding or shaking the experimental apparatus. The animals clearly recognized the inequity of the food division, but, when given the next opportunity to make the choice (typically within 40 to 60 seconds of the last trial), they consistently chose the larger array again, time after time. The difficulties forced a rethinking in trying to teach them this particular set of contingencies and instead we tried to better understand their apparent limitations.

There were some powerful interfering mechanisms at work, and the chimpanzees showed no signs of acquiring any understanding of the rules, despite 48 training sessions and more than 700 trials. At this point, we wanted to know why they were consistently wrong, from a reward-payoff perspective. As anyone who works closely with captive chimpanzees knows, altruism is not their strength, and thus we surmised that the chimpanzees were not intentionally choosing to provide their partners with the greater share of the reward. Recall their behavioral distress following the food division that provided more candy to their partners. They genuinely appeared to be unable to inhibit choosing the larger amount of candy. It was the basis for this failure to inhibit that now intrigued us.

THE EXPONENTIAL POWER OF A REPRESENTATION

A simple change in the nature of the reward stimuli changed everything. In an effort to alter the immediacy of the candy arrays, we replaced the two quantities of candies with their Arabic numeral counterpart (Boysen and Berntson 1995). The results were dramatic and immediate, as both of the chimpanzees who were involved in the pilot study of quantity judgment (Sarah and Sheba) readily selected the number symbol that represented the smaller quantity and thus consistently garnered the larger, remaining reward, represented by the other number. Subsequent testing was conducted with all five chimpanzees in the project who had the requisite counting repertoire of Arabic numerals, including animals of different ages, from differing rearing histories, cognitive training histories, and captive and wild birth origins. None of these factors seemed to have any effect on either the expression of the interference effect if candy arrays were used or the symbolic facilitation effect if numerals were used. The five animals' performances were virtually identical despite numerous other differences among them, and fell between a narrow range of choosing the smaller quantity on 26 to 31 percent of the time if candy was used and 80 to 95 percent if number symbols were used (Boysen et al. 1996). Thus we hypothesized that a powerful, highly predisposed interference effect was being expressed if candy arrays were used in the task.

To test for interfering effects of the literal physical immediacy of the candy rewards, we ran a second study during which choice comparisons were made between arrays of candies of different quantities or arrays composed of small rocks (Boysen et al. 1996). The rock arrays replicated some of the perceptual features of the candy arrays, without the inherent high valence associated with a high-incentive candy reward. If the chimpanzees chose, for example, an array composed of five rocks over two rocks, their partners received five candies, and the selecting chimpanzees received only two. The chimpanzees chose the larger number of rocks, too, and thus earned fewer rewards because the rock arrays served as a type of meta-representation between candy arrays and abstract Arabic symbols. Once again, the animals were unable to inhibit selecting the perceptually larger (and correspondingly the larger numerosity), whether the arrays were of candies or rocks (Boysen et al. 1996). Although there may have been some acquired associative properties of the rocks that interfered with the chimpanzees' ability to inhibit choosing the larger rock array as well, the perceptual, mass-related features of both arrays clearly and dramatically affected the animals' ability to respond optimally. A return to number symbols immediately reinstated the chimpanzees' high performance levels, and a return to candy arrays quickly resulted in a reestablishment of poor performance seen previously when candy or rock arrays were compared. We are continuing to explore the parameters that may be contributing to quantity judgments by chimpanzees, because, despite more than five years of testing with the task, the animals continue to choose the larger array if collections of edibles or objects are used. But as long as symbolic representations or tokens serve as the choice stimuli, the chimpanzees are able to invoke the optimal decision rules that garner them the larger payoff. These findings are in accord with earlier reports from Menzel (1960) and Menzel and Draper (1965), in which chimpanzees were offered a variety of foods differing in size, and their choices were always for the largest pieces, even though size differences were sometimes minimal. The general finding has also been supported by a replication of the candy array choices with rhesus monkeys, *Macaca mulatta* (Silberberg and Fujita 1996), with the same persistent response toward the larger array, and more recently Emmerton (1998) has demonstrated a similar phenomenon with pigeons.

A LONG HISTORY: PAST AND PRESENT CONCEPTUAL STUDIES WITH CHIMPANZEES

Historically, chimpanzees have been subjects for the exploration of a wide range of conceptual and problem-solving experiments, such as the classic box-stacking tasks by Kohler (1925) and decades of innovative studies by Yerkes and his colleagues (e.g., Crawford 1937; Spence 1938; Yerkes 1916, 1943). In the past 30 years there have been classic studies of artificial language acquisition and comprehension, beginning with the work of Gardner and Gardner (Gardner, Gardner, and Cantfort 1989), early work by Rumbaugh (1977; see also chapter 13, this volume) and Premack (1971, 1986), and continued work by Fouts (e.g., 1972; see also chapter 10, this volume), and Savage-Rumbaugh (1986) and colleagues (Savage-Rumbaugh, Brakke, and Hutchins 1992; Savage-Rumbaugh et al. 1989). More recently, chimpanzees have been the subject of empirical studies of imitation (Custance and Bard 1994; Custance, Whiten, and Bard 1995; Tomasello, Savage-Rumbaugh, and Kruger 1993; Whiten et al. 1996), as well as tool use in the wild and captivity (e.g., Matsuzawa, 1997; Nagell, Olguin, and Tomasello 1993), causality (Limongelli et al. 1995), pointing (Krause and Fouts 1997; Leavens and Hopkins 1999; Leavens, Hopkins, and Bard 1996), mirror self-recognition based on the classic studies of Gallup (1970; Lin et al. 1992; Povinelli 1987; Povinelli et al. 1993), and numerical skills (Boysen 1997; Matsuzawa 1985). What has emerged is a picture of the conceptual abilities of the chimpanzee that is a rich tapestry of inferential abilities, deft motor and cognitive skills, exquisite observational learning capacities, self-awareness, and an enormous capacity to learn and to be taught, representing myriad different information-processing capabilities and mechanisms. The envelope for capacity is best pushed under conditions of captive enculturation.

This enormous potential, under conditions of minimal tutelage by a human teacher, most clearly suggests how dramatic and exponential the impact of purposeful transmission of acquired skills and information would have been to an early hominid ancestor. Almost surely there were innovators, those few individuals who emerged among the great variability in burgeoning early hominid populations, those who had creative and sudden insights into conflict resolution, cooperative food sharing, rudimentary representations, enhanced communicative skills, primitive tool use, and eventually tool making (representing but a few of the capabilities that were emerging among these species). In the company of other group members who had significant behavioral plasticity and intellectual potential, and the requisite cognitive flexibility, these new skills were rapidly acquired through social learning mechanisms of observation and imitation at the hands of gifted enculturators or teachers. Indeed, such a process is readily observed in any chimpanzee facility today, where chimpanzees and humans work in close collaboration and as scientific partners (e.g., Boysen 1992; Matsuzawa 1997).

Although the archival literature on chimpanzee conceptual abilities is considerable, new areas of investigation remain. One such question involves the ability of the chimpanzee to understand perspective-taking. We were interested specifically in their understanding of scale models as representations for real-world space. This topic had been addressed previously by Premack and Premack (1983). The four juvenile chimpanzees in the Premacks' study were unable to recognize the relationship between a scale model of a room and the full-size room itself. Given the demonstrated expertise of chimpanzees in the many seemingly more sophisticated cognitive areas described earlier, the failure of Premack's research participants raised several questions. To address these issues, several recent experiments modeled after the unique studies of DeLoache and her colleagues on children's understanding of scale models (DeLoache 1987, 1991, 1995) were undertaken with the chimpanzees. In her studies, DeLoache asked children to observe an experimenter hide a miniature stuffed dog in a small-scale model of an adjacent test room; they found that younger children (ages 2.5 years) were unable to find a larger size stuffed toy that had been hidden in the analogous location in the real room. Three-year-olds, however, were successful, and thus appeared to understand the representational relationship between the scale model and the real room. DeLoache accounted for the younger research participants' failures based on a dual representation hypothesis (DeLoache 1987, 1991), suggesting that the older children were able to recognize the model as a toy-like object in and of itself, as well as a type of symbolic representation of its real-world referent. The younger children, DeLoache hypothesized, were unable to form such dual representations, and thus failed to interpret the model symbolically. When photographs of the room were substituted for the model, however, the 2.5-year-olds were able to find the toy in the actual room, demonstrating what the investigators referred to as "picture superiority effect" (DeLoache 1987). Using DeLoache's methodology fairly closely, including setting the experiment in a playroom, we tested two adult chimpanzees who were familiar with the testing room and were still tractable enough to move safely within the facility, accompanied by the first author, with whom they had a long-term relationship. A 1 to 7 scale model, the same proportion used in the DeLoache studies, was constructed and included the identical wall decorations, carpeting, and furnishings as the real playroom. The four furnishings served as hiding sites for a can of soda, which also became the chimpanzee's reward after she located the can during each test trial. The sites included a metal file cabinet, a brightly colored foam chair covered with fabric, a blue plastic tub, and a large artificial tree in a small basket. The chimpanzees' task was to observe the experimenter hide a miniature soda can at one of the four sites in the model, after which the experimenter went into the real room alone and hid the real soda can while the chimpanzee waited outside in the adjacent hallway. Cameras placed in front of the model and in the hallway outside the test room recorded the chimpanzees' choices during both retrieval phases of the task. Retrieval 1 occurred when the chimpanzee was allowed to enter the playroom after the experimenter had hidden the real can. Once the can was found, the chimpanzee was encouraged to return to the scale model in the hallway and indicate where among the four sites the experimenter had originally hidden the miniature can (Retrieval 2). Thus the second retrieval request was to ensure that the animals were able to remember the site in the model and that any failure they had in searching for the can in the real room was not a result of memory limitations.

Both chimpanzees tested, a 16-year-old female and 10-year-old male, were enthusiastic participants, although only the female was successful in finding the soda can at the correct site in the playroom. This included significant performance with the scale model, with the individual items representing the hiding sites (without the entire model), color photographs of the hiding sites, and a panoramic color photograph of the entire room. The male failed to reach statistically significant performance levels during all versions of Retrieval 1 (full model, individual items, or any photographs), although he was often successful on Retrieval 2, when he was required to return to the model and indicate that he remembered where the experimenter initially had hidden the miniature can. However, once he entered the playroom, his attempts to find the real soda can were below chance, and his apparent mapping of the room into four sites that were consistently visited clockwise led to suboptimal performance. This highly rigid and stereotyped response pattern was exhibited regardless of the mode in which the hiding sites were demonstrated, and thus, unlike the younger children who had failed the task with scale models but performed successfully once photographs were used, the younger male chimpanzee was unable to demonstrate any understanding of the relationship between the model or photographs and the actual playroom.

Given the limited number of chimpanzees tested and the demonstrated individual differences in chimpanzee cognitive abilities (Boysen 1992, 1994), it was not possible to say definitively why one chimpanzee succeeded and the other failed. It was intriguing that the chimpanzee who passed the test was female and the abysmal performance came from the male. However, we were already clearly aware of dramatic differences in male and female chimpanzee behaviors along a number of dimensions, including attention, and therefore we were eager to test the other five adult chimpanzees who did not typically work outside of their housing areas. To accomplish this second phase of testing, we constructed another model fashioned after the chimpanzees' outdoor play area (Kuhlmeier et al. 1999). Each chimpanzee was tested individually and watched as the experimenter "hid" a miniature plastic juice bottle in one of four model hiding sites that corresponded to four large play objects (large blue barrel, green plastic alligator-shaped teeter-totter, a large red plastic sandbox, and a small black garden tractor tire) in the outdoor play yard. The chimpanzee was then given access to the real play yard, where an actual fruit juice bottle had been hidden. Because of the size of the cage mesh, it was not possible to devise a way for the chimpanzees to complete Retrieval 2 in this experiment, so only Retrieval 1 performance was analyzed. All seven chimpanzees were tested, including a retesting of Sheba and Bobby, the two from the original study. The testing with the outdoor scale model also included Abby, the 25-year-old ex-pet chimpanzee who had lived in a human household for 20 years and had literally no cognitive task training whatsoever. She also had retinal damage from Type 2 diabetes, and thus we were unsure how well she could actually see. Nevertheless, Abby was highly motivated to obtain the juice bottles, and so she was given the opportunity to participate.

In the first phase of testing, with the four sites remaining in the same location on each trial and the juice bottle hidden in a different site on each, three of the seven chimpanzees performed successfully in finding the full-size juice bottle in the outdoor enclosure. These included Sheba, Sarah, and Darrell. Two adolescent males (Digger and Bobby, ages 8 and 10), a second 19-year-old male, Kermit, and Abby failed to reach significance. Just as Bobby had persisted in using a clockwise, rigid response pattern in the initial scale-model study, the four chimpanzees who failed with the outdoor scale model also searched in the same perseverative clockwise pattern among the four sites.

In an attempt to disrupt this highly stereotyped and inadequate response pattern, we ran a second phase of the outdoor model study, during which the hiding sites in the model and the actual enclosure were moved every trial, although both sets of objects remained in a corresponding arrangement. Under these conditions, if the chimpanzees opted to move to Site 1 as their initial choice, they encountered a different object each time, which allowed us to determine if the spatial arrangement or persistence at selection of a specific object as the initial choice was driving the clockwise pattern. The results were remarkable in that moving the objects on each trial enhanced performance for the two adult males (though not enough for significance) and resulted in Abby's performance being identical to Sheba's and Sarah's, thus making the performance of all three adult females highly significant. Both younger males, however, performed at chance. There was evidence, particularly for the younger males, of continued movement in a clockwise pattern among the four sites, beginning with Site 1, regardless of which object was placed there. Thus the overall spatial arrangement of the sites had more impact on the unsuccessful chimpanzees than the object itself, whereas the successful female chimpanzees could readily move from the miniature representation of an object in the model directly to the full-size object–site in the outdoor enclosure. Future studies with the scale model paradigm offer an intriguing and challenging direction for understanding the impact of the task demands on both individual and possible sex-related performance.

CHIMPANZEE COGNITIVE ABILITIES

We know a great deal about the range and depth of chimpanzee cognitive capacities. To say that they are extraordinary and considerable is a woeful understatement. Having alluded to a roster of demonstrated skills and conceptual understanding, including more detailed discussion of recent work from our own laboratory, we are left with a description of a nonhuman organism that shares much more than the anatomical model and framework on which our earliest introductions to behavior and reasoning abilities of the chimpanzee were based (e.g., Yerkes 1943). True, they resemble us morphologically, share a significant proportion of genetic material with us, and have a complex social structure based on reciprocity and alliance formation, including deep kinship bonds (de Waal 1982; de Waal and van Roosmalen 1979; Goodall 1986). Thus they are capable of information processing like us, reasoning like us, have the capacity for representational communication like us, and appear to be affected emotionally in situations and contexts that provoke similar responses in humans (e.g., de Waal 1982). Taken together, the full impact of their cognitive awareness and intelligence brings us to difficult questions of how and why we maintain chimpanzees in captivity. Putting aside for the moment the horrific impact of the bushmeat trade on wild chimpanzees and gorillas, and the dramatically increasing habitat loss for chimpanzees (see chapters 1, 3, 4, 12, this volume), the question of the legitimacy of the biomedical use of chimpanzees moves to the forefront. By all standard empirical measures of pain and distress, chimpanzees clearly respond to such situations in a manner that allows empathetic interpretation by a human observer. Thus chimpanzees may experience distress, pain, fear, even what might be appropriately characterized as, in some cases, the dehumanizing conditions of captivity much the same as you or I would. Chimpanzees also continue to be represented in advertising, television, and the film industry with little more respect than a century ago. In addition, the entertainment business also promotes through their purchase of baby chimpanzees a market for breeding chimpanzee infants who are often taken from their mothers shortly after birth, hand-reared, and subsequently sold for $25,000 to $35,000. Indeed, a recent advertisement in a publication geared toward exotic animal dealers offered a "baby boy chimpanzee" for $30,000—billed as "a chimp for every occasion." However, once a chimpanzee reaches 6 or 7 years of age, when male chimpanzees in particular are not as cute and cooperative as they had been, they are often disposed of.

One method is resale into the private sector as a pet, usually to unsuspecting buyers who have little idea of the potential size of their new little darling. Typically, such naive buyers also do not have an understanding of the necessary housing requirements, dietary and veterinary needs, or the dramatic behavioral changes that will likely overwhelm them when the chimpanzee reaches sexual maturity. At age 8 or 9, chimpanzees raised in species isolation are socially inexperienced, and as a consequence are often behaviorally aberrant, having little or no opportunity to interact normally with other chimpanzees. The result is that they can be difficult or impossible to rehabilitate for social living with other chimpanzees, even if there is possibility for placement. In the past, these types of behavioral difficulties made it problematic for isolate- or hand-reared chimpanzees to be relocated to accredited zoos. Further, zoos accredited by the American Zoo and Aquarium Association (AZA) also have reached capacity for housing the total number of chimpanzees for the United States zoo population, based on the chimpanzee Species Survival Plan (SSP; see chapter 18, this volume). Other placement options include surrender to biomedical laboratories and placement in nonaccredited zoos or smaller animal parks ("roadside zoos"), which might still have space. The few sanctuaries that are equipped to care for adult chimpanzees were long ago filled to capacity. Thus viable options for placement of surplus biomedical chimpanzees, ex-pets, or former entertainment animals are extremely limited. There must be a concerted and coordinated effort among the biomedical community, animal welfare organizations, the zoo community, and the public to press for federal legislation to more vigorously control the movement of chimpanzees (and other primates) to private owners, regardless of their intentions, and a major lobbying effort for further legislation that would provide federal funding and support for the long-term care and retirement of chimpanzees, regardless of their history as research participants or entertainment animals (chapter 7, this volume).

With projected life spans of 45 to 60 years in captivity, and most interest for the entertainment business and some types of invasive research in chimpanzees at 5 years of age or younger, the need for a long-term plan, with endowment-funding mechanisms in place, is critical. This is especially the case because the U.S. government has funded the purposeful breeding of hundreds of chimpanzees for the express purpose of having a supply of subjects for biomedical research. Now that this population of chimpanzees is older and includes a portion of animals exposed to a range of viruses during vaccine development and other similar protocols, these animals are in need of even more specialized long-term care. The scientific study of the cognitive and conceptual capacities of the chimpanzee, also funded by the U.S. federal government, has determined that this is a species dramatically similar to our own. This demands of us, as purportedly the more "intelligent" species, a moral obligation to care for them with empathy and grace, much as we are morally, ethically, and socially obligated to care for our own children, aging parents, and companion animals.

REFERENCES

Boysen, S. T. 1992. Pongid pedagogy: The contribution of human/chimpanzee interaction to the study of ape cognition. In H. Davis and D. Balfour (eds.), *The inevitable bond: Examining the scientist–animal interaction* (pp. 205–17). New York: Cambridge University Press.

———. 1993. Counting in chimpanzees: Nonhuman principles and emergent properties of number. In S. T. Boysen and E. J. Capaldi (eds.), *The development of numerical competence: Animal and human models* (pp. 39–59). Hillsdale, NJ: Erlbaum.

———. 1994. Individual differences in cognitive abilities in chimpanzees (*Pan troglodytes*). In R. W. Wrangham, W. C. McGrew, F. B. M. de Waal, and P. Heltne (eds.), *Chimpanzee cultures* (pp. 335–50). Cambridge, MA: Harvard University Press.

———. 1997. Representation of quantities by apes. In P. J. B. Slater, J. S. Rosenblatt, C. T. Snowdon, and M. Milinski (eds.), *Advances in the study of behavior* (pp. 435–62). New York: Academic Press.

Boysen, S. T., and Berntson, G. G. 1986. Cardiac correlates of individual recognition in the chimpanzee (*Pan troglodytes*). *Journal of Comparative Psychology* 100: 321–24.

———. 1989. Numerical competence in the chimpanzee (*Pan troglodytes*). *Journal of Comparative Psychology* 103: 23–31.

———. 1995. Responses to quantity: Perceptual versus cognitive mechanisms in chimpanzees (*Pan troglodytes*). *Journal of Experimental Psychology: Animal Behavior Processes* 21: 82–86.

Boysen, S. T., Berntson, G. G., Hannan, M. B., and Cacioppo, J. T. 1996. Quantity-based interference and symbolic representations in chimpanzees (*Pan troglodytes*). *Journal of Experimental Psychology: Animal Behavior Processes* 22: 76–86.

Crawford, M. P. 1937. The cooperative problem-solving by chimpanzees of problems requiring serial responses to color cues. *Journal of Social Psychology* 13: 259–80.

Custance, D. M., and Bard, K. A. 1994. The comparative and developmental study of self-recognition and imitation: The importance of social factors. In S. T. Parker, R. W. Mitchell, and M. L. Boccia (eds.), *Self-awareness in animals and humans: Developmental perspectives* (pp. 207–25). New York: Cambridge University Press.

Custance, D. M., Whiten, A., and Bard, K. A. 1995. Can young chimpanzees imitate arbitrary actions? Hayes and Hayes (1952) revisited. *Behavior* 132: 839–58.

DeLoache, J. 1987. Rapid change in the symbolic functioning of very young children. *Science* 238: 1556–57.

———. 1991. Symbolic functioning in very young children: Understanding of pictures and models. *Child Development* 62: 736–52.

———. 1995. Early symbol understanding in very young children: Understanding of pictures and models. In D. L. Medin (ed.), *The psychology of learning and motivation* (pp. 65–114). San Diego, CA: Academic Press.

de Waal, F. 1982. *Chimpanzee politics.* New York: Academic Press.

de Waal, F., and van Roosmalen, A. 1979. Reconciliation and consolation among chimpanzees. *Behavioral Ecology and Sociobiology* 5: 55–66.

Emmerton, J. 1998. *Biases in pigeons' responses to numerosity: Discrimination of "more" vs. "less."* Paper presented at the annual meeting of the Psychonomic Society, Dallas, TX, Nov.

Fouts, R. S. 1972. Use of guidance in teaching sign language to a chimpanzee (*Pan troglodytes*). *Journal of Comparative Psychology* 80: 515–22.

Gallup, G. G, Jr. 1970. Chimpanzees: Self-recognition. *Science* 167: 86–87.

Gardner, R. A., Gardner, B. T., and Cantfort, T. E. 1989. *Teaching sign language to chimpanzees.* New York: State University of New York Press.

Goodall, J. 1986. *The chimpanzees of Gombe: Patterns of behavior.* Cambridge, MA: Harvard University Press.

Kohler, W. 1925. *The mentality of apes.* New York: Liveright.

Krause, M. A., and Fouts, R. S. 1997. Chimpanzee (*Pan troglodytes*) pointing hand shapes, accuracy, and the role of eye gaze. *Journal of Comparative Psychology* 111: 330–36.

Kuhlmeier, V. A., Mukobi, K., and Boysen, S. T. 1999. Comprehension of scale models by chimpanzees (*Pan troglodytes*). *Journal of Comparative Psychology* 113: 396–402.

Leavens, D. A., and Hopkins, W. D. in press. Implications of pointing for social cognition in apes. *Journal of Comparative Psychology.*

Leavens, D. A., Hopkins, W. D., and Bard, K. 1996. Indexical and referential pointing in chimpanzees (*Pan troglodytes*). *Journal of Comparative Psychology* 110: 346–53.

Limongelli, L., Boysen, S. T., and Visalberghi, E. 1995. Comprehension of cause-effect relations in a tool-using task by chimpanzees (*Pan troglodytes*). *Journal of Comparative Psychology* 109: 18–26.

Lin, A. C., Bard, K. A., and Anderson, J. R. 1992. Development of self-recognition in chimpanzees (*Pan troglodytes*). *Journal of Comparative Psychology* 106: 120–27.

Matsuzawa, T. 1985. Use of numbers by a chimpanzee. *Nature* 315: 57–59.

———. 1997. *Chimpanzee field studies of intelligence.* Paper presented at the Smithsonian Institution/National Zoo Symposium, Exploring the Primate Mind (B. Beck, chair), Washington, D.C., April.

Menzel, E. W. 1960. Selection of food size in the chimpanzee and in comparison with human judgments. *Science* 131: 1527–28.

Menzel, E. W., and Draper, W. A. 1965. Primate selection of food by size: Visible versus invisible rewards. *Journal of Comparative Physiological Psychology* 59: 231–39.

Nagell, K., Olguin, R. S., and Tomasello, M. 1993. Processes of social learning in the tool use of chimpanzees (*Pan troglodytes*) and human children (*Homo sapiens*). *Journal of Comparative Psychology* 107: 174–86.

Povinelli, D. J. 1987. Monkeys, apes, mirrors and minds: The evolution of self-awareness in primates. *Journal of Human Evolution* 2: 493–509.

Povinelli, D. J., Rulf, A. B., Landau, K. R., and Bierschwale, D. T. 1993. Self-recognition in chimpanzees (*Pan troglodytes*): Distribution, ontogeny and patterns of emergence. *Journal of Comparative Psychology* 107: 347–72.

Premack, D. 1971. Language in the chimpanzee? *Science* 167: 86–87.

———. 1986. *Gavagai.* Cambridge: MIT Press.

Premack, D., and Premack, A. J. 1983. *The mind of an ape.* New York: W. W. Norton.

Rosenthal, R. 1965. *Clever Hans* (Carl L. Rahn, Trans.). New York: Holt, Rinehart and Winston. (Original work published 1911)

Rumbaugh, D. M. 1977. *Language learning in a chimpanzee: The LANA project.* New York: Academic Press.

Savage-Rumbaugh, E. S. 1986. *Ape language: From conditioned response to symbol.* New York: Columbia University Press.

Savage-Rumbaugh, E. S., Brakke, K. E., and Hutchins, S. S. 1992. Linguistic development: Contrasts between co-reared *Pan troglodytes* and *Pan paniscus.* In T. Nishida, W. C. McGrew, P. Marler, M. Pickford, and F. B. M. de Waal (eds.), *Topics in primatology: Human origin* (pp. 51–66). Tokyo: Tokyo University Press.

Savage-Rumbaugh, E. S., Romski, M. A., Hopkins, W. D., and Sevcik, R. 1989. Symbol acquisition and use by *Pan troglodytes, Pan paniscus, Homo sapiens.* In P. Heltne and L. Marquardt (eds.), *Understanding chimpanzees* (pp. 266–95). Cambridge, MA: Harvard University Press.

Silberberg, A., and Fujita, K. 1996. Pointing at smaller food amounts in an analogue of Boysen & Berntson's (1995) procedures. *Journal of Experimental Analysis of Behavior* 66: 143–47.

Spence, K. W. 1938. The solution of multiple choice problems by chimpanzees. *Comparative Psychology Monographs* 15: 1–54.

Tomasello, M., Savage-Rumbaugh, E. S., and Kruger, A. C. 1993. Imitative learning of actions on objects by children, chimpanzees, and enculturated chimpanzees. *Child Development* 64: 1688–1705.

Whiten, A., Custance, D. M., Gomez, J. C., Teixidor, P., and Bard, K. A. 1996. Imitative learning of artificial fruit processing in children (*Homo sapiens*) and chimpanzees (*Pan troglodytes*). *Journal of Comparative Psychology* 110: 3–14.

Yerkes, R. M. 1916. *The mental life of monkeys and apes.* New York: Scholars' Facsimiles and Reprints.

———. 1943. *Chimpanzees: A laboratory colony.* New Haven, CT: Yale University Press.

RICHARD W. WRANGHAM

12
MORAL DECISIONS ABOUT WILD CHIMPANZEES

For many of us, chimpanzees and other great apes have a particularly high moral value. This often helps conservation efforts, for example by increasing the commitment to protecting populations. However, strategies for upholding individual rights can also interfere with efforts to maximize population welfare, creating awkward questions for managers. To what extent, for example, should individual chimpanzees in the wild be forced to suffer the stress of being visited by tourists, given that tourism benefits the population as a whole? Under what circumstances does a chimpanzee deserve to die because he or she harms humans? When should we rescue individuals who are unlikely to contribute to the breeding pool in the wild? A moralist concerned with the purest form of rights might insist that in such dilemmas, we ought always to protect or care for individuals even if we thereby jeopardize a larger population. On the other hand, a conservationist who thinks only in terms of demographic success would say that individual rights should sometimes give way to population benefits.

The problem is further complicated by the need to take larger perspectives into account. Humans suffer in various ways from their chimpanzee neighbors. The survival of chimpanzees may therefore depend on moral support from local advocates who see beyond the short-term costs. Morally enlightened views accordingly may promote population survival by fostering long-term support for apes.

To show the kinds of problems faced in practice, in this chapter I present a case study of the relationship between humans and chimpanzees around Kibale National Park, Uganda. After introducing the site, I consider the conflicts and conservation problems created by humans and chimpanzees, respectively. I discuss the role that tourism and research can play and consider some of the moral dilemmas raised by on-site conservation problems.

KIBALE NATIONAL PARK AS A MODEL

Kibale is a useful example because in many ways it is an ideal kind of reserve. It is in a country where eating chimpanzees (and other primates) is generally regarded as wrong, where wildlife conservation laws are strict and well-supported by the government, and where the current president (Yoweri Museveni) strongly supports forest conservation. The park has been a protected area since the 1920s. It is large enough (306 sq. mi; 766 sq. km) to have a self-sustaining population of chimpanzees, conservatively estimated as 550 (Distribution and Habitat Working Group Report 1997), but there may be more than 1,000 (see the legend to Figure 12-1). Furthermore, Kibale has been a long-term site of primate and chimpanzee research (since 1970 and 1979, respectively), which has led to substantial local employment both at a university field station and, since 1991, in an ecotourism project. Uganda is also noteworthy for a strong press that is critical of wildlife abuse and active specifically in promoting the care and conservation of chimpanzees. A variety of governmental and nongovernmental efforts have been effective over the past decade in virtually eliminating both international trade in and private keeping of chimpanzees and in developing world-class sanctuaries. Thus the circumstances around Kibale are in many ways especially favorable for chimpanzee–human relationships.

Many kinds of conflict occur, however, even around this almost ideal site. Conflicts occur primarily near the forest edge, which is visited by 60 to 80 percent of the park's chimpanzees. As Figure 12-1 shows, even in protected areas as big as 1,200 sq. mi (3,000 sq. km) or more, 50 percent of chimpanzees may visit the edge. In smaller and irregularly shaped protected areas, the proportion of chimpanzees that visit the edges rises sharply, up to 100 percent even for some areas larger than 200 sq. mi (500 sq. km).

Edge-living chimpanzees move between Kibale and the surrounding farmlands, using strips of valley forest and patches such as crater forests as travel routes and food sources. They also use these forest patches as bases from which to enter village areas, to harvest crops such as banana and sugarcane. Some chimpanzees appear to live wholly outside the park, migrating among such forest patches. Within a decade or two, all such forest patches will likely have been converted to fields. However, even farm bush can provide park chimpanzees with sufficient cover that they can visit fields and villages within 1 to 2 mi beyond the park boundary. Con-

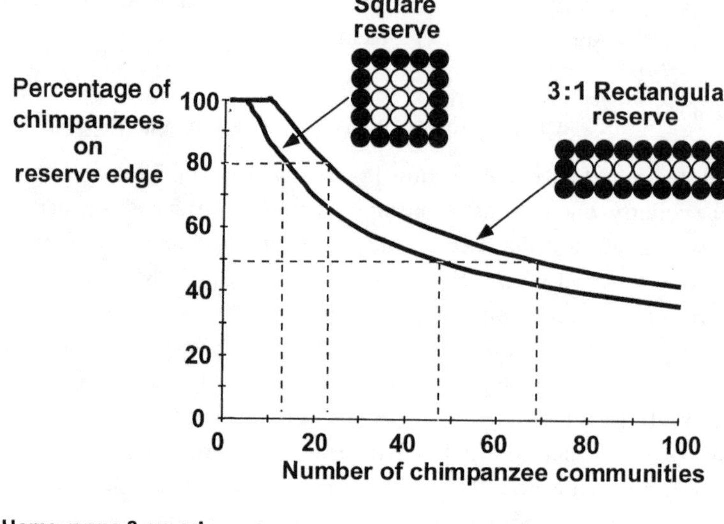

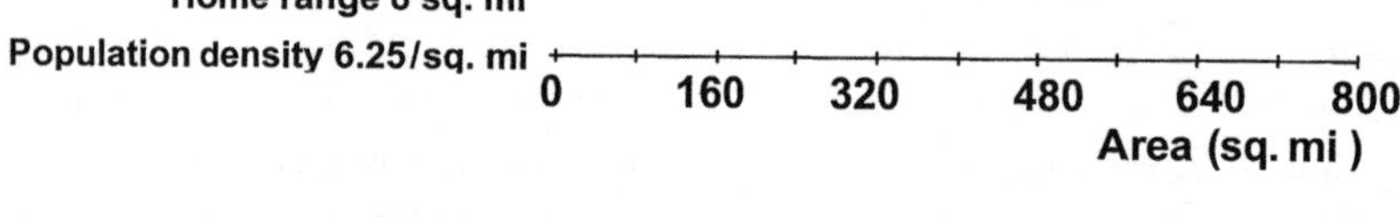

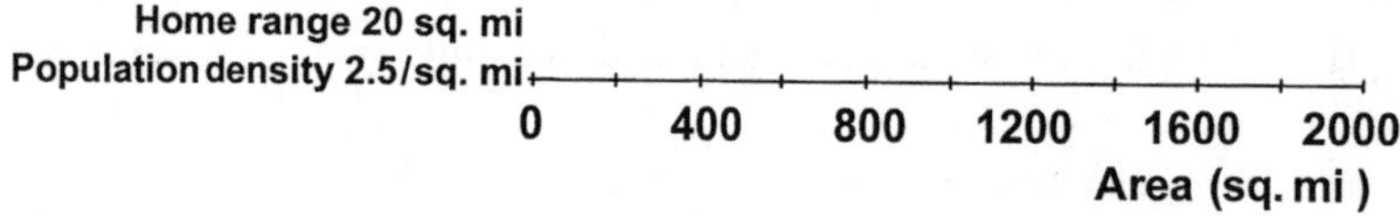

Figure 12-1. The proportion of chimpanzees on reserve edges in relation to reserve size and shape. The percentage of individuals visiting the forest edge is calculated from regular geometrical models. For example, in a square protected area containing 25 communities, 16 communities are peripheral, with 9 communities buffered from the edge by another community. Lower line: square protected area. Upper line: protected area is a regular rectangle, with a 3-to-1 length-to-width ratio. More than 80 percent of chimpanzees visit edges when a 3-to-1 rectangular protected area contains 22 or fewer communities. (In practice, protected area shapes are assumed to be closer to 3-to-1 than 1-to-1 rectangles.) Chimpanzees in Kibale National Park occur at an estimated population density of .4 to .8 per sq. mi (1 to 2 per sq. km), in about 280 sq. mi (700 sq. km). This figure, therefore, suggests that an estimated 60 to 80 percent of Kibale chimpanzees visit the edges. Based on communities of 50 individuals, occupying ranges of 8 sq. mi (20 sq. km), the figure suggests that 3-to-1 protected areas need to be at least 544 sq. mi (1,360 sq. km; holding 3,400 chimpanzees) to ensure that at least 50 percent of chimpanzees are protected from edge effects.

flicts between chimpanzees and humans can therefore be expected far into the future.

PROBLEMS CREATED BY HUMANS

Conflicts between humans and chimpanzees result partly from cultural attitudes. Because cultures vary, it is important to remember that the following examples may not be typical of all chimpanzee habitats.

Humans Deliberately Killing or Capturing Chimpanzees

Although the ape bushmeat crisis continues unabated in much of West and Central Africa (Fa et al. 1995; Fitzgibbon, Mogaka, and Fanshawe 1995; Wilkie and Carpenter 1999; chapters 1, 3, and 4, this volume), the traditional East African disinterest in eating chimpanzees has implied that chimpanzees might be safe from hunting in Uganda. Four points suggest that this perspective is naive. All evidence suggests that individual humans will continue to behave ruthlessly until the moral commitment to ape welfare is enhanced beyond that typical in a poor, rural community.

First, chimpanzees are vulnerable to local migrations of workers without the Ugandan food taboo—for example, those from the Congo who take up work as sawyers in forests containing chimpanzees. Second, even where people do not eat primates, chimpanzees are killed for profit. In 1998, for example, a man living about a half mile from Kibale National Park was found guilty of having killed five or six adults and captured five or six juveniles, apparently for sale to buyers in the Congo (Brian Hare, personal communication). He sold body parts of the adults, including genitals (for aphrodisiacs) and skulls. Third, chimpanzees are killed by farmers defending their crops. This most likely occurs when farmers anticipate crop-raiding at a particularly attractive resource. For example, in May 1997 a farmer living on the boundary of Kibale speared to death a lone adult female who fed from a large pile of bananas being ripened for beer production. This female, Ngonya, was a resident of the Kanyawara research community, known since 1991. The farmer's story was that he killed to defend his possessions and that he was forced to do so because Ngonya attacked him. Detective work by Brian Hare, a research student with the Kibale Chimpanzee Project, and others suggested an alternative interpretation: The farmer probably lay in wait for Ngonya (who had also visited on the night before her death) and attacked her. Certainly, in addition to protecting his bananas, the farmer gained other benefits. He allowed his hunting dogs to eat from her chest, because dogs that eat chimpanzees are thereby supposed to gain the qualities of chimpanzees—namely strength, intelligence, and courage. Thus the innocent explanation of killing in defense of crops may not always be accurate. Nevertheless, farmers who kill chimpanzees in the context of crop-raiding are unlikely to be prosecuted. Fourth, young chimpanzees are kept surreptitiously. In 1994, John Babiiha (a guide working at the Kibale National Park ecotourism project) led authorities to a juvenile female, about 5 years old, in a farmer's home about 2.5 mi from Kibale. An attempt was made to reintroduce this female (Bahati) into the wild. Although she was socially accepted by the Kanyawara chimpanzee community, the reintroduction attempt failed when Bahati showed a preference for agricultural foods during a period of low food availability in the wild (Treves and Naughton-Treves 1997). Bahati is now living on Ngamba Island in Lake Victoria, in a chimpanzee sanctuary.

These examples show that chimpanzees are not safe merely because they live in a well-protected reserve, even in a country with good laws and generally positive ethics with respect to apes.

Humans Accidentally Killing or Wounding Chimpanzees

Around Kibale, as in much of forested Africa, people set snares to catch ungulates such as pigs and small antelopes. Animals are captured both to acquire meat and to protect crops from being eaten. In recent years, the nooses of snares set in and around Kibale have been made mostly of multistrand wires rather than the traditional woody vines that apes could break or bite through (Noss 1997). Wire snares often trap chimpanzees. Deaths occur occasionally when juveniles are unable to tear the noose from its ground pole. Injuries from snares in Kanyawara occur at a rate of 3.7 percent of the population per year (Wrangham and Goldberg 1997). Snare rates are higher for males than females, probably because males more often lead groups. Kanyawara chimpanzees appear to recognize snares and are probably caught mainly when their attention is distracted (e.g., when running or during play). Injuries can be permanent (e.g., crippling or loss of hands, feet, fingers, and toes) or temporary (e.g., lameness). They affect about 20 to 30 percent of chimpanzees at Kanyawara in the Budongo Forest Reserve (Sonso Community; V. Reynolds, personal communication; Wrangham and Goldberg 1997).

Throughout ape-inhabited regions snares are a familiar sight on the limbs of chimpanzees (e.g., Budongo Forest Reserve, personal observation), gorillas (e.g., Virunga, Bwindi, Mgahinga, and Kahuzi-Biega National Parks; personal observation) and bonobos (e.g., Lomako Forest; G. Hohmann, personal communication). If the Kanyawara and Budongo data are representative of apes living near agriculture and if most apes are close to humans, it seems that across Africa tens of thousands of great apes are likely to be suffering crippling wounds.

Unfortunately no effective solution has yet been found for preventing or caring for snare wounds. Immobilization of chimpanzees to remove a snare is possible in theory. In practice, darting a chimpanzee to address a snare brings its own dangers. For example Mkono, a chimpanzee living in the Budongo Forest, Uganda, died from a fall when U.S. researchers darted him in 1969. Twice, when snares attached to individuals appeared to be potentially life-threatening, the Kibale Chimpanzee Project has made attempts to immobilize chimpanzees. Both times, the attempts failed without a shot being fired, because there were no safe opportunities. Thus even the removal of snares from the limbs of habituated individuals is a severe problem.

Preventing apes from being snared is harder still. Currently the Kibale Chimpanzee Project's main approach is to organize a small team of experienced rangers to search for snares that have been set in the forest. Because all hunting in the park is illegal, the Uganda Wildlife Authority permits our team to remove all snares that they find (about 100 per month). In the long term, however, this is a costly and inefficient procedure. Our ultimate aim is to reduce the rate at which snares are set, which will take education and an understanding of the sociology of snare hunting. Moral arguments may help, but unfortunately greater sympathy for chimpanzees will sometimes clash with desires for ungulate meat. For Kibale chimpanzees and for most populations of African apes, this is a problem in desperate need of a humane solution.

Habitat Loss

Africa is in general still experiencing a major increase in population (Cohen 1995). Although urbanization is increasing, population growth is creating intensified pressure to convert forests to agriculture. Population growth is a good predictor of wars (Mesquida and Wiener 1996). Wars lead to poor regulation of protected areas.

Kibale exemplifies these trends. Along the western boundary of the park, a growing human population has reduced the forest patches to farmland substantially in the past decade, causing undoubted losses of chimpanzees dependent on these patches. During the civil-war years from the late 1970s to 1986, thousands of migrant farmers settled inside the park (then a forest reserve). They were removed by presidential decree in the early 1990s. People now live and farm up to the edge of the border along much of the park, and they are cutting trees for charcoal and clearing both along the boundary and even occasionally inside the park. Uganda's population growth has been slowed by war and AIDS. As it recovers, and as the interstitial forest patches are eliminated, political pressures to sanction invasion of the protected forests will likely increase.

Uganda's managerial attitude to forest conservation is exceptionally positive.

For example, a debate at a national seminar of the Forest Department in 1991, involving most of the country's officers, produced a near-unanimous vote of support for conservation in preference to logging as a long-term managerial goal (Howard 1991). Yet this ethic is vulnerable to the appeals of starving families of farmers. We should not deceive ourselves into thinking that benign or proactive government strategies will succeed in maintaining or increasing the land available for ape populations, even in a country like Uganda. Local pressures will likely only intensify in the future.

Incidental Disease Transmission

Edge chimpanzees, in addition to being vulnerable to hunting, incidental snaring, and habitat loss, are exposed to potential sources of human infection (Wallis and Lee 1998; chapter 1, this volume). For example, around Kanyawara some of the poorer farming families often defecate in their fields. Chimpanzees commonly harvest agricultural crops in these fields, and are therefore likely to pick up human intestinal parasites. Because in many protected areas the proportion of chimpanzees visiting the edges reaches 50 percent or more (Figure 12-1), the potential for pathogen transmission is high.

PROBLEMS RESULTING FROM CHIMPANZEE ACTIVITY

Chimpanzees are generally fearful of people, and often have very little interaction with their human neighbors. Unfortunately, however, when conflicts do arise the consequence can be serious. Like humans, chimpanzee behavior varies among individuals and between local cultures, so the following examples merely show some of the types of interaction possible in different areas.

Chimpanzees Attacking or Killing Humans

Attacks on human children by chimpanzees are uncommon, but have been reported in both Tanzania (Goodall 1986) and Uganda (L. Naughton-Treves, personal communication). In the four years between August 1994 and September 1998, nine children were victims of chimpanzee attacks in village areas on the west side of Kibale National Park, within 6 mi of the park headquarters and research station of Kanyawara. Three of the children (all younger than 1 year old) were killed and partially eaten. Two others were permanently maimed.

The motivation for the attacks was uncertain, but most evidence suggests they were carried out by a single male living largely alone in the affected area. We named this putative killer "Saddam." There was sufficient forest in valley strips to enable Saddam to move secretively among villages. The first attack was made on a poorly defended baby lying only 59 ft from the forest. Subsequent attacks were as far as 600 ft from the forest edge, including on infants lying in a house doorway. Most of the attacks suggested the chimpanzee was hunting for meat.

Naturally, the people in the afflicted areas were afraid and angry, and they called on those of us involved in chimpanzee research and conservation to stop the attacks. We were sympathetic to the human victims and their families and were eager to end the state of fear. However, it was hard to know how many chimpanzees were responsible or how to identify or catch them (or him). We devised various ways of assessing how many adult male chimpanzees visited the area. We hoped to prove that there was a single Saddam, in which case we would try to catch and remove him. For nine months, two field assistants working for the Kibale Chimpanzee Project became full-time detectives in search of Saddam. Data slowly emerged to support the hypothesis that a single adult male was responsible for the attacks. But although we sought information, we were accused of not trying to stop him and of caring more for chimpanzees than for children. There was talk of revenge on chimpanzees in general.

It is interesting to consider what we would have done if we had been able to catch Saddam. It was clearly necessary to remove him. However, it seemed unlikely that he could achieve a pleasant life anywhere else in the wild, because no suitable habitat without resident chimpanzees could be found. On the other hand, living alone in a cage would be deeply unpleasant for him. So we determined, reluctantly, that it was best that he be killed, but we did not want to encourage the killing of innocent chimpanzees. Our chosen method was therefore to dart him, take him away in a cage, and euthanize him quietly in private. This strategy created considerable ambivalence among chimpanzee supporters, because in some senses he was innocent—he was engaging in a natural feeding behavior. To make matters worse, it would be difficult to be sure of the chimpanzee's guilt.

In the end, such considerations were irrelevant. Saddam killed for the last time late one evening in September 1998. At dawn the next morning, an angry mob hunted him down with the aid of dogs. By the time field assistants from the Kibale Chimpanzee Project arrived, Saddam had already been badly injured. Saddam was then shot dead to save him from further agony. Most people were excited at Saddam's death, and a foreigner who appeared upset at the sight of his dead body was roundly criticized for her inappropriate affection.

Initially there had been considerable local empathy for chimpanzees. After the first baby was killed, a rumor spread that the killer was a mother chimpanzee whose baby had been killed by a human. Revenge was therefore supposed to be "her" motive for killing a human infant. But this empathy for the plight of chimpanzees quickly fell victim to fear and anger. Following the death of Saddam, there were rumors of other chimpanzees being secretly killed. I have no doubt that if Saddam had survived to kill again, hostility toward chimpanzees in general would have grown dangerously, threatening the lives of many other chimpanzees. In this unfortunate case, therefore, too much protection for the killer chimpanzee would have endangered the lives of many others.

Agricultural Harvesting ("Crop-Raiding")

Chimpanzees forage or "crop-raid" regularly in fields around Kibale National Park, particularly for bananas and sugarcane (more than for maize, cassava, sweet potatoes, or groundnuts; Naughton-Treves 1997, 1998). Crop-raiding happens most during periods of fruit scarcity in the park (Naughton-Treves et al. 1998).

Chimpanzees are less damaging to crops than are elephants, baboons, bush pigs, and monkeys, which in total consume 4 to 7 percent of annual agricultural production around Kibale (Naughton-Treves 1998). The competitive effect of chimpanzees is therefore small and becomes less in fields up to 1 to 2 mi (2 to 3 km) from the park edge (Naughton-Treves 1998). Nevertheless, crop-raiding by chimpanzees creates a negative perception among farmers, which is amplified by legal prohibitions against killing the offenders. This is exacerbated by fear of violence from chimpanzees. Even in the absence of any direct physical threats from chimpanzees, therefore, crop-raiding makes chimpanzees unpopular.

Disease Transmission

Around Kibale many people say that the presence of chimpanzees may increase the risks of humans contracting disease. This is a reasonable fear because the close relationship between chimpanzees and humans makes chimpanzees vulnerable to human diseases as well as a possible source of infection of dangerous pathogens (Gao et al. 1999; Morell 1995; chapter 1, this volume). Of course, unless chimpanzees are eaten or captured, the risks of transmission are probably small. Nevertheless, as the results of biomedical research become more widely published, such fears may become stronger.

ON-SITE SOLUTIONS

These examples show that, partly because chimpanzees are difficult neighbors, humans persistently threaten their welfare. Legislation alone cannot be expected to provide adequate protection for chimpanzee populations. Accordingly, active on-site problem-solving by wildlife authorities is necessary. In addition, two kinds of support activity occur in Kibale: ecotourism and research. Both contribute to conservation and occur widely in other ape habitats. Both incur moral questions.

Ecotourism

Those concerned about the future of protected areas for apes share far more agreements than disagreements. We all would like to see apes have more land and be better protected from killing, kidnapping, disease, and stress.

But there are also disagreements. In 1991, a chimpanzee ecotourism project (the Kanyanchu Project) was initiated in Kibale National Park with help from the Wildlife Conservation Society. But since 1988, when some of us first advocated such a project, there have been objections from scientists worried about negative consequences. These objections include disease risk to apes (Wallis and Lee 1998), psychological stress to apes (about which too little is known), and trampling of the habitat. Concerns have also included more convoluted possibilities, such as profits attracting corrupt or cynical officials or tourists being disappointed in what they see and therefore losing interest in supporting conservation, which could lead to loss of government support of the protected area. As a result, despite substantial governmental and nongovernmental interest in supporting Kibale ecotourism, development was several times slowed, halted, or even reversed during its first decade. Despite this, by 2001 the project was flourishing with chimpanzees well habituated and offering excellent viewing to tourists. The project currently attracts substantial attention and appears to be contributing importantly to local and governmental support for conservation of the forest.

The phrase "chimpanzee ecotourism" refers to a program of taking tourists to visit chimpanzees in close proximity (e.g., as close as 16 ft), much as tourists visit mountain gorillas. This program has benefits and costs (see chapter 1, this volume, for a discussion of costs). Some costs were listed earlier in this chapter.

Benefits include the flow of profits to local people, government organizations, the national treasury, and foreign tour operators. The constituency of interest in chimpanzee welfare is enlarged and deepened both within the host country and internationally.

The relative importance of costs and benefits will vary between contexts. Purely from an ape ethics perspective, we might want to protect chimpanzees and other apes from the disturbance, stress, and risks of daily visits by tourists (Butynski and Kalina 1998). But banning ecotourism involves the gamble that apes will fare well without paying for themselves. Given the conservation problems outlined previously, even in one of the best of contexts, apes desperately need allies, even if those allies are in it for the money.

Around Kibale there appears to be substantial appreciation of the potential value of chimpanzee ecotourism. In seminars we conduct in villages to discuss conflicts between chimpanzees and humans, we often hear people talking positively about our program. It therefore seems likely that local people's tolerance for the problems imposed by chimpanzees, or for the boundaries of the national park, is enhanced. Data from ape tourism projects are therefore needed to test the idea that they increase the net welfare of their ape populations (e.g., Weber 1993). This is hard given the difficulty of obtaining control data, but it is important (Weber 1987).

The practical issue is the level of tax the apes must pay. Chimpanzees and the people who live as their neighbors have conflicts that we would ignore at the apes' peril. Is it too much to require 10 percent of the ape population to be endlessly visited by tourists, even at the risk of occasional epidemics? What about 20 percent or 50 percent? An argument can be made that although we should do all we can to minimize the disease risks, we should promote a substantial flow of tourists wherever the market can sustain it. If every ape-range country ends up with several ape tourism facilities, we can expect the concern for apes to rise.

Research

Though there seems to have been no formal analysis, research projects normally seem to improve the conservation status of reserves. Research creates "stories" that interest the public, it often draws tourists, and it provides information and emotional commitment that can influence government policy. In addition, like tourism, a long-term research project means that a population and its habitat are at least partially monitored. Experience suggests that this is important. For example, the only time when the Kanyawara community of chimpanzees in Kibale was left unwatched since 1983 was for 21 months in 1986 to 1987 between the end of G. Isabirye-Basuta's study and the initiation of the Kibale Chimpanzee Project. During this period almost half the community disappeared for unknown reasons, whereas since then no more than two adults have died in any year (G. Isabirye-Basuta, personal communication).

Research also appears beneficial to reserve status. Several areas have become protected after a decade or more of being home to an ape project (e.g., Gombe, Mahale, and Kibale National Parks). Whether "protected" status is good for a reserve depends on the particular country, department, and area involved, but at least it implies that an effort is being made in the right direction. Research projects are cheap to the conservation community because they are often funded from scientific sources, and they are cost-effective because researchers tend to be committed conservationists. They can complement other efforts such as human community development, habitat conservation projects, or working for forest conservation at the global level (Wolfensohn and Fuller 1998).

Future ape field studies will be particularly beneficial if they involve or are run by host-country field scientists, as is beginning to happen. Host-country scientists can have high visibility and authority for persuading governments and protected-area neighbors about appropriate action, tend to understand ape–human conflicts particularly well, and are well-placed to take a long-term perspective. In an optimistic future in which the primary problems of food supply, disease, war, and government are brought under control, national interests in secondary issues can be expected to rise. The fostering of host-country scientists and conservationists is therefore a critical part of a larger strategy. Ultimately, this means that novel sources of funds may be needed for both research and training.

Even if field research tends to benefit apes, it might have costs as well. For example, the long-term consequences for population health and demography are unknown and could be negative. Nevertheless, suppose we find that individuals tend to die young in study communities, or that study communities tend to die out faster than those left unobserved. What would our attitude be? We would be faced with the same problem as with ecotourism: How much tax should one community pay so that others can survive? A morally balanced answer should consider not only the risks to certain individuals but also the long-term benefits that research can bring.

SYNTHESIS

Past experience shows that it is impossible to predict when human population growth (and habitat conversion) will stabilize (Cohen 1995). In central Africa, we must hope that within the next century, populations level off before the remaining wild areas of great ape habitat are too small to support self-sustaining populations.

Even with the most optimistic assumptions, however, it seems likely that great apes will soon be largely restricted to national parks and similar reserves (Barnes 1990). Traditionally there would have been high rates of interchange among such areas (Goldberg 1998), but parks and reserves in the future will mostly be biological islands (Wilkie et al. 1998) in which local extinctions will be predictably high (cf. Newmark 1995). Reserves therefore need to be large to allow for adequate population size. As reserves become smaller, their edges will become relatively large (Figure 12-1; Malcolm 1994). The potential for direct conflicts can be expected to grow, and humans will normally win.

Ultimately all protected areas may need to be fenced, but fencing will be impossible until economies in Africa are enormously stronger than they are now. Until then, chimpanzees are unlikely to survive even in protected areas without active, on-site advocates. For the future, such advocates will rarely be found in the local population of farmers. A reasonable vision of the middle of the twenty-first century, therefore, has on-site activity by conservationists in virtually every ape reserve.

Some people dislike the intrusion and risk that research and tourism bring. The romantic "ape advocates" of urban Western civilization, of whom I am one, feel enormous respect for apes. It would be wonderful if this respect could be translated into mutually tolerant relations among apes and people living as neighbors in Africa. In the distant future, it may be. But in much of Africa we must surely move through several decades of harsh rural poverty, continued population growth, and civil and national wars before neighborly relations will be even as good as those between bears and people in the United States. Until that time, we cannot afford to forego benefits for ape populations for the sake of excessive ethical standards. If apes were people, intrusive ecotourism would be wrong. But apes are not people. They are other minds with their own agendas. Even in countries without an ape bushmeat crisis, their future depends on local attention, strong advocates, and economic realism.

On the other hand, wild ape populations will benefit from future generations in range countries cherishing apes as special beings. We should therefore be working to encourage a change in attitudes within range countries.

If the high moral value of apes is a useful weapon in the conservationist armory, it depends on a consistent standard. Our arguments are weakened if, in developed countries, apes are seen to be maltreated or if conservation efforts are allowed to fail. If captive apes in the developed world are kept solitarily in cages, used as subjects in invasive experiments, or treated badly in the entertainment industry, we lose the moral high ground. The same is true if manatees or wolves are allowed to go extinct. Because chimpanzees and other apes will continue to need all the help we can give them, a high moral standard in the developed world is a key component of the strategy for their long-term survival in the wild.

The moral questions thus remain confused. Though we must cherish individuals, I believe we must sometimes forsake individual or community benefits for population welfare. It is to be hoped, however, that improved management techniques will mean that in the future, conflicts between humans and wild chimpanzees will be reduced. The moral choices will then be easier.

ACKNOWLEDGMENTS

The Uganda Forest Department, Uganda National Parks, and Uganda National Council for Science and Technology kindly granted permission for fieldwork, and have been consistently supportive. Activities of the Kibale Chimpanzee Project were facilitated by Dr. Gilbert Isabiriye-Basuta and Dr. John Kasenene on behalf of the Makerere

University Biological Field Station. Thanks to them and to members of the Kibale Chimpanzee Project for important contributions to data and conservation, including especially Adam Arcadi, Colin Chapman, Brian Hare, Christopher Katongole, Aloysius Makuru, Samuel Mugume, Lisa Naughton-Treves, John Okwilo, and Michael Wilson. KCP research has been funded through grants from the Getty Foundation, Leakey Foundation, National Geographic Society, and National Science Foundation. Conservation activities have been supported by those organizations, as well as by the Columbus Zoo (Ohio, USA), Jane Goodall Institute (USA), and Jane Goodall Institute (Uganda).

REFERENCES

Barnes, R. F. W. 1990. Deforestation trends in tropical Africa. *African Journal of Ecology* 28: 161–73.

Butynski, T. M., and Kalina, J. 1998. Gorilla tourism: A critical review. In E. J. Milner-Gullard and R. Mace (eds.), *Conservation of biological resources* (pp. 280–300). Oxford: Blackwell.

Cohen, J. E. 1995. *How many people can the earth support?* New York: W. W. Norton.

Distribution and Habitat Working Group Report. 1997. Ugandan chimpanzee distribution, status, threats and conservation priorities: Criteria and definitions for data fields. In E. Edroma, N. Rosen and P. Miller (eds.), *Conserving the chimpanzees of Uganda: Population and habitat viability assessment for* Pan troglodytes schweinfurthii (pp. 33–46). Apple Valley, MN: SSC/IUCN Conservation Breeding Specialist Group.

Fa, J. E., Juste, J., Perez Del Val, J., and Castroviejo, J. 1995. Impact of market hunting on mammal species in Equatorial Guinea. *Conservation Biology* 9: 1107–15.

Fitzgibbon, C. D., Mogaka, H., and Fanshawe, J. H. 1995. Subsistence hunting in Arabuko-Sokoke Forest, Kenya and its effects on mammal populations. *Conservation Biology* 9: 1116–26.

Gao F., Bailes, E., Robertson, D. L., Chen, Y., Rodenburg, C. M., Michael, S. F., Cummins, L. B., Arthur, L. O., Peeters, M., Shaw, G. M., Sharp, P. M., and Hahn, B. H. 1999. Origin of HIV-1 in the chimpanzee *Pan troglodytes troglodytes. Nature* 397: 436–41.

Goldberg, T. 1998. Biogeographic predictions of genetic diversity in populations of Eastern African chimpanzees (*Pan troglodytes schweinfurthii*). *International Journal of Primatology* 19: 237–54.

Goodall, J. 1986. *The chimpanzees of Gombe: Patterns of behavior.* Cambridge, MA: Harvard University Press.

Howard, P. C. (ed.). 1991. Nature conservation in tropical forests: Principles and practice. In *Proceedings of a symposium for District and Regional Forest Officers, Makerere University, Kampala 6–7 May 1991.* Kampala, Uganda: Forest Department.

Malcolm, J. R. 1994. Edge effects in Central Amazonian forest fragments. *Ecology* 75: 2438–45.

Mesquida, C. G., and Wiener, N. I. 1996. Human collective aggression: A behavioral ecology perspective. *Ethology and Sociobiology* 17: 247–62.

Morell, V. 1995. Chimpanzee outbreak heats up search for Ebola Origin. *Science* 268: 974–75.

Naughton-Treves, L. 1997. Farming the forest edge: Vulnerable places and people around Kibale National Park, Uganda. *The Geographical Review* 87(1): 27–46.

———. 1998. Predicting patterns of crop damage by wildlife around Kibale National Park, Uganda. *Conservation Biology* 12: 156–68.

Naughton-Treves, L., Treves, A., Chapman, C. A., and Wrangham, R. W. 1998. Temporal patterns of crop raiding by primates: Linking food availability in croplands and adjacent forest. *Journal of Applied Ecology* 35: 596–606.

Newmark, W. D. 1995. Extinction of mammal populations in western North American National Parks. *Conservation Biology* 9: 512–26.

Noss, A. J. 1997. The impacts of cable snare hunting on wildlife populations in the forests of the Central African Republic. *Conservation Biology* 12: 390–98.

Treves, A., and Naughton-Treves, L. 1997. Case study of a chimpanzee recovered from poachers and temporarily released with wild conspecifics. *Primates* 38: 315–24.

Wallis, J., and Lee, J. R. 1998. Primate conservation: The prevention of disease transmission. *International Journal of Primatology* 20: 803–26.

Weber, A. W. 1987. Socioecologic factors in the conservation of Afromontane forest reserves. In C. Marsh and R. Mittermeier (eds.), *Primate conservation in the tropical rain forests* (pp. 205–29). New York: Alan R. Liss.

———. 1993. Primate conservation and eco-tourism in Africa. In C. S. Potter, J. I. Cohen, and D. Janczewski (eds.), *Perspectives on biodiversity: Case studies of genetic resource conservation and development* (pp. 129–50). Washington, D.C.: American Association for the Advancement of Science Press.

Wilkie, D. S., Curran, B., Tshombe, R., and Morelli, G. A. 1998. Modeling the sustainability of subsistence farming and hunting in the Ituri Forest of Zaire. *Conservation Biology* 12: 137–47.

Wilkie, D. S., and Carpenter, J. 1999. Bushmeat hunting in the Congo Basin: An assessment of impacts and options. *Biodiversity and Conservation* 8: 927–55.

Wolfensohn, J. D., and Fuller, K. S. 1998. Making common cause: Seeing the forest for the trees. *International Herald Tribune,* May 27, p. 11.

Wrangham, R. W., and Goldberg, T. 1997. An overview of chimpanzee conservation and management strategies. In E. Edroma, N. Rosen, and P. Miller (eds.), *Conserving the chimpanzees of Uganda: Population and habitat viability assessment for* Pan troglodytes schweinfurthii (pp. 156–62). Apple Valley, MN: SSC/IUCN Conservation Breeding Specialist Group.

DUANE M. RUMBAUGH, E. SUE SAVAGE-RUMBAUGH, AND MICHAEL J. BERAN

13
THE GRAND APES

In this chapter we choose to call the orangutan, *Pongo* sp., the gorilla, *Gorilla* sp., the chimpanzee, *Pan troglodytes*, and the bonobo, *Pan paniscus*, "grand apes," not just great apes, in an effort to emphasize their unique plight both in their history and probable future. In some manner, the origins of our own species are shared with theirs. On one or more occasions about five million years ago, evolutionary divisions produced new lineages within the order Primates, one of which would spawn a new species of ape distinguished in time by its several colors, its mode of locomotion (bipedalism), and its seemingly endless ability to exploit its surroundings. Taxonomists would name it *Homo sapiens,* and it is us. The division was far from complete, however, because we still share with two of the grand apes (*P. troglodytes* and *P. paniscus*) more than 98 percent of our DNA (Sarich 1983; Sibley and Ahlquist 1987). It is interesting to note that the DNA of *Pan* and *Homo* is more similar than the DNA of *Pan* and *Gorilla* (Andrews and Martin 1987; Sibley and Ahlquist 1987).

Research of the past quarter century has made it abundantly clear that environmental contexts, particularly those of infancy, play a large role in determining what the grand apes, and we, become. Apes and humans are highly susceptible to the impact that their respective cultures make on them as they develop. Apes are not as different from humans in terms of potential as one might think. The development and potential of both are highly plastic in that early environment has major effects on their social, cognitive, and language skills (Bard and Gardner 1996; Bruner 1972, 1983; Davenport, Rogers, and Rumbaugh 1973; Davenport 1979; Gallup et al. 1971; Rumbaugh and Savage-Rumbaugh 1994; Rumbaugh, Savage-Rumbaugh, and Sevcik 1994; Savage-Rumbaugh 1986, 1991; Savage-Rumbaugh and Lewin 1994; Savage-Rumbaugh et al. 1986; Tomasello, 1992, 1996; Tomasello and Call 1997).

GRAND APES AND THE NUCLEAR-TOOTHED APE

Our own species has become so intelligent, as it came to stand and walk erect, that it has become a threat to the rest of life on our planet. While the grand apes kept their razor-sharp canines, sheathed in hard enamel, we abandoned ours as our first line of defense and offense. In lieu of sharp canines, we shaped flint into points to serve our dexterous hands as knives and points to be thrown and eventually shot from bows. Stone on shafts of wood substituted for our diminished canines. Spears and arrows gave way to explosion-propelled missiles, called balls, bullets, and shells. They, in turn, gave way to self-propelled guided missiles, tipped with nuclear devices that can be delivered anywhere on the planet. Metaphorically, we have become the nuclear-toothed ape, not by the design of nature but rather as the consequence of our unnaturally emphasized intelligence.

Not so with the grand apes. Even to this day they more closely reflect the essence of what we were when we were "the way we were," when the first of several lineages emerged that eventually gave rise to us. We are at a point in history where, in the interest of survival of our world as we know it, our species needs to revise the long-held perspective of itself in relation to nature. This is particularly true for the developed nations of the world. Knowledge of the grand apes can help us immeasurably in this effort.

The revision of our perspective of ourselves advanced in this chapter is based on evidence that there is psychological as well as biological continuity among all primates—and that includes us. The degree of continuity is a primary function of genetic relatedness. Hence the continuity is more pronounced between the grand apes and us than between, for example, the rhesus macaque, *Macaca mulatta,* and us. As important as genetics surely are, it has become particularly clear that early environmental effects on intelligence and social competencies can be pervasive and long-lived. During the earliest years of life, genetics and environment have major interactive effects on the development of the psychologies of the grand apes and of our own species. Those effects can serve either to elaborate or to stultify our intelligence and social functioning (Bard and Gardner 1996; Davenport 1979; Davenport et al. 1973). Accordingly, it follows either that the grand apes are not necessarily the beast–machines that Descartes (1637/1956) declared animals to be, or it follows that both they and we might run shared risks of becoming such.

We argue that apes are sentient, feeling, sensitive-to-pain, intelligent creatures in ways not foreign to us. They have symbolic thought, basic dimensions of language, elemental numeric skills, impressive memory and planning capabilities, and other cognitive skills (Beran, Rumbaugh, and Savage-Rumbaugh 1998; Beran et al. 2000; Boysen and Berntson 1989; Byrne 1996; Matsuzawa 1985; McGrew, Marchant, and Nishida 1996; Menzel 1999; Premack and Premack 1983; Rumbaugh 1977; Rumbaugh et al. 1989; Rumbaugh and Savage-Rumbaugh 1994; Russon, Bard, and Parker 1996; Savage-Rumbaugh 1986; Savage-Rumbaugh and Lewin 1994; Savage-Rumbaugh, Rumbaugh, and Boysen 1978; Savage-Rumbaugh, Shanker, and Taylor 1998; Savage-Rumbaugh et al. 1986, 1990; Tomasello and Call 1997; chapters 10 and 11, this volume). They, along with other nonhuman primate species, have personalities that are subject to being profiled by methods and patterns remarkably similar to those appropriate to humans (Bolig et al. 1992; Buirski, Plutchik, and Kellerman 1978; King, Rumbaugh, and Savage-Rumbaugh 1998; McGuire, Raleigh, and Pollack 1994; Stevenson-Hinde and Zunz 1978). In these and other attributes we share common ground with our primate relatives.

This is not to deny distinctive species markers and differences, however, as we both look and behave differently than the grand apes. Those differences are a function both of genetics and cultural differences. The societies into which they and we are born contribute relatively more, of course, to behavioral than to morphological differences. Given shared environments, however, it seems probable that both their and our psychologies and senses of being and knowing can become so similar as to permit a greater understanding of each other. To the degree that their and our offspring are reared in similar cultural contexts, they and we come to manifest similar intellectual skills and to coordinate social and communicative behaviors in effective ways (Bard and Gardner 1996; Call and Tomasello 1996; Gardner and Gardner 1969; Hayes and Hayes 1952; King and Mellen 1994; Premack and Premack 1983; Savage-Rumbaugh et al. 1989, 1993; Tomasello, Savage-Rumbaugh, and Kruger 1993).

In his early evolutionary theory, Darwin (1860, 1871) posited both psychological and biological continuities between animals and humans. Although evidence for biological continuity has been strong for decades, we had to await the end of the twentieth century for evidence to affirm psychological continuity. The previous absence of strong evidence for the processes of intelligence and language in animals permitted the earlier proclamations of Descartes (1637/1956) in the mid-1600s to spawn the belief of discontinuity. Because animals could not talk sensibly, it was easy for people to conclude that animals were, indeed, not sensible. Animals came to be viewed as "subhuman" or "infrahuman"—that is, as creatures that sadly had just not made it to the exalted status of humanhood. They were held to be without reason, thought, affect, intelligence, and language. This view was not reserved only for members of species other than *H. sapiens.* It also was extended to members of our own species because of differences in skin color that somehow offset and overrode in the minds of many Europeans the myriad similarities between all "races" of humans in terms of morphology and behavior. Only in more recent centuries (and in some cases decades) have these views become remedied and dismissed by the majority of the scientific community as foolishness. Today, most of us cannot believe that those views were ever considered scientifically legitimate. However, they were, and they provide good evidence of what happens when the "thinking" ape chooses to address differences instead of similarities between the "us" group and a "them" group.

Those members of the species *H. sapiens* who were granted status as humans were believed able to think because they had rational souls. Accordingly, they could be held accountable by God for their actions. Their unacceptable behaviors were sins. And when they sinned, God inflicted pain. But because animals were without rational souls, they could not sin. Accordingly, it was maintained that God, being just, surely would spare them from experiencing any and all pain. They were, therefore, used in manners similar to that of inanimate objects.

Despite the fact that the rationale for Descartes's beast–machine concept of animals has faded with the twentieth century, the concept of discontinuity between even the grand apes and humans is still very much alive. The frequency and vehemence with which humans argue for demarcations that clearly set themselves apart from animals remains strong. As a consequence, animals are subject to denigration and to having their lives wasted. Indeed, the basic strategy of the schools of behaviorism in this century has been to formulate models that would render *both* animals and humans as nothing but behaving beast–machines. For behaviorism, implicit adherence to the principle of continuity has served substantively to simplify animals and humans alike. One might argue, in a benevolent moment, that to its credit behaviorism has done so even-handedly. Indeed, behaviorism has been working at cross-purposes with the more dominant historic trend, the principle of discontinuity. However, behaviorism has failed as a complete explanation of the complexity of the behavior of living organisms. Something more is needed to understand this complexity, and we propose the idea that behaviors can emerge with no explanation in past performance, past reinforcement, or in species-specific innate patterns of behavior.

EMERGENT BEHAVIOR

The field has come a long way since, only a half-century ago, behavior theorists were confident that essentially all of the psychology necessary for understanding human behavior could be garnered by research on respondent and operant behaviors. In the behavior of even the rat, as it executed its choices in a maze or pressed a lever in an operant chamber, all that would be needed to understand human development, perception, and language was thought to be present. But that perspective now seems nothing short of incredible to most of us.

Respondent and operant behaviors clearly are important to survival. However, they are not the entire picture. Rather, they provide the foundations for still other processes and mechanisms, some with properties not evident in the components but only evident in the outcome. We call these entities *emergents* (Rumbaugh, Washburn, and Hillix 1996). Emergents have four distinctive attributes, none of which are characteristic of respondents and operants: (1) Emergents are forms of silent learning, by which it is meant that learning or acquisition of new response patterns might progress with no obvious manifestation until several months or even years later. Emergents are acquired silently through social observation. (2) Emergent behaviors are not, and cannot be, specifically reinforced via training regimens as are those that condition specific responses to specific stimuli. (3) Emergent behaviors are established through induction by the organism. (4) Emergent behaviors are noted for their appropriateness to novel situations. Examples of emergent behaviors include language acquisition, numerical competence, transfer of learning, imitation, and tool use.

Though based in respondents and operants, emergents are distinctive in their origins, properties, and functions. Their origins are not in the specifics of reinforcement of responses to specific stimuli, and they are more likely to be obtained in organisms noted for having relatively complex brains and enriched rearing in logically structured environments. Emergents reflect relationships imputed between other relationships. Their function is similar to that of speciation: Emergents are the fountainhead of new ways of doing things and creating better solutions to old problems and ingenious solutions to new ones in an efficient and timely manner.

What do emergents tell us? In short, that there is no easy definition of behavior, and there is no easy explanation for the way the living world operates. Emergents do share something in common with the physical, inanimate world. However, the common ground is not something predicted by behaviorism, for behaviorism tried to equate behavior with a deterministic view of the physical world. Today, in physics and chemistry, there is a new emphasis on the unpredictable and possibly indeterminate processes and mechanisms of the natural world. Terms like *complexity* and *chaos theory* now abound in the search for a "grand theory of everything." Although we humbly acknowledge our limits in the understanding of these exciting new fields, we note that the drive within them is to explain natural and human-made systems not in terms of their parts but in terms of the whole that "emerges" above and beyond the parts of the system. Initial states are no longer enough to determine outcomes, as we all know intuitively when we watch the stock market, the weather report, and dozens of other daily activities that affect our lives. At the molecular and submolecular level, the story remains the same. There is a growing recognition that there are processes in action that go beyond the entities from which they are made. From this, we feel even more confident in proposing that the same view be taken of living processes. When we question where these emergent processes should be most evident, we turn to none other than the grand apes.

GRAND APE BRAINS

So where might we begin in our quest for a revised perspective of ourselves and nature? As we have already noted, studies of the grand apes are particularly relevant. What, or rather *who,* are the grand apes? They are the pongids and the natural residents of the hearts of the remaining forests of our world. They reside as tattered, remnant populations of the most exotic species of primates ever known. They are also repositories of some of the most complex brains in the animal kingdom.

Brain functions have important roles in defining dimensions of psychological continuity. To Le Gros Clark and others, the course of evolution of the human brain is outlined in brains of the some 200 species that make up the order Primates (Le Gros Clark 1959; Martin 1990). Primate species' brains tend to encephalize, to become larger than what is predicted for a mammal of their body sizes and weights (Jerison 1985). Thus brains are larger as the body weights of primate species increase because of mammalian allometric brain–body relationships, and primate brains are relatively larger than the brains of other orders of mammals because of encephalization. From this perspective, it is no coincidence that the larger primates, and notably the grand apes, are the more intelligent and more facile learners. They have more brain than monkeys and other mammals, both relatively and absolutely. With their enhanced brains, the grand apes and humans are able to leverage even small amounts of learning to advantage. They are also more likely to use relational forms of learning than to use only stimulus–response associative learning (Rumbaugh 1970, 1995, 1997; Rumbaugh and Pate 1984; Rumbaugh, Savage-Rumbaugh, and Washburn 1996; Rumbaugh, Washburn, et al. 1996). It is relational learning, in turn, that probably is the foundation of observational or vicarious learning, insight, and creativity—and all of the other dimensions of competence that, in history, have been held to be unique to humans.

THE GRAND APES, LANGUAGE, AND TOOL USE

It is significant with regard to a revised perspective of ourselves to be reminded that the infant ape acquires optimal cognitive competence only by being reared in enriched environments. If reared in impoverished physical and social environments during even only the first two years of its infancy, the evidence is that neither cognitively nor socially will its competence ever be restored to normal (Davenport et al. 1973). So it also is with the human infant. The human infant and the ape infant are both hypersensitive to early environmental effects. The larger and more complex the brain of a species, the more sensitive the infants of that species are to the effects of early environment. This sensitivity is not just a general one, however. It is specific as well. For instance, it is through rearing in a language-structured environment that the infant ape most efficiently acquires many of its competencies in language (Rumbaugh and Savage-Rumbaugh 1996; Savage-Rumbaugh 1991; Savage-Rumbaugh et al. 1986, 1998). That competence can include the ability to decode even the syntax of novel sentences of request, a skill never instated through more conventional projects designed to teach language to apes (Greenfield and Savage-Rumbaugh 1990, 1991). Apes so reared also learn spontaneously the meanings and the appropriate use of their word lexigrams, symbols that for them function as words because they cannot speak. The grand apes are capable of learning the meanings and representational use of sign language and arbitrary symbols that for them have all the functional properties of words (Miles 1990; Patterson and Linden 1981; Premack and Premack 1983; Rumbaugh 1977; Savage-Rumbaugh 1984, 1986, 1993; Savage-Rumbaugh et al. 1980; Terrace 1979). The symbols stand for things that are not necessarily present in time and space, for activities, for the properties of things (i.e., temperatures of drinks, ambient noise levels), for the individual's state (i.e., hunger, thirst, sleepiness), for other animates (either ape or human or canine), for places to which they would like to go, and for making comments on activities and recent happenings in the laboratory. For the chimpanzees and bonobos at our laboratory, the Language Research Center, these assertions are based on data obtained from controlled scientific tests, replications of studies with different individuals, and are affirmed by social communication between humans and apes across decades (Rumbaugh and Savage-Rumbaugh 1994, 1996; Savage-Rumbaugh 1986; Savage-Rumbaugh et al. 1986, 1990, 1993). The processes whereby symbols, signs, and gestures optimally acquire these *semantic* properties are cultivated during early rearing of the ape, just as with the human child. Formal, discrete trial training of language skills is relatively ineffective and does not establish the ability to comprehend substantial amounts of human speech (Rumbaugh 1977; Rumbaugh and Savage-Rumbaugh 1994; Savage-Rumbaugh 1986, 1991). As with the human child's pattern of language acquisition, the ape reared in a linguistically complex environment first comes to comprehend speech and lexigrams and then, later, spontaneously begins to produce language, through competent use of its word lexigrams (Beran, Savage-Rumbaugh, et al. 1998b; Brakke and Savage-Rumbaugh 1995, 1996; Savage-

Rumbaugh 1991; Savage-Rumbaugh, Brakke, and Hutchins 1992; Savage-Rumbaugh and Lewin 1994). Early rearing of the bonobo can establish the ability to comprehend language in formal tests similar to that of a $2\frac{1}{2}$ to 3-year-old child and the ability to use grammar similar to the 1- to $1\frac{1}{2}$-year-old child (Greenfield and Savage-Rumbaugh 1990, 1991; Savage-Rumbaugh et al. 1993).

As exciting as it has been to document the apes' language skills, even more important is the principle derived from these results: It is in the logic structure of the infants' environments that their complex abilities, competencies, and dimensions of intelligence and expression are formed. Their formation is behaviorally "silent" in that their expression might not occur until the age of 2 years or older (Savage-Rumbaugh 1991). It is during infancy that important basic vectors of competence are formed. Both ape and child seemingly learn the most complex things regarding social and cognitive competence not through executing motor responses and receiving pellets or other units of extrinsic reinforcement but rather by observing. It is through observation that they learn of the relationships between the things of their world. They learn not just how to make specific responses to specific stimuli. Rather, they accrue information about the world and about behavior, including their own behavior (Bruner 1972, 1983; Savage-Rumbaugh and Rumbaugh 1993; Savage-Rumbaugh et al. 1989).

It is in the evolution of the primate brain's design, structure, and function that organization is brought to "making sense" of the experiences of being alive. The information base instated by the structured experiences of infancy may remain "silent" for several months or even years. But in due course it becomes active and useful. Infant apes can take the principles of complex behaviors extracted by observations of others' behaviors and eventually recast them into patterns that are their own. Thus they can even learn how to knap flint so as to make tools for use in extricating incentives from puzzle boxes. The bonobo Kanzi uses quite sophisticated techniques as he makes chips of flint for use as knives. He turns the cobble to present the best edge for producing a large and sharp chip, and assesses each for sharpness and size. The thicker the cable of rope to be cut, the larger the chip he produces—letting the smaller ones fall by the way (Savage-Rumbaugh and Lewin 1994; Schick et al. 1999; Toth et al. 1993).

GRAND APES AND THE NUCLEAR-TOOTHED APE IN THE FUTURE

As argued by Deborah Blum (1994) in her recent acclaimed book, *The Monkey Wars,* given what we know now about apes and what it is that they are capable of learning and doing in highly complex tasks, it is no longer clear that we and they are qualitatively different in terms of intelligence. Different we are, grand apes and humans, but not so totally different as some would think. Abundant evidence now exists for rejecting the Cartesian beast–machine concept of animals and for supporting Darwin's postulations of continuity. Such evidence has been slow in coming because, within science, Lloyd Morgan's (1894) well-intended canon for parsimony has been used in a manner that has rendered an impoverished and incomplete perspective of animals and behavior. In brief, it produced an "empty-organism psychology," one that denied sentience, rationality, and feelings, even pain, to animals. Its use served to obscure the fact that, unlike the case with physics and chemistry, the data of both psychology and philosophy come *exclusively* from entities that are distinguished by the fact that they are alive. We recognize the shift in this perspective that is taking place, but we desire a quickening of that shift.

Somehow, in the sciences of behavior we need to put the qualities of life and being alive into our database, both for ourselves and for animals. One might object on the grounds that to do so for animals is blatant anthropomorphism, the ascription of "human" characteristics to animals (Kennedy 1992). Notwithstanding, because of our close genetic relationship, we should not disallow from the study of the grand apes any attribute or process or trait that we include as a proper one in the study of humankind. Indeed, we should bear in mind that many "human" qualities are not, in fact, restricted to humans. Rather, they can be general states and processes shared notably by the grand apes. Who is to deny that the mechanisms that induce us as humans to sense common feelings, intentions, and goals among ourselves might not have at least some validity as we formulate impressions about apes and even other animals?

The real cost as we continue to deny all sense of being and knowing to animals is to ensure that they will not be highly valued, other than as chattel and resources for food and materials. If animals are so poorly viewed, why should nature itself be valued, other than as chattel and a resource for food and materials? Indeed, such a view can foster the devaluation of even human lives.

A well-defended, revised perspective regarding our relationship to the grand apes, one based on the "null hypothesis" that there is "no significant difference" between ourselves and nonhuman animals, can serve the survival interests of future generations and the habitability of our planet for the following reason. As we place value on the apes, we are more likely to identify our kind with nature and its systems and be motivated to conserve rather than consume them. Let us take a long view of ourselves. We *H. sapiens* are not lacking in the ability to do things on a grand scale. We have cut the forests, drained swamps, raised crops, produced children by the billions, built cities and universities, created new technologies, and visited the moon.

We celebrate our accomplishments. We enjoy being impressed with ourselves. But lest we be too proud, too impressed with being the *sapiens* ("thinking") species of the genus *Homo,* we need to be sensitive to our several other-than-laudable accomplishments that include the devastation of nature throughout the broad reaches of land masses and oceans of the world by wanton consumption and pollution and the wasting of tens of millions of persons through senseless conflagrations, repetitions of which promise to continue. We have accomplished both the elegantly beautiful and unbelievably ugly throughout our imperfect history as a species. We have done so because of our hyper-smartness for the accrual of knowledge, on the one hand, and because of our hyper-ignorance, on the other hand, as to its impact on both nature and ourselves. We are proving to be a poor fit with the natural world that we need to sustain ourselves and the living organisms around us. Just how long nature will tolerate our kind so beneficently is in question, but nature's patience is not without limit. We should be instructed by the fact that 99 percent of all species from the history of this planet are now extinct. We are the crucial species at the most crucial point in the history of life as we know it, because at no point in history have species disappeared as quickly and in as large numbers as today, in our lifetimes. The majority of the blame for this is our relationship with the planet, a relationship based on exploitation and short-sightedness. Our genus, *Homo,* has existed for approximately two million years on this planet, but in only five one-thousandths of that time have we started, and nearly completed, the devastation of the planet that has produced us.

Even as we seek answers to our problems, rhetoric about sustainable economic development and family planning, though attractive, likely will fail to forestall the devastating crises of our population's relentless expansion into what little remains of virgin nature. Through the course of decades of living and thriving in consumer-based, adversarial (if not warring) societies, we are now caught short by the worry that we are consuming our planet—our one and only planet. Our planet can lose its fecundity, its ability to create and nurture life as we now know it. The planet itself will survive, but nothing of the lifestyle we and other mammals experience today will remain. However, what took only five one-thousandths of our existence as a species to produce can be reversed. A return to the Garden of Eden is impossible, but a return to a reciprocal relationship with nature, and especially our grand ape nature, is not.

About 32 years ago, the first author had the rare privilege of sitting by Adolph Schultz, an accomplished physical anthropologist, as we traveled by bus on an outing for those who attended one of the earliest meetings of the International Primatological Society. During that time he made clear his distress with our species. "We never should have happened," he said. He viewed us as being too intelligent, that our level of intelligence was unnatural, a dangerous "over-run." In his view, for the first time there was a life form that had a characteristic so powerful that it could successfully confound the controlling influences that natural forces otherwise exercise over all other species' populations and distributions. That life form was us, and the attribute of reference was our intelligence.

Schultz died many years ago, but his fears were not ill-founded. His concerns regarding the impact of *H. sapiens* on the planet sadly have proven to be underestimations in both rate and scope. But one might protest and say that there are strong conservation movements afoot. Sadly, conservation, in the genuine sense of it being the protection and maintenance of systems of this world in their virgin states, is not a popular concept. Indeed, it is openly scoffed at politically as being antigrowth, antiemployment, and antiimprovement. *Conservation* today too frequently means putting endangered animals in zoos or parks through which we drive and in which we camp and mine and cut timber and build highways.

The main threat to nature is not the struggling citizens of the developing world who strive to provide food, however unwisely obtained, to their families. They eat what they can access, and we would do the same in their stead. Rather, the threat is our money and our appetites that drive the markets of the world. What we do with our money controls the future of the natural world. For instance, it is the investment of money in the lumber and mining operations of Africa that is both devastating the rain forests and driving the heinous bushmeat market. The needs of this market have led to the killing of hundreds of apes each passing year to obtain protein for loggers and gourmet dishes for the wealthy of the cities (chapters 1, 3, 4, this volume). Thanks to the bold leadership of Karl Ammann of Kenya and the World Society for the Protection of Animals, the ugly bushmeat market might be effectively limited through multinational agreements recently negotiated and announced a few years ago. If it is not, our grand ape cousins will no longer inhabit the land of their, and our, birth as a species.

The grand apes are so closely related to us that it is reasonable that they might share with us several basic elements of intelligence and language. Descartes's beast–machine model of animals is clearly wrong. Accordingly, its influence on our science and society should now end. To become responsible managers of our planet, we must come to perceive ourselves in relation to other life. Are we only a transient tenant species of this planet and owner of all of its resources—licensed to consume, to alter, to engineer, and even to despoil whatever we will? We believe that the definition of the natural and continuous biological and psychological relationship between the grand apes and us will bridge our logic to new perspectives that will serve well our collective survival.

ACKNOWLEDGMENTS

This research was supported by National Institutes of Health grant HD-38051 and HD-06016 to the Language Research Center of Georgia State University and by the College of Arts and Sciences, Georgia State University. The authors thank all the people who have cared for the apes at the Language Research Center throughout the years.

REFERENCES

Andrews, P., and Martin, L. 1987. Cladistic relationships of extant and fossil hominoids. *Journal of Human Evolution* 16: 101–108.

Bard, K. A., and Gardner, K. H. 1996. Influences on development in infant chimpanzees: Enculturation, temperament, and cognition. In A. E. Russon, K. A. Bard, and S. T. Parker (eds.), *Reaching into thought: The minds of the great apes* (pp. 235–56). New York: Cambridge University Press.

Beran, M. J., Pate, J. L., Richardson, W. K., and Rumbaugh, D. M. 2000. A chimpanzee's (*Pan troglodytes*) long-term retention of lexigrams. *Animal Learning and Behavior* 28: 201–207.

Beran, M. J., Rumbaugh, D. M., and Savage-Rumbaugh, E. S. 1998. Chimpanzee (*Pan troglodytes*) counting in a computerized testing paradigm. *The Psychological Record* 48: 3–19.

Beran, M. J., Savage-Rumbaugh, E. S., Brakke, K. E., Kelley, J. W., and Rumbaugh, D. M. 1998. Symbol comprehension and learning: A "vocabulary" test of three chimpanzees (*Pan troglodytes*). *Evolution of Communication* 2: 171–88.

Blum, D. 1994. *The monkey wars.* New York: Oxford University Press.

Bolig, R., Price, C. S., O'Neil, P. L., and Suomi, S. J. 1992. Subjective assessment of reactivity level and personality traits of monkeys. *International Journal of Primatology* 13: 287–306.

Boysen, S. T., and Berntson, G. G. 1989. Numerical competence in a chimpanzee (*Pan troglodytes*). *Journal of Comparative Psychology* 103: 23–31.

Brakke, K., and Savage-Rumbaugh, E. S. 1995. The development of language skills in bonobo and chimpanzees: Comprehension. *Language and Communication* 15: 121–48.

———. 1996. The development of language skills in *Pan*—Production. *Language and Communication* 16: 361–80.

Bruner, J. 1972. Nature and uses of immaturity. *American Psychologist* 27: 687–708.

———. 1983. *Child's talk: Learning to use language.* New York: Norton.

Buirski, P., Plutchik, R., and Kellerman, H. 1978. Sex differences, dominance, and personality in the chimpanzee. *Animal Behaviour* 26: 123–29.

Byrne, R. W. 1996. The misunderstood ape: Cognitive skills of the gorilla. In A. E. Russon, K. A. Bard, and S. T. Parker (eds.), *Reaching into thought: The minds of the great apes* (pp. 111–30). New York: Cambridge University Press.

Call, J., and Tomasello, M. 1996. The effects of humans on the cognitive development of apes. In A. E. Russon, K. A. Bard, and S. T. Parker (eds.), *Reaching into thought: The minds of the great apes* (pp. 111–30). New York: Cambridge University Press.

Darwin, C. 1860. *The origin of species.* New York: Hurst.

———. 1871. *The descent of man—And selection in relation to sex.* London: Murray.

Davenport, R. K. 1979. Some behavioral disturbances of great apes in captivity. In D. Hamburg and E. R. McCown (eds.), *The great apes* (pp. 341–56). Menlo Park, CA: Benjamin/Cummings.

Davenport, R. K., Rogers, C. M., and Rumbaugh, D. M. 1973. Long-term cognitive deficits in chimpanzees associated with early impoverished rearing. *Developmental Psychology* 9: 343–47.

Descartes, R. 1956. *Discourse on method* (L. J. Lafleur, trans.). New York: Liberal Arts Press. (Original work published 1637)

Gallup, G., McClure, M. K., Hill, S. D., and Bundy, R. A. 1971. Capacity for self-recognition in differentially reared chimpanzees. *Psychological Record* 21: 69–74.

Gardner, R. A., and Gardner, B. T. 1969. Teaching sign language to a chimpanzee. *Science* 165: 664–72.

Greenfield, P. M., and Savage-Rumbaugh, E. S. 1990. Grammatical combination in *Pan paniscus:* Processes of learning and invention in the evolution and development of language. In S. T. Parker and K. R. Gibson (eds.), *"Language" and intelligence in monkeys and apes: Comparative developmental perspectives* (pp. 540–78). New York: Cambridge University Press.

———. 1991. Imitation, grammatical development, and the invention of protogrammar by an ape. In N. A. Krasnegor, D. M. Rumbaugh, R. L. Schiefelbusch, and M. Studdert-Kennedy (eds.), *Biological and behavioral determinants of language development* (pp. 235–58). Hillsdale, NJ: Erlbaum.

Hayes, K. J., and Hayes, C. 1952. Imitation in a home raised chimpanzee. *Journal of Comparative and Physiological Psychology* 45: 450–59.

Jerison, H. J. 1985. On the evolution of mind. In D. A. Oakley (eds.), *Brain and mind* (pp. 1–31). London: Methuen.

Kennedy, J. S. 1992. *The new anthropomorphism.* Cambridge: Cambridge University Press.

King, J. E., Rumbaugh, D. M., and Savage-Rumbaugh, E. S. 1998. Perception of personality traits and semantic learning in evolving hominids. In M. C. Corballis and S. E. G. Lea (eds.), *The descent of mind: Psychological perspectives on hominid evolution* (pp. 98–115). New York: Oxford University Press.

King, N. E., and Mellen, J. D. 1994. The effects of early experience on adult copulatory behavior in zoo-born chimpanzees (*Pan troglodytes*). *Zoo Biology* 13: 51–59.

Le Gros Clark, W. E. 1959. *The antecedents of man.* Edinburgh: Edinburgh University Press.

Martin, R. D. 1990. *Primate origins and evolution: A phylogenetic reconstruction.* London: Chapman and Hall.

Matsuzawa, T. 1985. Use of numbers by a chimpanzee. *Nature* 315: 57–59.

McGrew, W. C., Marchant, L. F., and Nishida, T. (eds.). 1996. *Great ape societies.* London: Cambridge University Press.

McGuire, M. T., Raleigh, M. J., and Pollack, D. B. 1994. Personality features in vervet monkeys: The effects of sex, age, social status, and group composition. *American Journal of Primatology* 33: 1–13.

Menzel, C. R. 1999. Unprompted recall and reporting of hidden objects by a chimpanzee *(Pan troglodytes)* after extended delays. *Journal of Comparative Psychology* 113: 426–34.

Miles, H. L. 1990. The cognitive foundations for reference in a signing orangutan. In S. T. Parker and K. R. Gibson (eds.), *"Language" and intelligence in monkeys and apes: Comparative developmental perspectives* (pp. 451–68). New York: Cambridge University Press.

Morgan, C. 1894. *An introduction to comparative psychology.* London: Walter Scott.

Patterson, F. L., and Linden, E. 1981. *The education of Koko.* New York: Holt, Rinehart, and Winston.

Premack, D., and Premack, A. J. 1983. *The mind of an ape.* New York: Norton.

Rumbaugh, D. M. 1970. Learning skills of anthropoids. In L. Rosenblum (eds.), *Primate behavior: Developments in field and laboratory research* (pp. 231–45). New York: Academic Press.

———. 1977. *Language learning by a chimpanzee: The LANA project.* New York: Academic Press.

———. 1995. Primate language and cognition: Common ground. *Social Research* 62: 711–30.

———. 1997. Competence, cortex and animal models: A comparative primate perspective. In N. Krasnegor, R. Lyon, and P. Goldman-Rakic (eds.), *Development of the prefrontal cortex: Evolution, neurobiology and behavior* (pp. 117–39). Baltimore: Paul H. Brookes.

Rumbaugh, D. M., Hopkins, W. D., Washburn, D. A., and Savage-Rumbaugh, E. S. 1989. Lana chimpanzee learns to count by "NUMATH": A summary of a videotaped experimental report. *The Psychological Record* 39: 459–70.

Rumbaugh, D. M., and Pate, J. L. 1984. The evolution of cognition in primates: A comparative perspective. In H. L. Roitblat, T. G. Bever, and H. S. Terrace (eds.), *Animal cognition* (pp. 569–87). Hillsdale, NJ: Erlbaum.

Rumbaugh, D. M., and Savage-Rumbaugh, E. S. 1994. Language in a comparative perspective. In N. J. Mackintosh (ed.), *Animal learning and cognition* (pp. 307–33). San Diego, CA: Academic Press.

———. 1996. Biobehavioral roots of language: Words, apes, and a child. In B. M. Velichkovsky and D. M. Rumbaugh (eds.), *Communicating meaning: The evolution and development of language* (pp. 257–74). Mahwah, NJ: Erlbaum.

Rumbaugh, D. M., Savage-Rumbaugh, E. S., and Sevcik, R. A. 1994. Biobehavioral roots of language: A comparative perspective of chimpanzee, child, and culture. In R. W. Wrangham, W. C. McGrew, F. B. M. de Waal, and P. G. Heltne (eds.), *Chimpanzee cultures* (pp. 319–34). Cambridge, MA: Harvard University Press.

Rumbaugh, D. M., Savage-Rumbaugh, E. S., and Washburn, D. A. 1996. Toward a new outlook on primate learning and behavior: Complex learning and emergent processes in comparative perspective. *Japanese Psychological Research* 38: 113–25.

Rumbaugh, D. M., Washburn, D. A., and Hillix, W. A.. 1996. Respondents, operants, and emergents: Toward an integrated perspective on behavior. In K. Pribram and J. King (eds.), *Learning as a self-organizing principle* (pp. 57–73). Hillsdale, NJ: Erlbaum.

Russon, A. E., Bard, K. A., and Parker, S. T (eds.). 1996. *Reaching into thought: The minds of the great apes.* New York: Cambridge University Press.

Sarich, V. 1983. Retrospective on hominid macromolecular systematics. In R. L. Ciochon and R. S. Corruccini (eds.), *New interpretations of ape and human ancestry* (pp. 137–50). New York: Plenum Press.

Savage-Rumbaugh, E. S. 1984. Acquisition of functional symbol usage in apes and children. In H. L. Roitblat, T. G. Bever, and H. S. Terrace (eds.), *Animal cognition* (pp. 291–310). Hillsdale, NJ: Erlbaum.

———. 1986. *Ape language: From conditioned response to symbol.* New York: Columbia University Press.

———. 1991. Language learning in the bonobo: How and why they learn. In N. A. Krasnegor, D. M. Rumbaugh, R. L. Schiefelbusch, and M. Studdert-Kennedy (eds.), *Biological and behavioral determinants of language development* (pp. 209–33). Hillsdale, NJ: Erlbaum.

———. 1993. Language learnability in man, ape, and dolphin. In H. L. Roitblat, L. M. Herman, and P. E. Nachtigall (eds.), *Language and communication: Comparative perspectives. Comparative cognition and neuroscience* (pp. 457–84). Hillsdale, NJ: Erlbaum.

Savage-Rumbaugh, E. S., Brakke, K., and Hutchins, S. 1992. Linguistic development: Contrasts between co-reared *Pan troglodytes* and *Pan paniscus.* In T. Nishida (ed.), *Proceedings of the 13th international congress of primatology* (pp. 293–304). Tokyo: University of Tokyo Press.

Savage-Rumbaugh, E. S., and Lewin, R. 1994. *Kanzi: The ape at the brink of the human mind.* New York: Wiley.

Savage-Rumbaugh, E. S., McDonald, K., Sevcik, R. A., Hopkins, W. D., and Rubert, E. 1986. Spontaneous symbol acquisition and communicative use by pygmy chimpanzees (*Pan paniscus*). *Journal of Experimental Psychology* 115: 211–35.

Savage-Rumbaugh, E. S., Murphy, J., Sevcik, R. A., Brakke, K. E., Williams, S. L., and Rumbaugh, D. M. 1993. Language comprehension in ape and child. *Monographs for the Society for Research in Child Development* 1: 1–221.

Savage-Rumbaugh, E. S., Romski, M. A., Hopkins, W. D., and Sevcik, R. A. 1989. Symbol acquisition and use by *Pan troglodytes, Pan paniscus, Homo sapiens.* In P. G. Heltne and L. A. Marquardt (eds.), *Understanding chimpanzees* (pp. 266–95). Cambridge, MA: Harvard University Press.

Savage-Rumbaugh, E. S., and Rumbaugh, D. M. 1993. The emergence of language. In K. R. Gibson and T. Ingold (eds.), *Tools, language and cognition in human evolution* (pp. 86–108). Cambridge: Cambridge University Press.

Savage-Rumbaugh, E. S., Rumbaugh, D. M., and Boysen, S. T. 1978. Symbolic communication between two chimpanzees (*Pan troglodytes*). *Science* 210: 922–24.

Savage-Rumbaugh, E. S., Rumbaugh, D. M., Smith, S. T., and Lawson, J. 1980. Reference: The linguistic essential. *Science* 210: 922–25.

Savage-Rumbaugh, E. S., Sevcik, R. A., Brakke, K. E., Rumbaugh, D. M., and Greenfield, P. 1990. Symbols: Their communicative use, comprehension, and combination by bonobos *(Pan paniscus)*. In C. Rovee-Collier and L. P. Lipsitt (eds.), *Advances in infancy research, Vol. 6* (pp. 221–71). Norwood, NJ: Ablex.

Savage-Rumbaugh, E. S., Shanker, S., and Taylor, T. J. 1998. *Apes, language and the human mind.* New York: Oxford University Press.

Schick, K. D., Toth, N., Garufi, G., Savage-Rumbaugh, E. S., Rumbaugh, D., and Sevcik, R. 1999. Continuing investigations into the stone tool-making and tool-using capabilities of a bonobo (*Pan paniscus*). *Journal of Archaeological Science* 26: 821–32.

Sibley, C. G., and Ahlquist, J. E. 1987. DNA hybridization evidence of hominoid phylogeny: Results from an expanded data set. *Journal of Molecular Evolution* 26: 99–121.

Stevenson-Hinde, J., and Zunz, M. 1978. Subjective assessment of individual rhesus monkeys. *Primates* 19: 473–82.

Terrace, H. S. 1979. *Nim.* New York: Knopf.

Tomasello, M. 1992. The social bases of language acquisition. *Social Development* 1: 67–87.

———. 1996. The cultural roots of language. In B. M. Velichkovsky and D. M. Rumbaugh (eds.), *Communicating meaning: The evolution and development of language* (pp. 275–308). Mahwah, NJ: Erlbaum.

Tomasello, M., and Call, J. 1997. *Primate cognition.* New York: Oxford University Press.

Tomasello, M., Savage-Rumbaugh, E. S., and Kruger, A. C. 1993. Imitative learning of actions on objects by children, chimpanzees, and enculterated chimpanzees. *Child Development* 64: 1688–1705.

Toth, N., Schick, K. D., Savage-Rumbaugh, E. S., Sevcik, R. E., and Rumbaugh, D. M. 1993. *Pan* the tool maker: Investigations into the stone tool-making and tool-using capabilities of a bonobo (*Pan paniscus*). *Journal of Archaeological Science* 20: 81–91.

COLIN ALLEN

14
COGNITIVE RELATIVES AND MORAL RELATIONS

The close kinship between humans, chimpanzees, gorillas, and orangutans is a central theme among participants in the debate about human treatment of the other apes, and is emphasized by numerous contributors to *The Great Ape Project*. Richard Dawkins (1994) has made the point vivid by imagining two chains of daughters holding their mothers' hands, starting with a human and a chimpanzee side by side on the African coast. Barely 300 mi inland, at the head of both chains, would be an individual ancestor of both the human and the chimpanzee. In a subsequent issue of *Etica & Animali* dedicated to *The Great Ape Project,* Maxine Sheets-Johnstone (1996, 125) chided philosophers for failing to recognize the importance of evolutionary history for a proper understanding of who we humans are and where we have come from. What we need, she said, is "not a different conception of nonhuman animals . . . but a different conception of ourselves."

By delivering the message that we humans are apes, these authors and others seek to establish an empathetic bond between our primate cousins and ourselves. Empathy is probably the single most important determinant of actual human moral behavior, including the treatment of nonhuman animals. Given the applied nature of questions about the treatment of captive apes, it is entirely appropriate that the close relationship between us should be highlighted. But the role that relatedness should play in ethical theory is less clear (a point Dawkins has acknowledged). To the extent that legal and regulatory challenges to keeping apes in captivity are likely to be based on principles of theory, it is important to understand what roles evolutionary theory can play in deriving such principles.

In the ethical literature on animal rights, phylogenetic relatedness plays no direct role in determining the moral status of animals. Rather, various capacities such as the ability to experience pain, to suffer, to be an intentional agent, to participate intentionally in reciprocal social arrangements, and to be self-aware have been put forward as the relevant factors for moral consideration (chapter 17, this volume). From this perspective, the relationship of humans to chimpanzees and the other great apes is indirectly of interest to ethical questions insofar as it is directly relevant to the likelihood that we humans share certain basic mental capacities with our ape cousins (Crisp 1996). But the capacities themselves are assumed to be attributable on the basis of behavioral and neurological properties of the animals themselves, independent of their historical or phylogenetic provenance. Indeed, for a truly nonanthropocentric, primatocentric, or taxocentric ethic, it is essential that such determinations be made independently of membership in any particular phyletic group because of the possibility of convergent evolution, the independent evolution of similar traits in distantly related species.

Relatedness is thus secondary to the attribution of mental characteristics in the development of an adequate theoretical framework for resolving ethical questions about animals. Nonetheless, evolutionary theory has a very significant role to play in any adequate conceptualization of the mind itself. An important development has been the development of what Donald Griffin (1978) has called "cognitive ethology": the attempt to graft Darwinian ideas about mental continuity onto comparative methods of classical ethology (see Allen and Bekoff 1997, especially chapter 1). Ruth Millikan's work in the philosophy of mind (Millikan 1984, 1993) provides an example of the potential for evolutionary ideas to alter our very conception of mind. Millikan has developed a theory of mental states that has its roots in the general biological treatment of adaptive characters and that allows her to treat the inherent meaning or content of mental states in terms of biological function. The result is a theory that can properly accommodate the enormous variety of mental capacities that we find in the animal kingdom without forcing theories of mind into an all-or-nothing approach that sometimes seems to be the consequence of other philosophical views (see Allen and Bekoff 1997, chaps. 6 and 9). For instance, whereas Dennett (1987) seems to regard his intentional stance as a phenomenon that is illusory in the absence of completely general rationality, Millikan's approach provides a realist conception of intentionality linked to a conception of biological functions that allows intentional attributions to be predicated on the basis of domain-specific adaptations.

The pioneers of ethology, Konrad Lorenz and Niko Tinbergen, were themselves skeptical of the possibility of engaging in a serious scientific study of animal mentality. Given that their efforts to establish ethology as a scientific discipline took place during the height of behaviorism in psychology, it is hardly surprising that they did not emphasize the notion of mental continuity that Charles Darwin had promoted in his book *The Descent of Man* (1871/1936). Darwin, of course, had strong reasons for wanting to emphasize continuity in all respects between humans and the rest of nature, as illustrated in the following passage:

> If no organic being excepting man had possessed any mental power, or if his powers had been of a wholly different nature from those of the lower animals, then we should never have been able to convince ourselves that our high faculties had been gradually developed. But it can be shown that there is no fundamental difference of this kind. We must also admit that there is a much wider interval in mental power between one of the lowest fishes, as a lamprey or lancelet, and one of the higher apes, than between an ape and a man; yet this interval is filled up by numberless gradations. (Darwin 1871/1936, 445)

Contrary to what this passage might suggest, there is not likely to be any single scale along which we can rank lampreys to humans with respect to mental capacities. For a start, all living organisms are the tips of the so-called tree of life. Humans are not descended from lampreys, and even though we share a common ancestor there is no route from that ancestor to humans through, for example, rabbits, or even through chimpanzees. Thus there is no evolutionary reason to expect that rabbits or chimpanzees should lie somewhere midway between humans and any other organism on any particular scale, whether it is of intelligence or ear shape.

If we have learned anything at all from contemporary cognitive science, it is that the human mind is modular: Our minds are designed and implemented around a number of specific domains and operations, and it is quite possible for these modules to operate in a relatively isolated fashion. Thus, for instance, a recent discovery in evolutionary psychology is that our ability to reason about social rules, evolved in the cauldron of primate sociality, may be a modular system that is distinct from our ability to reason about abstract logical principles (Cummins 1998). Minds, despite how they may seem from the inside, are not general-purpose devices. The result of such discoveries is a realization that it is not appropriate to ask "Can animals reason?" as did Descartes, and before him Aristotle. Instead we should be asking whether they can reason about some specific thing or another. Again, this puts pressure on the idea of a single scale of intelligence or mind on which all organisms might be arrayed.

This attack on the idea of a single scale cannot be construed as an attack on the idea of continuity in general. Although any trait of any organism may be the novel feature that sets that lineage apart from others, the general lesson from evolutionary biology is that evolution is both conservative and convergent. It is conservative in the sense that useful characteristics, once evolved, tend to be preserved. It is convergent in the sense that common solutions to common problems are frequently reached by different paths. These principles should lead us to expect elements of mind to be found across a variety of species. But each case must be considered on its own merits. Early defenders of Darwin's views about mental continuity, such as George Romanes (1883), were well aware of these issues, and their discussions led to specific methodological recommendations, such as C. Lloyd Morgan's (1894) canon: Never attribute a higher psychical cause when a lower one is available.

Why, then, did the investigation of mental capacities in nonhumans effectively disappear from science during the early part of the twentieth century? The rise of behaviorism coincided not just with the realization that greater objectivity was required in the description of animal behavior, but also with the positivist movement in the philosophy of science, with its emphasis on directly observable events and strict empirical testing of scientific hypotheses. In the face of this apparent onslaught of hard-headed logicality, it was hard to see how to defend investigations of such ephemeral notions as mind and consciousness as science. The philosophical problem of other minds had come to roost.

Positivist philosophy of science was, and continues to be, I am convinced, a good and essential force in the maturation of science. But it was not the final word, and the revival of the ideas of turn-of-the-century physicist Pierre Duhem by the philosopher Willard Quine during the mid-twentieth century brought this into focus (Quine 1953). Duhem's thesis was that no scientific hypothesis is ever tested in isolation, but only against the background of numerous other hypotheses and factual claims. Quine went on to argue that although experience remains the court in which our beliefs are tested, there is no tenable distinction to be drawn between science on the one hand or mathematics and philosophy on the other. Theories from all the disciplines constitute an interconnected network that can be modified to provide the highest overall coherence with our observations. But (and here I depart perhaps from Quine's view) if our best explanation of nonhuman animal behavior postulates mental capacities, as it does for human behavior, then it is appropriately scientific to speak in such terms.

So what we have is a picture of scientific investigation where observable facts may be used to defend theoretic claims about processes and entities that may themselves be far from directly observable. Current physics has, of course, been quick to make use of inferences of this kind to posit and investigate all manner of otherwise recalcitrant phenomena. But psychologists and ethologists have not been so keen to embrace the lesson and return to mentalistic hypotheses about animal behavior.

This reluctance is somewhat understandable, because what comparative psychology lacks, modern physics has in abundance in the form of sophisticated mathematical analysis linking theory to observation. In the present state of play it is hard to imagine what a similarly mathematical theory of mind might be like, and there may even be reasons to suspect that there will be no such thing forthcoming.

Is this fatal to the attempt to develop cognitive ethology? I think not, although it remains to be seen whether cognitive ethologists can meet the charges of all their critics. There are, however, some grounds for optimism, and what I want to do next is to identify some specific areas where such optimism may be justified.

There are two very hard topics with which all attempts to study animal behavior and cognition must come to grips. One is the topic of conscious awareness. Without a doubt this particular topic has been the single biggest bludgeon used by critics to bash cognitive ethology, and especially by critics of Griffin's attempts to make scientists face up to the task of giving an adequate scientific account of consciousness in nonhuman animals. Griffin himself is regarded by some as a defector, for his early reputation was based on exquisite physical analyses of the echolocation abilities of bats, biological "hard" science at its best. A pair of quotes that illustrate the common attitudes follow.

> A Griffin bat is a miniature physics lab. So imagine the consternation among behavioristic ethologists when Mr. Griffin came out a decade ago, with *The Question of Animal Awareness,* as a sentimental softy. . . . For Mr. Griffin, all this [cleverness] suggests consciousness. He is wrong. If such cleverness were enough to demonstrate consciousness, scientists could do the job over coffee and philosophers could have packed up their scholarly apparatus years ago. (Cronin 1992, 14)

> We submit that it is this very goal of investigating animal consciousness that, although grand and romantic, falls far outside the scope of a scientific psychology that has struggled for the better part of the past century to eschew such tantalizing, but ultimately unsubstantiable, analyses of subjective mental experience. (Blumberg and Wasserman 1995, 133)

In fact, Blumberg and Wasserman, like some other critics of cognitive ethology, conflate the difficult issue of consciousness with the perhaps more general issue of meaning in animal psychology. As evidence for this I take their dismissive comments about my own work with primatologist Marc Hauser (Blumberg and Wasserman 1995) to show that they have not carefully drawn the distinction between consciousness, a subject we made no mention of, and animal concepts, the sole focus of our work at that point (Allen and Hauser 1991).

Philosophers, and some psychologists, in general have been more careful to consider questions about the meaning or *content* of mental states separately from questions about their *phenomenology* or subjective feel. Both are difficult issues, but some, although not all, philosophers think that a divide-and-conquer strategy has the best hope of progress.

Phenomenological consciousness is, without doubt, the more recalcitrant of the two. The subject of nonhuman phenomenology was brought to the fore by philosopher Thomas Nagel's famous paper, "What Is It Like to Be a Bat?" (Nagel 1974). Indeed Griffin credits Nagel for stirring his own interest in this topic leading to *The Question of Animal Awareness* (Griffin 1976).

Nagel, however, starts off by assuming that there is something that it is like to be a bat—that bats do have a phenomenology. But this is an assumption that the sternest critics are not even willing to grant. Some of these critics are motivated by the generic problem of other minds, the question of how knowledge of other minds is possible, to deny even the possibility of a scientific study of human consciousness. By focusing their attacks on those who would study consciousness of nonhuman animals, I believe that these critics are being disingenuous. Their complaints would not get much of a hearing, say, at the meetings of the Society for Neuroscience, where attempts to understand the neural bases of human consciousness are well underway. Nor should general skeptical worries about knowledge of other minds carry much weight, any more, say, than general skepticism about the nature of the external world should carry any weight among physicists.

But perhaps there is a more specific issue that pertains only to nonhuman minds; call it the problem of other species of mind. If language convinces us that other humans are conscious, what can languageless creatures do to convince us? Some philosophers (Jamieson 1998; Searle 1999) believe that this question presupposes a general strategy of inferring mentality from behavior, which is doomed to failure. They argue that this strategy should be abandoned in favor of more commonsense methods for directly recognizing or perceiving that nonhuman animals are conscious. I believe, however, that the inferential strategy should not be so readily abandoned, and that it is possible to improve the strategy.

One approach has, of course, been the attempt to teach human languages to members of other species: chimpanzees, gorillas, dolphins, and parrots, among others. The specific interpretations of these experiments depend on their details. But I think the same general worry can be launched at all of them, and it is one of the worries that Cronin launched at Griffin in the earlier quotation. When it comes to specific claims about the conscious phenomenology of animals, what is missing is a general theoretical reason for thinking that the particular behaviors are relevant evidence for conscious awareness.

Arguments from analogy to human behavior and experience can only take us part of the way because the strategy is inherently weak. For any target property *P* that we take to be a correlate of consciousness in humans, it is possible to find some other property *Q* that humans possess and other animals lack, and one can then use the absence of *Q* to reject the analogy. It is necessary to justify a tighter connection between the target property *P* and consciousness than mere cooccurrence of *P* with consciousness in humans. (See Allen 1998 for a more detailed discussion of the analogy argument.) I wish to give a brief indication of how considerations about the evolution of consciousness can be used to reorient this debate.

As mentioned, the focus in Nagel's paper was the question of what it is like to be a bat, but I have been suggesting that the more fundamental question is whether it is like anything at all to be a bat. This suggests the possibility that bats are really little zombies, flying around catching insects and avoiding obstacles without any awareness. This possibility of zombiehood has also been raised by philosophers concerned with human consciousness in thought experiments involving entities whose behavior is completely indistinguishable from regular human beings but lacking entirely any subjective awareness. Quite apart from questions about the coherence of such a thought experiment, the conception of consciousness that is underlying it leaves us, I believe, with a complete mystery about what the point of being conscious would be. If, after all, a nonconscious being could behave identically to a conscious one, then there would seem to be no obvious biological function for consciousness itself.

What good is consciousness? Too simplistic a view of this is often promoted via what I call the "lightbulb model" of consciousness. In this model some physical stimulus, say a pinprick, causes the light to come on, a conscious pain, which in turn causes the reaction of withdrawing the affected part of the body. The trouble with this picture is that the lightbulb seems completely superfluous; the stimulus might as well be wired directly to the response. Indeed the neurological evidence seems to support this. Recently decapitated alligators will, for instance, swipe very precisely with their limbs at the point of a scalpel incision despite the fact, presumably, that we can safely say "no brain, no pain!"

More sophisticated ideas about the functions of consciousness need to be developed, and Marc Bekoff and I try to do this in chapter 8 of our book *Species of Mind* (Allen and Bekoff 1997). The idea that we pursue is that phenomenological consciousness enables an organism to learn and adjust to errors derived from its sensory perceptions. To illustrate this point briefly, consider an organism that follows a signal that normally signifies the presence of food, and imagine that the organism reaches the source of the signal and finds no food. Either of two things may have happened: Either there never was any food in the first place, it was a false signal, or the organism got there too late and the food had already been snapped up by a faster competitor. The adaptive responses to these alternatives are different. In the first case, the organism would do better to respond less vigorously to similar presentations of the signal. In the second case, it would do better to respond more vigorously to similar presentations. To distinguish these two cases the organism requires some way of comparing the meaning of the signal to the facts of the case. Or, to put it slightly differently, it would benefit if it could distinguish the sensory appearance ("food over there") from the reality, where there is a difference.

This suggests an important function for perceptual consciousness: the discrimination of particular appearances from the corresponding reality. Indeed, on this view of function the philosophers' description of conscious states as "appearance states" is apropos. Adaptation to perceptual error may not be the only function of such states, but it is a good place, we think, to start looking for evidence of consciousness in nonhuman animals. Furthermore, we can associate this function with a class of ecological problems that it would behoove any organism to solve if it lives long enough for learning to be adaptive, hence putting the problem into an evolutionary framework.

I want now to turn to the other difficult issue facing cognitive ethologists, that of the meaning or semantic content of thoughts. The major interest of philosophers developing accounts of semantic content has been to give a theory of the content of human thought. As such, progress in this area is encouraging. But there remain philosophers and scientists who think that there are special problems facing attempts to attribute semantic content to animal thoughts. These special problems can be brought into focus by considering the notion of a concept and its applicability to nonhuman animals.

There has been much work by comparative psychologists purporting to show that animals categorize objects (and photographs of objects) in ways comparable to humans. But a persistent problem with all such studies is that mere categorization does not seem to entail conceptual abilities. For instance, it is known that many species of ant will remove dead nest mates from the nest, and that they do so on the basis of a chemical cue, the production of an acid that is a by-product of the decomposition of the dead ant's body. The fact that ants will remove pieces of paper and even other live ants that have been artificially marked with this chemical cue argues against any conceptual recognition of death in these animals. The worry about ants, pigeons, or apes successfully trained on a discrimination task is that the discrimination is being made on the basis of a purely perceptual cue that reveals nothing about the animal's cognitive or conceptual structure.

Even though we may be convinced that, for example, the pigeon's discrimination capacity is more sophisticated than an ant's, there still remains a problem of interpretation. Pigeons, for instance, have been trained to discriminate between photographs of trees and photographs that do not contain trees, and to discriminate photographs of human faces from those containing none. But even if we think the pigeon really can make more or less the same judgment as a human about the categories of the photographs, would it be correct to attribute a concept of tree or person? Critics have suggested that such a move is unwarranted because the concept identified in English with the word "tree" is closely related to numerous other concepts—for example, plant, trunk, root, leaf—that the pigeon may lack; it is even associated to some technical notions, perhaps xylem and phloem, that the pigeon certainly lacks. Yet if the pigeon does not think of trees as we do, how can it be correct to say that it has the concept of tree?

These and other problems have led some critics of cognitive ethology to recommend against the further pursuit of studies of animal mind. Are things truly hopeless? I think not. It is worth looking in a little more detail at the argument, such as it is, that a number of philosophers, including Daniel Dennett, Steven Stich, and Alexander Rosenberg, have been tempted to give against the possibility of using notions of mental content for scientific purposes. It goes as follows.

1. Because nonhuman animals do not have the same associations between concepts as we do (e.g., "bone" is associated with "skeleton"), it is imprecise to use human language to describe the contents of their thoughts (can Fido really think of a bone if he does not have the concept of a skeleton?).
2. Precision—that is, precise content specification—is required for the scientific objective of predicting behavior.
3. (Hence) The use of human language to describe animal thought contents is unsuitable for scientific purposes.

Bekoff and I have argued that both premises of this argument are questionable (Allen and Bekoff 1997, chap. 5). Although it may be true that we cannot simply map our English dictionary into the thoughts of a gorilla, it does not follow that we cannot adequately construct ways to describe the gorilla's thought contents. It may be hard, but that is still a long way from being impossible. I would add that in the philosophical discussion of these matters there is generally a lot of loose talk about concepts such as "the concept of tree" and "our concept of tree," as if humans all shared one concept of trees. More sensitivity to human variability on these matters might lead to more sophistication about nonhuman animal concepts (Allen 2000).

The second premise is also questionable because it assumes that precise prediction is the proper measure of scientific respectability. In fact, modeling of complex systems is another goal that may be independent of precise prediction. We may, for example, understand the general principles of tornado dynamics well before we can predict the course of any particular tornado. Likewise, modeling of animal behavior at the level of concepts may provide understanding but not precise predictive power.

Where does this leave the topic of concepts? As with consciousness what is needed is a more theory-based approach to concept attribution. In 1991, Marc Hauser and I suggested that we should seek evidence of a level of representation that is somewhat independent of perceptual information and representation (Allen and Hauser 1991).

For example, suppose I were to pick on someone in your group and, without the rest of the group knowing that I had done this, give this person a drug that suppresses his vital signs, such as pulse and breathing, making the person appear to be dead. I could then sit back and watch while you all go through the motions of starting resuscitation and calling for an ambulance. Suppose I then step forward and provide the antidote to the drug. Assuming you get over your annoyance with me, if I were to play the same trick the next day I wager that you would all look to me rather than going through the same motions. You are not bound to treat something as dead just because it provides perceptual evidence for being dead. Your judgments about death can transcend the particular perceptual cues.

If this is correct, what such results would reveal is a theoretical reason for attributing a two-track representational system of what we can call *percepts* and *concepts.* Why might such a two-track system have evolved? To facilitate certain kinds of learning, I have suggested. Not every species faces the same learning problems, so we might expect varying abilities to represent the world conceptually, just as we find great variation in the ability, for instance, to fly.

Finally, I would like to bring the discussion back to the ethics of keeping great apes in captivity. I would reiterate the point that ethical and regulatory issues will necessarily be based on principles other than evolutionary relatedness. Evolutionary theory is very important for conceptualizing the nature of all species of mind. Although there are no direct arguments from mental capacity to moral status, moral judgments must be informed by the facts about animal cognition and consciousness. Furthermore, consistency demands that whatever connections between mentality and moral status we apply to humans must be applied equally to other apes.

No simple blanket statements about cognition or consciousness can cover all apes. Captivity involves confinement in a variety of conditions, ranging from small cages to entire wildlife reserves. Different apes have evolved under different conditions and may, for example, consequently require different amounts of freedom to range and have evolved different capacities for perceiving and reasoning about captivity and confinement. Prehistoric humans were extremely peripatetic, colonizing almost every conceivable environment on the planet, from arctic tundra to tropical deserts, and from coastal marshes to the high mountains. Whether this wanderlust is part of our genetic or cultural heritage, it helps to account for the high value we place on freedom to roam. Individuals of other species might be satisfied by freedom to range within smaller areas than the entire planet. There will surely be individual differences, too, but these differences must be investigated empirically.

Captivity also denies choice to the captive organisms. As with cognition of space, the cognitive capacities that animals bring to bear on the choices they make are also likely to vary widely between species. A fundamental tenet of liberal political theory is the right of individuals to choose their own destinies. One of the most basic rights in any liberal democracy is the freedom to emigrate. No one's participation in a society should be coerced, because participants must be willing contractors, if a very plausible version of social contract theory is to have any force. This line of thought is represented in the contribution to the *Etica & Animali* special issue by Galdikas and Shapiro (1996), where choice is the central feature of their approach to orangutan ethics.

No zoo would last 10 minutes without physical barriers to the animals' escape. Zoos deny the opportunity to choose to emigrate. Indeed, as the planet becomes increasingly affected by human population growth, even large wildlife reserves are becoming more zoo-like in that they do not allow free migration of nominally wild animals across park boundaries. Some psychologists argue that basic movements of animals, such as approaching food, are not intentionally chosen (Heyes and Dickinson 1990), but these conclusions are based on laboratory studies that may have limited application to a proper understanding of animal cognition (Allen and Bekoff 1995). If, indeed, members of other species choose where to go, then we are faced with the possibility that we should respect those choices.

The development of ethically correct policies for captivity of animals will depend on taking into account both species-specific and individual differences in the ways that individuals perceive and conceptualize the spaces in which they live and the choices with which they are presented. A fully evolutionary approach to cognition, a cognitive ethology, that is not just limited to the great apes or to primates is the best hope we have for understanding such perceptions and conceptions. Such an approach is an important component of the development of Sheets-Johnstone's (1996) "different conception of ourselves," a conception of ourselves as primates, linked by evolution to other primates and to nature. It is also a necessary step toward the development of what Dawkins (1994) called a "continuously distributed morality," the natural destination for our evolving sense of morality.

ACKNOWLEDGMENTS

I wish to thank Marc Bekoff, Dale Jamieson, Bryan Norton, and Paola Cavalieri for helpful comments on this chapter, which is based on my presentation to the Great Apes and Humans at an Ethical Frontier workshop held in Orlando in June 1998. I would also like to thank the organizers and sponsors of that workshop for a most stimulating experience.

REFERENCES

Allen, C. 1998. The discovery of animal consciousness: An optimistic assessment. *Journal of Agricultural and Environmental Ethics* 10: 217–25.

———. 2000. Concepts revisited: The use of self-monitoring as an empirical approach. *Erkenntnis* 50: 1–12.

Allen, C., and Bekoff, M. 1995. Cognitive ethology and the intentionality of animal behavior. *Mind and Language* 10: 313–28.

———. 1997. *Species of mind.* Cambridge: MIT Press.

Allen, C., and Hauser, M. D. 1991. Concept attribution in nonhuman animals: Theoretical and methodological problems in ascribing complex mental processes. *Philosophy of Science* 58: 221–40.

Blumberg, M., and Wasserman, E. A. 1995. Animal mind and the argument from design. *American Psychologist* 50: 133–44.

Crisp, R. 1996. Evolution and psychological unity. In M. Bekoff and D. Jamieson (eds.), *Readings in animal cognition* (pp. 309–21). Cambridge: MIT Press.

Cronin, H. 1992. Review of Griffin, *Animal minds. New York Times Book Review,* Nov. 1, p. 14.

Cummins, D. 1998. Social norms and other minds: The evolutionary roots of higher cognition. In D. Cummins and C. Allen (eds.), *The evolution of mind* (pp. 30–50). New York: Oxford University Press.

Darwin, C. 1936. *The descent of man and selection in relation to sex.* New York: Random House/Modern Library. (Original work published 1871)

Dawkins, R. 1994. Gaps in the mind. In P. Cavalieri and P. Singer (eds.), *The Great Ape Project: Equality beyond humanity* (pp. 80–87). New York: St. Martin's Press.

Dennett, D. C. 1987. *The intentional stance.* Cambridge: MIT Press.

Galdikas, B. M. F., and Shapiro, G. L. 1996. Orangutan ethics. *Etica & Animali* 8: 50–67.

Griffin, D. R. 1976. *The question of animal awareness: Evolutionary continuity of mental experience.* New York: Rockefeller University Press.

———. 1978. Prospects for a cognitive ethology. *Behavioral and Brain Sciences* 4: 527–38.

Heyes, C., and Dickinson, A. 1990. The intentionality of animal action. *Mind and Language* 5: 87–104.

Jamieson, D. 1998. Science, knowledge, and animal minds. *Proceedings of the Aristotelian Society* 98: 79–102.

Millikan, R. 1984. *Language, thought, and other biological categories: New foundations for realism.* Cambridge: MIT Press.

———. 1993. *White Queen psychology and other essays for Alice.* Cambridge: MIT Press.

Morgan, C. L. 1894. *An introduction to comparative psychology.* London: Walter Scott.

Nagel, T. 1974. What is it like to be a bat? *Philosophical Review* 83: 435–45.

Quine, W. V. O. 1953. *From a logical point of view.* Cambridge, MA: Harvard University Press.

Romanes, G. 1883. Animal intelligence. New York: D. Appleton.

Searle, J. R. 1999. Animal minds. *Etica & Animali* 9: 37–50.

Sheets-Johnstone, M. 1996. Taking evolution seriously: A matter of primate intelligence. *Etica & Animali* 8: 115–30.

STEVEN M. WISE

15
A GREAT SHOUT: LEGAL RIGHTS FOR GREAT APES

The earliest known law is preserved in cuneiform on Sumerian clay tablets. These Mesopotamian law codes, 4,000 years old, the Laws of Ur-Nammu, the Lipit-Ishtar Lawcode, the Laws of Eshunna, and the Laws of Hammurabi, assumed that humans could own both nonhuman animals and slaves (Wise 1996). It took most of the next 4,000 years for subjective legal rights to develop. Even in Republican and Imperial Rome, legal rights were understood to exist only in the objective sense of being "the right thing to do" (Wise 1996, 799). Subjective legal rights, claims that one person could make on another, first glimmered in twelfth-century writings. It was only in the fourteenth century that the notion that one's legal rights were one's property began to root (Wise 1996).

Not until the nineteenth century was slavery abolished in the West and every human formally cloaked with the legal personhood that signifies eligibility for fundamental legal rights. So the final brick of a great legal wall, begun millennia ago, was cemented into place. Today, on one side of this legal wall reside all the natural legal persons, all the members of a single species, *Homo sapiens.* We have assigned ourselves, alone among the millions of animal species, the exalted status of legal persons, entitled to all the rights, privileges, powers, and immunities of "legal personhood" (Wise 1996).

On the other side of this wall lies every other animal. They are not legal persons but legal things. During the American Civil War, President Abraham Lincoln was said to have spurned South Carolina's peace commissioners with the statement, "As President, I have no eyes but Constitutional eyes; I cannot see you" *(Oxford Dictionary of Quotations* 1979, 313). In this way, their "legal thinghood" makes nonhuman animals invisible to the civil law. Civil judges have no eyes for anyone but legal persons. But to any one who has had the privilege to observe the work or writings of such primatologists as Jane Goodall, Sue Savage-Rumbaugh, Duane Rumbaugh, Sally Boysen, Roger Fouts, Birute Galdikas, Richard Wrangham, or Geza Teleki, the overbreadth of this legal rule, as applied to the great apes, should naturally stimulate an inquiry into its justice.

The legal thinghood of nonhuman animals has a unique history. An understanding of this history is instrumental to what Oliver Wendell Holmes Jr. called the "deliberate reconsideration" to which every legal rule must eventually fall subject (Holmes 1897). Alan Watson has concluded from his studies of comparative law that "to a truly astounding degree the law is rooted in the past" (Watson 1993, 95). The most common sources from which we quarry our law are the legal rules of earlier times. But when we borrow past law, we borrow the past. Legal rules that may have made good sense when they were fashioned may make good sense no longer. Raised by age to the status of self-evident truths they may perpetrate ancient ignorance, ancient prejudices, and ancient injustices that may once have been less unjust because we knew no better. Like Theseus in the palace of the Minotaur, we must follow the thread of the legal thinghood of nonhuman animals through the shadows of legal history. It will lead us to the most ancient legal systems known.

"All law," said the third-century Roman jurist Hermogenianus, "was established for men's sake" (Mommsen, Krueger, and Watson 1985, 1.5.2). Why should law not have been established solely for the sake of men? Everything else was. This the Greeks and Hebrews knew well. Justifying the massacre of guiltless nonhuman animals by a flood meant to punish an evil humanity, Rashi, the medieval Jewish scholar, explained that "since animals exist for the sake of man, their survival without man would be pointless" (Katz 1992, 274). So it was that the ancient Greek, Roman, and Hebrew worlds fully embraced the idea that the universe had been divinely designed for a single end: the benefit of human beings. It was not just that humans were different from every other animal, but that their inherent value was radically incommensurable with anything else (Wise 1995).

The world that spawned the legal thinghood of every nonhuman animal is not our world. It is not *the* world. We know today that the universe in which they believed was imaginary. It has since collapsed beneath a staggering weight of evidence provided by a process of which they knew almost nothing: science. As a result, my 10-year-old daughter, when she graduated from the fifth grade, understood nature more truly than did the authors of the Five Books of Moses, or Socrates, or Plato, or Aristotle. Today, the great majority of biologists understand that the universe was not divinely designed for the benefit of human beings. More than 90 percent of the members of the elite National Academy of Science and 95 percent of its biologist members do not even believe in the divine, being either proclaimed atheists or agnostics (Larson and Witham 1999). Yet although philosophy and science have recanted, the law has not (Wise 1995).

These hoary ideas play a critical role in perpetuating the legal thinghood of nonhuman animals, because Hermogenianus's teaching, that all law was made for humans, implicit throughout the Covenant and other Old Testament codes and other ancient law, was incorporated by the Byzantine Emperor Justinian into his immensely influential sixth-century *Institutes* and *Digest*. From there it was absorbed into the writings of the Glossators of Continental Europe, where it echoed throughout the works of the great lawyers, judges, and commentators on the English common law, Bracton to Britton to Fleta to Coke and Blackstone, and was received nearly whole by their American descendants, Kent, Holmes, and the supreme courts of every jurisdiction in the United States (Wise 1996). Today, the heart of this curious and imaginary physical world of the Ancients lies beating within the breasts of common law judges, animating the law that regulates the modern relationships between human and every nonhuman animal.

On encountering this legal wall, one is initially awed by its thickness, its height, and its history of success at all levels of law in maintaining the legal apartheid between the human and every other animal. It is not surprising because they draw from a common well: International law, constitutional law, statutory law, the common law, and the civil law all treat nonhuman animals in nearly the same ways. But an expert inspection of this wall will eventually yield up its more important, if less obvious, qualities.

TO THINK ABOUT IT WAS TO CONDEMN IT

The wall's foundations have rotted. Because its intellectual foundations are unprincipled and arbitrary, unfair and unjust, its greatest vulnerability, at least in the English-speaking countries, is to the unceasing tendency of the common law "to work itself pure," to borrow a phrase from Lord Mansfield, the great eighteenth-century English judge.[1] Once a great injustice is brought to their attention, common law judges have the duty to place the legal rules that are its source alongside those great overarching principles that have been integral to Western law and justice for hundreds of years—equality, liberty, fairness, and reasoned judicial decision making—to determine if, in light of what are believed to be true facts and modern values, those rules should be found wanting.

Unlike the law of international treaties, national constitutions, and municipal statutes, the common law is developed by judges through their use of reasoned judgment. To be sure, the common law values consistency and certainty, so that its subjects may order their lives in harmony with rights and obligations that they know. But the common law also values reason, fairness, and flexibility. As judges make the common law, judges may unmake it if they come to believe they have erred. Legislatures that disagree with the decisions of common law judges can overrule them. Judges, of course, know that they can unmake what they have made and that legislatures may revise their decisions. But the very fact that their mistakes can be undone in relatively painless and swift ways encourages common law judges to be innovative and sensitive to those arguments about what is reasonable, fair, right, and just that shine through the light of a kaleidoscopic world of changing facts and values.

The writer Edith Hamilton reminded us of the plight of human slaves before the time of the Greek Stoics of the second century BC. The words she chose to describe their plight today describes the effects of the legal rule that excludes all nonhuman animals, even the great apes, from eligibility for even the most fundamental legal rights.

> Everywhere . . . the way of life depended upon them. One cannot say that they were accepted as such, for there was no acceptance. Everyone used them; no one paid attention to them . . . what must be remembered is that the Greeks were the first who thought about slavery. To think about it was to condemn it. (Hamilton 1957, 23)

Recall that the abomination of human slavery was finally abolished in the West little more than 100 years ago. It continues in a few countries to this day. The first thinking about the justice of the legal thinghood of nonhuman animals occurred just as slavery was flickering in the West. To date it has resulted mostly in the enactment of pathetically inadequate anticruelty statutes. But as the scientific evidence of the true natures of such nonhuman animals as chimpanzees continues to mount, that thinking will be its undoing. Because to think about the legal thinghood of such creatures as the great apes will be finally to condemn such a notion.

This process has begun. Modern law has begun slowly to disassemble the radical incommensurability said to exist between all human and all nonhuman animals from both the top down and the bottom up. The intrinsic value of human beings is now seen in law as commensurable with other legal values. This was reflected, for example, in the enactment in the English-speaking countries of wrongful death statutes in the middle of the nineteenth century. These statutes were intended to alter the ancient, and unfair, common law rule that the loss of human life, understood to be incommensurable with anything else, could never be compensated by money (Wise 1998b). The lives of at least some nonhuman animals have begun to be infused with a degree of intrinsic and not merely instrumental value. The preamble to the United Nations World Charter for Nature states that "every form of life is unique, warranting respect regardless of its worth to man" (World Charter 1982, 992). Respected international law commentators have argued that the legal right of individual whales to life may be becoming a part of binding international law (D'Amato and Chopra 1991). While interpreting the federal Endangered Species Act, the U.S. Supreme Court was guided in its decision by the declaration of the American Congress that endangered species were of "incalculable value."[2]

As we have seen, the legal thinghood of all nonhuman animals is as ancient and as deeply woven into the fabric of our law as is any legal rule. But I suggest that this ancient legal rule, as applied to the great apes, is so unfair and so outright contradictory to the overarching and, if I may say, sacred principles of liberty, equality, and justice that reasoned judicial decision making must reject it.

LIBERTY: THE SUPREME VALUE OF THE WESTERN WORLD

Today liberty "stands unchallenged as the supreme value of the Western world" (Patterson 1991, ix). Out of the more than 200 recorded senses of "liberty," Sir Isaiah Berlin famously identified two central senses: negative and positive (Berlin 1969). One's negative liberty, with which we are concerned, is often described as "freedom from" and depends on being able to do what one wishes without human interference (Berlin 1998). On the other hand, one's positive liberty may be described as "freedom to" (Berlin 1969, lvi; McPherson 1990, 61). Berlin explained that "if my negative freedom is specified by answering the question 'How far am I controlled?' the question for the second sense of freedom is "Who controls me?'" (Berlin 1998, 58). Lincoln explained the difference in a speech he made the year before the Civil War ended:

> We all declare for liberty; but in using the same word we do not all mean the same thing. . . . The shepherd drives the wolf from the sheep's throat, for which the sheep thanks the shepherd as a liberator, while the wolf denounces him for the same act as the destroyer of liberty. . . . Plainly the sheep and the wolf are not agreed upon a definition of the word liberty. (quoted in McPherson 1994, 10; see also Basler 1953–1955, 301)

Berlin put it as "freedom for the pike is death for the minnow" (Berlin 1969, 124). Fundamental liberty rights protect fundamental anthropological or intrapersonal human interests that we derive from objectively ascertainable qualities of the human body and personality, such as our sentience and consciousness. Most fundamental, they derive from that autonomy and self-determination that are said to produce the dignity that in turn engenders the fundamental legal rights to bodily integrity and bodily liberty (Correia 1994; van Vyer 1991). That is why I refer to these fundamental negative liberty rights as "dignity-rights." International and domestic courts and legislatures around the world recognize that for a human being to have a minimal opportunity to flourish, such dignity-rights as bodily integrity and bodily liberty must be protected by sturdy barriers of negative liberty rights that form a protective legal perimeter around our bodies and personalities (Berlin 1969; Dworkin 1977; Feinberg 1966).

Anglo-American common law recognizes a general negative liberty right. The constitutions of most modern nations protect fundamental negative liberty rights (Allan 1991).[3] Chastened by the past century's destruction of human liberty from Nazi Germany to Stalinist Russia to Cambodia to Bosnia to Rwanda to Kosovo, the international community has adopted numerous international treaties and agreements that expressly emphasize the link between dignity-rights and the qualities of human personality, and routinely and repeatedly affirm the fundamental and inherent dignity and worth of human beings (Wise 1998a).

A few of many possible examples follows. The Nuremberg Charter, under whose authority the prosecution of Nazi leaders proceeded, authorized the prosecution of such crimes against humanity as murder and enslavement "whether or not in violation of the domestic law of the country where perpetrated."[4] That is why in his opening speech at the Nuremberg trials, Justice Jackson, the chief prosecutor, argued that the "real complaining party . . . (was) Civilization."[5] The Supplementary Convention on the Abolition of Slavery, the Slave Trade, and Institutions and Practices Similar to Slavery states that "freedom is the birthright of every human being" and emphasizes "the dignity and worth of the human person."[6] The Preamble of the International Convention on the Elimination of All Forms of Racial Discrimination affirms that the United Nations Charter "is based on the principles of the dignity . . . inherent in all human beings" and that the Universal Declaration of Human Rights "proclaims that all human beings are born free and equal in dignity and rights."[7] The preamble of the Convention against Torture and Other Cruel, Inhuman or Degrading Treatment or Punishment states that "the . . . inalienable rights of all members of the human family . . . derive from the inherent dignity of the human person."[8] The preamble of the International Covenant on Civil and Political Rights recognizes principles proclaimed in the United Nations Charter concerning "the inherent dignity and of the . . . inalienable rights of all members of the human family" and that "these rights derive from the inherent dignity of the human person."[9] No torture is allowed; no slavery is allowed. These international norms can never be waived, their violation can never be excused. This is the teaching of the fundamental value of liberty.

FUNDAMENTAL RIGHTS DERIVE FROM A PRACTICAL AUTONOMY

It is true, as the Kansas Supreme Court has said, that "Anglo American law starts with the premise of thorough-going self determination."[10] But what kind of autonomy is required? Philosophers often understand autonomy to mean what the German philosopher Immanuel Kant intended it to mean 200 years ago. We will call Kant's notion of autonomy "full autonomy." Though whole books have been written about what Kant meant, I will try to catch much of his core meaning in a single sentence: I have autonomy if, in determining what I ought to do in any situation, I have the ability to understand what others can and ought to do, I can rationally analyze whether it would be right for me to act in some way or another, keeping in mind that I should act only as I would want others to act and as they can act, and then I can do what I have decided is right. My ability to perform something like this calculus is what makes me autonomous, gives me dignity, and requires that I be treated as a person. If I cannot do this, I lack autonomy and dignity and can justly be treated as a thing, according to Kant.

Whether I have summarized Kant's idea perfectly or not is irrelevant. What is important is that anything that resembles this analysis demands an ability to reason at an almost inhumanly high level. Perhaps our Aristotles, Kants, Freuds, and Einsteins achieved it some of the time. But it is only a glimmering possibility for infants and children, most normal adults never reach it, and the severely mentally limited and the permanently vegetative do not even begin. How did Kant deal with them? Well, he did not, and his "deep silence" on the moral status of children and nonrational adults has not gone unnoticed (Herman 1993). Even Aristotle and company pass significant portions of their lives on automatic pilot or often act out of desire, and not reason, which is precisely how Kant argued nonhuman animals act (Herman 1993; Langer 1997). Were judges to demand full autonomy as a prerequisite for dignity, they would exclude most of us, themselves included, from eligibility for dignity-rights.

Beings may possess a much simpler ability that allows them to act to fulfill their intended purposes. They may have varying capacities for mental flexibility and responsiveness. Autonomy can encompass a range of capacities for consciousness from the most simple awareness of one's present experience to a much broader and deeper self-awareness, self-reflection, and an awareness of the past, present, and future. The full autonomy of, say, Plato might be said to approximate the high end of full Kantian autonomy, whereas the consciousness of a typical preschooler might approximate the low end of a practical autonomy.

A full Kantian autonomy is too narrow a prerequisite for dignity-rights. Not all humans possess it to any degree. Most possess it only in varying degrees. No humans possess it all the time and no one expects them to. Many more humans, though still not anencephalic or even normal infants, the most severely retarded adults, or adults in persistent vegetative comas, possess a practical autonomy than possess full Kantian autonomy (Russell 1996; Wright 1993). A practical autonomy merely recognizes that a being has a somewhat "less than perfect ability to choose appropriate actions" (Cherniak 1985, 5). Any being capable of desires and beliefs has a practical autonomy if she can have beliefs and desires and is able to make "some, but not necessarily all of the sound inferences from the belief set that are apparently appropriate" (Cherniak 1985, 10; Rachels 1990; Regan 1983; Wright 1993). A practical autonomy, therefore, much more closely coincides with the way in which human beings are normally understood to be autonomous.

Perhaps most important for our purposes, fundamental common law and constitutional rights were not designed to protect only the fully autonomous. Courts are exquisitely sensitive to autonomy's practical sense.[11] Once some minimum capacity is attained, courts generally respect the choices made within exceedingly wide parameters, at least with respect to human adults. Practical autonomy acts as a trip wire for dignity-rights. This is because a choice emanating from even a flickering autonomy is more highly valued, regardless of whether the actions are rational, reasonable, or even inimical to one's own best interests, than is any specific choice.[12] That is why the judges of the California Court of Appeals said that "respect for the dignity and autonomy of the individual is a value universally celebrated in free societies. . . . Out of fidelity to that value defendant's choice must be honored even if he opts foolishly to go to hell in a handbasket."[13]

American courts routinely hold that incompetent human beings are entitled to the same dignity-rights as competent human beings. For example, the U.S. Supreme Court held that a man with an I.Q. below 10 and the mental capacity of an 18-month-old child had an inextinguishable liberty right to personal security.[14] Every American jurisdiction recognizes a cause of action in tort, almost always under the common law, for a fetus born alive who suffered prenatal injury after reaching viability (Kader 1980).

Courts sometimes limit the dignity-rights of those who possess the less-than-normal degree of cognition required for full Kantian autonomy to those appropriate to the capacity for autonomy and self-determination that they do possess. This allows normal older children and all but the most seriously retarded human adults a realistic and nonarbitrary measure of dignity-rights.[15] But to avoid the arbitrariness, lack of integrity, and invidious discrimination antithetical to the overarching principles and values of traditional Western law, courts would have to recognize that nonhuman animals who possess capacities for autonomy and self-determination of similar degree, or their equivalents, also possess a dignity sufficient to trigger a realistic and nonarbitrary degree of dignity-rights as well.

Now comes the exceedingly odd part: Even humans who have always lacked autonomy and self-determination are said to possess the requisite dignity for legal personhood.[16] "Can it be doubted," state the judges of high courts rhetorically, "that the value of human dignity extends to both (competent and incompetent humans)?"[17] Undoubting courts grant competent, incompetent, and never-competent humans the same common law dignity-rights and the rights to protect their powers to use them as well.[18] The result is that humans with minimal, or even no, capacity for autonomy and self-determination, even terminally ill infants who lack all cognition, possess not just protected dignity-rights but the right to use their power to enforce them.[19]

When the guardian of a 10-month-old in a persistent vegetative state who was subjected to a do-not-resuscitate order argued that "the child had no dignity interest in being free from bodily invasions," as she "has no cognitive ability," the justices of the Massachusetts Supreme Judicial Court retorted that "*cognitive ability is not a prerequisite for enjoying basic liberties*" and insisted that the child had a "dignity" that could be offended by invasive resuscitation attempts.[20] "Any other view," said justices of the New Jersey Supreme Court, "would permit obliteration of an incompetent's panoply of rights merely because the patient could no longer sense the violation of those rights."[21] "Allowing *someone* to choose," said the California Court of Appeals, "is more respectful of an incompetent person than simply declaring that such a person has no more rights."[22] But how can someone exercise the dignity-rights of another when those dignity-rights were supposedly created by her autonomy? Thus we are left with the following choices. Either (1) dignity-rights are not derived from autonomy and self-determination, (2) it is a logical error to assign dignity-rights to those who lack autonomy and self-determination, or (3) courts arbitrarily assign dignity-rights to humans who lack autonomy and self-determination.[23]

Most courts prefer the third choice. They act as if even never-competent incompetents are competent.[24] So a do-not-resuscitate order for a 10-month-old in a persistent vegetative coma is affirmed by the Massachusetts Supreme Judicial Court on the ground that "the *child* would refuse resuscitative measures in the event of a cardiac or respiratory arrest."[25] The New Jersey Supreme Court holds that an incompetent woman "has the same constitutional right of privacy as anyone else to choose whether or not to undergo sterilization. Unfortunately, she lacks the ability to make that choice for herself. We do not pretend that the choice of (others) is her own choice. But it is a genuine choice nevertheless."[26]

The problem of ascertaining the choices of a once-competent adult who, while competent, has speculated on her preferences should she become incompetent or left insufficient evidence of what decision she would have made if competent is Sisyphean.[27] It is irrational to allow a substitute decision maker to exercise rights founded on the autonomy and self-determination of another (Minow 1985).[28] Such is the power of the rule that dignity-rights arise from autonomy and self-determination that its embrace may undermine the higher value of reasoned judgment in judicial decision making.

Courts recognize human dignity-rights in the complete absence of autonomy only by using an arbitrary legal fiction that controverts the empirical evidence that no such autonomy exists. Conversely, courts refuse to recognize dignity-rights of the great apes only by using a second arbitrary legal fiction in the teeth of empirical evidence that they possess it (Nino 1993; Rachels 1990). But legal fictions can only be justified when they harmonize with, or at least do not undermine, the overarching values and principles of a legal system. Thus the legal fiction that a human who actually lacks autonomy has it is benign, for at worst it extends legal rights to those who might not need them. At best it protects the bodily integrity of the most helpless humans alive. But the legal fiction that great apes are not autonomous when they actually are undermines every important principle and value of Western justice: liberty, equality, fairness, and reasoned judicial decision making. It is pernicious.

BEING HUMAN IS NOT NECESSARY FOR FUNDAMENTAL LEGAL RIGHTS

If being human were necessary for fundamental legal rights, the enormous judicial solicitude for such alleged universal human qualities as autonomy, self-determination, and dignity would be so much flimflam, mere smoke screens that covered that which was really necessary, being human, and for no good or discernible reason. Consider the remarkable 1837 debate in the House of Representatives over the proposed censure of John Quincy Adams for presenting a petition from slaves. South Carolinean Henry Laurens Pinckney mocked Adams. He "just as soon would have supposed that the gentleman from Massachusetts would have offered a memorial from a cow or horse—for he might as well be the organ of one species of property as another" (quoted in Miller 1996, 257). To them

Adams replied, "Sir . . . if a horse or a dog had the power of speech and of writing, and he should send (me) a petition, (I) would present it to the House" (quoted in Miller 1996, 268).

With the exception of the handful of scientists who explore the cognitive abilities of such species as the great apes, few of us have the opportunity to see nonhumans demonstrate sophisticated cognition, either in the wild or in the laboratory. Many of us are scarcely able to imagine that such creatures exist. But once we did not have to imagine them, because they probably outnumbered us.

On the traditional view, *Homo sapiens* is the only surviving member of the genus *Homo.* Some cladists have begun to argue that all great apes, humans included, fall within the same tribe, that gorillas, chimpanzees, bonobos, and humans lie within the same subtribe, and that chimpanzees, bonobos, and modern humans, as well as *Ardipithecus* and *Australopithecus,* share the same genus (Goodman et al. 1998). But between 27,000 and 53,000 years ago, two hominid species, *H. neanderthalensis* and the older *H. erectus,* may have coexisted with us (Swisher et al. 1996).[29] Evidence suggests that *H. erectus* walked erect, was large-brained, probably possessed some degree of reflexive consciousness, crafted technically complex hand axes and other tools, probably used fire, had a primitive culture, colonized several continents, and possessed a well-developed Broca's area, but probably only used a limited range of sounds to communicate that were too simple to be considered language (Eldredge and Tattersall 1982; Leakey 1994; Lewin 1993; Mithen 1996). The Neanderthals were even larger brained than we and may have been our neurological and mental equals. They probably possessed a more complex degree of reflexive consciousness, crafted sophisticated tools, organized group hunts, used fire, possibly—and even probably—spoke a rudimentary language, and developed a complex culture that likely included taking care of their sick and weak, burying their dead, creating primitive art, and possibly included playing the flute (Leakey 1994; Lewin 1993; Mithen 1996) Both Neanderthals and *Homo erectus* have vanished. If a member of either suddenly emerged in an unexplored corner of the world, why should her eligibility for dignity-rights turn on her taxonomic classification rather than her qualities? Would we not care *what* she was, rather than how taxonomists classified her?

In a legal system such as ours that so highly values both fairness and rationality in judicial decision making, the question of the dignity-rights of such nonhuman animals as the great apes must be resolved in the same way as dignity-rights for a *H. sapiens* or a Neanderthal or a *H. erectus.* This is not because this process forms perhaps the strongest ground for granting dignity-rights to nonhuman animals, but because it underpins human dignity-rights wherever they exist. Today, at least in theory, they exist nearly everywhere. It is arbitrary to continue to deny such fundamental legal rights as bodily integrity and bodily liberty to the great apes if their natures and interests would justify those rights if only they were us.

EQUALITY: LIKES SHOULD BE TREATED ALIKE

Equality is the axiom of Western justice that likes be treated alike. Equality prohibits arbitrary classifications that conflict with the reasoned judgment that Western law and justice demand. A claimant is generally entitled to an equality right if others similarly situated unfairly enjoy benefits that the claimant does not enjoy or unfairly escape burdens to which a claimant is subjected.[30] But the promise of equality often collides with the reality that classification is an inevitable feature of much legislation and many judicial decisions, because perfect equality is neither possible nor desirable. How can we determine when a classification violates equality? The "unfairness" of a discriminatory classification may be logical in that it arises from an improper relationship between ends and means. Or it may be normative in that it stems from the nature of the ends or means themselves. Equality demands that the relationship between means and ends must be rational and that both ends and means must be normatively acceptable.

Equality's logical component requires that dissimilar treatment rests on some relevant and objectively ascertainable difference between a favored and disfavored class with respect to the harm to be avoided or the benefit to be promoted (Simons 1989).[31] If a favored class actually poses a lesser harm or deserves a greater benefit, a discriminatory classification is probably rational and not arbitrary.[32] But if both classes pose the same harm or deserve the same benefit, a classification is probably irrational and arbitrary (Tribe 1988).[33] When, as happens on occasion, a favored class actually poses a greater harm or deserves a lesser benefit, the classification is not merely arbitrary and irrational but perverse (Simons 1989).

Equality's normative component means that no matter how perfect the relationship between ends and means may be, some means and some ends are unacceptable solely because they are arbitrary, irrelevant, invidious, or are otherwise normatively illegitimate.[34] Thus the legislative classification enacted by the state of Florida, only struck down in 1964, that punished the habitual cohabitation of men and women of different races was utterly rational in light of the legislative purpose of deterring interracial sexual relations (Kull 1992). However, it violated equality because such a purpose was invidious.

Together, the logical and normative elements of equality mean that only qualities that are objectively ascertainable and normatively acceptable should be compared. Actual and relevant likenesses that are examined in light of current knowledge and normative understandings that are not false, assumed, unproveable, or anachronistic assumptions or contain "fixed notions" about likeness should be the measure.[35] These are the teachings of the fundamental value of equality.

Closely related to equality rights are proportionality rights. Proportionality requires that unalikes be treated proportionately to their unalikeness (Simons 1989). Alan Gewirth's "principle of proportionality" attributes legal rights in proportion to the possession of qualities to less-than-normal degrees.

> When some quality Q justifies having certain rights R, and the possession of Q varies in degree in the respect that is relevant to Q's justifying the having of R, the degree to which R is had is proportional to or varies with the degree to which Q is had. . . . Thus, if x units of Q justify that one have x units of R, then y units of Q justify that one have y units of R. (Gewirth 1978, 121)

Gewirth's is not a perfectionist theory of justice, by which the justice to which one is due varies proportionately to the degree to which one possesses a certain characteristic, such as the quality Q. Anyone who possesses the quality Q possesses the right *R* in full measure. However, the degree to which one may *approach* having the quality Q can determine the degree to which one possesses the right *R*. But once the quality Q is attained, one possesses the full right *R*, no matter how much of the quality Q one may have (Gewirth 1978). As Gewirth explained,

> according to the principle of proportionality . . . although children, mentally-deficient persons, and animals do not have (fundamental) rights in the full-fledged way normal human adults have them, members of these groups *approach* having (fundamental) rights in varying degrees, depending on the degree to which they have the requisite abilities. (Gewirth 1978, 122)

Carl Wellman has postulated three dimensions along which one's legal rights might vary "in proportion to his possession of the justifying quality Q" (Wellman 1995, 129). First, one might possess *fewer* legal rights. A severely mentally limited human—or a bonobo—might not be able to participate in the political process, but he or she should still have the right freely to move about. Second, one might possess *narrower* legal rights. A severely mentally limited human, or a chimpanzee, might not have the right to move about freely in the world at large, but he or she should have the right to move within the confines of his or her home or habitat. Third, one might possess a *partial* right. A profoundly retarded human—or a gorilla—might possess a legal right to bodily integrity, but he or she might lack the power to waive it. Thus he or she might be unable to consent to a risky medical procedure that might benefit her or to a risky medical procedure that might not benefit her but another (Wellman 1995).

At least three independent equality or proportionality arguments support fundamental legal rights for the great apes (Wise 1998a). First, great apes who possess Kant's full autonomy should be entitled to dignity-rights *if* humans who possess full autonomy are entitled to them. To do otherwise would be to undermine the major principled arguments against racism and sexism.

Second, great apes who possess a practical autonomy should be entitled to dignity-rights in proportion to the degree to which they approach full Kantian autonomy *if* humans who possess a practical autonomy are entitled to dignity-rights in proportion to the degree to which *they* approach full Kantian autonomy. Thus if a human is entitled to fewer, narrower, or partial legal rights as their capabilities approach the quality Q, so should nonhuman animals whose capabilities also approach the quality Q.

Third, in perhaps the clearest argument for equality, great apes who possess either full Kantian autonomy or a practical autonomy should be entitled to the same fundamental rights to which humans who entirely lack autonomy are entitled. Placing the rightless legal thing, the bonobo Kanzi, beside an anencephalic 1-day-old human with the legal right to choose to consent or withhold consent to medical treatment highlights the legal aberration that is Kanzi's legal thinghood.

Probably the strongest argument that just being human is necessary for the possession of fundamental equality rights has been offered by Carl Cohen, who has argued that at least moral rights should be limited to all and only human beings. "The issue," said Cohen, "is one of kind" (Cohen 1986, 866). He acknowledged that some humans lack autonomy and the ability to make moral choices. However, because humans as a "kind" possess this ability, it should be imputed to all humans, regardless of their actual abilities. But Cohen's argument can succeed only if the species, *H. sapiens,* can nonarbitrarily be designated as the boundary of a relevant "kind." That is doubtful. Other classifications, some wider, such as animals, vertebrates, mammals, primates, and apes, and at least one narrower—normal adult humans—also contain every fully autonomous human.

As well as being logically flawed, Cohen's argument for group benefits is normatively flawed. It "assumes that we should determine how an individual is to be treated, not on the basis of *its* qualities but on the basis of *other* individuals' qualities" (Rachels 1990, 187). Rachels calls the opposing moral idea "moral individualism" and defines it to mean that "how an individual may be treated is to be determined, not by considering his group memberships, but by considering his own particular characteristics" (Rachels 1990, 173). It is individualism, and not group benefits, that is more consistent with the overarching principles and values of a liberal democracy and that has the firmer basis in present law. It is surprising that even Cohen has recognized this, but in another context. Elsewhere he has criticized racial affirmative action policies precisely because they sacrifice the fundamental rights of individuals on the altar of group rights. Therefore they are underinclusive with respect to the disfavored race and overinclusive with respect to the favored race, even though he acknowledged that some blacks have indeed suffered because of their race (Cohen 1995).

Laurence Tribe has acknowledged the "tenaciously-held principle . . . with undeniable constitutional roots . . . that each person should be treated as an individual rather than as a statistic or as a member of a group, particularly of a group the individual did not knowingly choose to join" (Tribe 1988, 1589). For example, the U.S. Supreme Court, in striking down the exclusion of women by a state military academy, remarked that the plaintiff

> does not challenge any expert witness testimony estimation on *average capacities* or preferences of men and women. Instead . . . state actors controlling gates to opportunity . . . may not exclude *qualified individuals* based on "fixed notions concerning the roles and abilities of males and females."[36]

Something like Cohen's notion of group benefits and burdens has, on rare occasions, been used to *correct* the effects of invidious discrimination, such as racial or gender affirmative action, but *never* for the purpose that he advocates: to further *oppress.* Concurring in the seminal case of *Regents of University of California* v. *Bakke,* Justice Powell identified the "principal evil" of the university's affirmative action program as the denial to Bakke of his "right to individualized consideration without regard to his race."[37]

This kind of argument against group rights explains much of the "considerable unease" that has accompanied even the Supreme Court's sharply divided, and often pluralistic, endorsements of racial affirmative action, despite its anchor in the laudable desire to achieve equality for a long-oppressed minority (Tribe 1988, 1523).[38] In contrast, drawing the line of legal personhood at *H. sapiens* is unreasonable, arbitrary, and unprincipled. It does not attempt to remedy earlier discrimination.[39] Instead, it is an odious and unprincipled equality-subverting line drawn solely for the purpose of exploiting a long-oppressed group.[40] This is an illegitimate end that produces indefensible results.

WILL WE AFFIRM OR UNDERMINE OUR COMMITMENT TO FUNDAMENTAL HUMAN RIGHTS?

The destruction of the legal thinghood even of the great apes, our closest cousins, will necessarily involve a long and difficult struggle. It is the nature of great change to stimulate great opposition. But the anachronistic legal thinghood of the great apes so contradicts outright the overarching principles of equality, liberty, fairness, and rationality in judicial decision making that it will eventually be denied only by those in whom a narrow self-interest predominates. Everywhere we look, the great principles of Western law that underlie the arguments for the fundamental legal rights of the great apes stand unchallenged by any opposing principles of remotely high stature. Instead, like Gulliver in Lilliput, they are temporarily bound by a myriad of narrow self-interests. These may allow the winning of temporary victories, but they are doomed in the long run.

Of course, everything depends on some knowledge of the inner worlds of the great apes. But anyone who has struggled with the protean nature of consciousness knows that we can never know for sure that other humans are conscious. As Martha Nussbaum has written, we are only left with "a choice only between a generous construction and a mean spirited construction" (Nussbaum 1995, 38). Our choice will not only have a profound impact on the great apes but also on ourselves, because it will either affirm or undermine our commitment to fundamental human rights (Wise 2000).

NOTES

1. *Omichund v. Barker,* 1 Atk. 21, 33 (K.B. 1744).
2. *Tennessee Valley Authority v. Hill,* 437 U.S. 153, 188 (1978). The Endangered Species Act, 16 U.S.C. §§ 1531–43 (1973).
3. *E.g., Youngberg v. Romeo,* 457 U.S. 307, 317 (1982); *Bowers v. Devito,* 686 F.2d 616, 618 (7th Cir. 1982).
4. 82 U.N.T.S. 279 (Aug. 8, 1945).
5. II *Trial of the Major War Criminals before the International Military Tribunal* 155 (1947). (Opening Speech of Justice Robert H. Jackson, Nov. 21, 1945). The Tribunal itself asserted that "humanity is the sovereignty which has been offended," *United States v. Ohlendorf* (Case No. 9), IV *Trials of War Criminals before the Nuernberg Military Tribunals under Control Council Law No. 10 497* (1950).
6. Sept. 7, 1956, 18 U.S.T. 3201, T.I.A.S. No. 6418, 266 U.N.T.S. 3 (entered into force April 30, 1957) (entered into force for United States, Dec. 6, 1967).
7. The International Convention on the Elimination of All Forms of Racial Discrimination, adopted Dec. 21, 1965, 660 U.N.T.S. 195 (entered into force January 4, 1969) (entered into force for United States 1994). Universal Declaration of Human Rights, adopted Dec. 19, 1948, G.A. Res. 217A (III), 3 U.N. GAOR (Resolutions pt. 1), at 71, U.N. Doc. A1810 (1948).
8. Adopted Dec. 10, 1984, G.A. Res. 39/46, 39 U.N. GAOR Supp. (No. 51) at 197, U.N. Doc. A/39/51 (1985) (entered into force June 26, 1987) (entered into force for United States 1994).

9. Adopted Dec. 16, 1966, G.A. Res. 2200, 21 U.N. GAOR Supp. (No. 16) at 52, 6999 U.N.T.S. 171 (entered into force March 23, 1976) (entered into force for United States 1992).
10. *Natanson v. Kline,* 350 P.2d 1093, 1104 (Kan. 1960). *See also Stamford Hospital v. Vega,* 674 A.2d 821, 831 (Conn. 1996).
11. *E.g., Rivers v. Katz,* 495 N.E.2d 337, 341, *reargument denied,* 498 N.E.2d 438 (N.Y. 1986); *Schmidt v. Schmidt,* 450 A.2d 421, 422–23 (Pa. Super. 1983) (a 26-year-old woman with Down's syndrome with the mental ability of a child between $4\frac{1}{2}$and 8 years can rationally decide whether to choose to visit a parent).
12. *E.g., Thornburgh v. American College of Obstetricians and Gynecologists,* 476 U.S. 747, 778 n.5 (Stevens, J., concurring); *Application of President & Directors of Georgetown College,* 331 F.2d 1010, 1017 (D.C. Cir. 1964); *State v. Wagner,* 752 P.2d 1136, 1178 (Ore. 1988).
13. *People v. Nauton,* 34 Cal. Rptr. 2d 861, 864 (Ct. App. 1994).
14. *Youngberg, supra* note 3, at 315–16.
15. *E.g., Thompson v. Oklahoma,* 487 U.S. 815, 824, 825 n.23 (1988); *Parham v. J.R.,* 442 U.S. 584 (1979).
16. *E.g., Gray v. Romeo,* 697 F. Supp. 580, 587 (D.R.I. 1987); *Conservatorship of Drabick,* 245 Cal. Rptr. 840, 855, *cert. denied sub nom., Drabick v. Drabick,* 488 U.S. 958 (1988); *Superintendent of Belchertown State School v. Saikewicz,* 370 N.E.2d 417, 427, 428 (Mass. 1977); *Eichner v. Dillon,* 426 N.Y.S.2d 517, 542 (App. Div.), *modified* 52 N.Y.2d 363 (1980); *see Conservatorship of Valerie N.,* 707 P.2d 760, 776 (Cal. 1985), citing *Matter of Moe,* 432 N.E.2d 712, 720 (Mass. 1982).
17. *Matter of Guardianship of L.W.,* 482 N.W.2d 60, 69 (Wis. 1992), quoting *Eichner, supra* note 16, at 542. *See also Gray, supra* note 16, at 587; *Matter of Moe, supra* note 16, at 719; *Delio v. Westchester County Medical Center,* 516 N.Y.S.2d 677, 686 (N.Y. App. Div. 1987).
18. *E.g., Gray, supra* note 16, at 587; *Conservatorship of Drabick, supra* note 16, at 855; *Foody v. Manchester Memorial Hospital,* 482 A.2d 713, 718 (Conn. Sup. Ct. 1984); *Matter of Tavel,* 661 A.2d 1061, 1069 (Del. 1995); *Severns v. Wilmington Medical Center, Inc.,* 421 A.2d 1334, 1347 (Del. 1980); *John F. Kennedy Memorial Hospital v. Bludworth,* 452 So.2d 921, 921, 923, 924 (Fla. 1985); *In re Guardianship of Barry,* 445 So.2d 365, 370 (Fla. Dist. Ct. App. 1984); *DeGrella by and through Parrent v. Elston,* 858 S.W.2d, 698, 709 (Ky. 1993); *In re L.H. R.,* 321 S.E.2d 716, 722 (Ga. 1984); *Care and Protection of Beth,* 587 N.E.2d 1377, 1382 (Mass. 1992); *Matter of Conroy,* 486 A.2d 1209, 1229 (N.J. 1985); *In re Grady,* 426 A.2d. 474–75 (N.J. 1981); *In re Quinlan,* 355 A.2d 647, 664 (N.J.), *cert. denied sub nom., Garger v. New Jersey,* 429 U.S. 922 (1976); *Eichner, supra,* note 16, at 546; *Matter of Guardianship of Hamlin,* 689 P.2d 1372, 1376 (Wash. 1984); *In re Colyer,* 660 P.2d 738, 774 (Wash. 1983); *Matter of Guardianship of L.W., supra* note 17, at 67, 68.
19. *E.g., In re L.H.R., supra* note 18 (4-month-old in chronic vegetative state); *Care and Protection of Beth, supra* note 18 (10-month-old in irreversible coma); *Strunk v. Strunk,*

445 S.W.2d 145 (Ky. 1969) (27-year-old with an I.Q. of 35 and a mental age of 6 years); *In re Grady, supra* note 18; *In re Penny N.*, 414 A.2d 541 (N.H. 1980); *Saikewicz, supra* note 16 (67-year-old with an I.Q of 10 and a mental age of 31 months); *Matter of Guardianship of L.W., supra* note 18, at 68; *In re Guardianship of Barry, supra* note 18 (anencephalic 10-month-old with no cognitive brain function).

20. *Care and Protection of Beth, supra,* note 18, at 1382 (emphasis added).
21. *Matter of Conroy, supra* note 18, at 1229. *See also DeGrella by and through Parrent, supra* note 18, at 706.
22. *Conservatorship of Drabick, supra* note 16, at 855 (emphasis added).
23. *E.g., In re Storar,* 420 N.E.2d 64, 72–73 (N.Y.) *cert. denied sub nom., Storar v. Storar,* 454 U.S. 858 (1981); *Saikewicz, supra* note 16, at 427–28; *In re Quinlan, supra* note 18, at 666; *In Matter of Guardianship of Eberhardy,* 307 N.W.2d 881, 893 (Wis. 1981).
24. *E.g., In re Guardianship of Barry, supra* note 18, at 372. *See Conservatorship of Valerie N., supra* note 16, at 777; *Matter of Tavel, supra* note 18, at 1068–69.
25. *Care and Protection of Beth, supra* note 18, at 1382 (emphasis added).
26. *In re Grady, supra* note 18, at 480–81.
27. *Matter of Conroy, supra,* note 18, at 1231.
28. *E.g., Conservatorship of Valerie N., supra* note 16, at 786 (Bird, C.J.) (dissenting); *Curran v. Bosze,* 566 N.E.2d. 1319, 1326 (Ill. 1990); *In re Storar, supra* note 23, at 72–73; *Matter of Westchester County Medical Center,* 531 N.E.2d 607, 612 (N.Y. 1988); *Matter of Guardianship of L.W., supra* note 17, at 69 n.11; *In Matter of Guardianship of Eberhardy, supra* note 23, at 893.
29. Though the issue of whether the Neanderthals were a subspecies of *H. sapiens* or a separate species remains controversial, recent evidence supports the hypothesis that they were a distinct species; *see* Johanson 1998; Krings et al. 1997; Lieberman 1998; Tattersall 1995, 1998.
30. *E.g., Samaad v. City of Dallas,* 940 F.2d 925, 941 (5th Cir. 1991). *See* Simons 1989, 465.
31. *E.g., Logan v. Zimmerman Brush Co.*, 455 U.S. 422, 442 (1982) (plurality opinion); *Rinaldi v. Yeager,* 384 U.S. 305, 308–309 (1966); *McLaughlin v. Florida,* 379 U.S. 184, 191 (1964).
32. *E.g., New York City Transit Authority v. Beazer,* 440 U.S. 568, 591 (1979).
33. *E.g., Skinner v. Oklahoma,* 316 U.S. 535 (1942).
34. *E.g., Romer v. Evans,* 116 S. Ct. 1620, 1627–29 (1996); *Skinner, supra* note 34. *See Thoreson v. Penthouse International, Ltd.*, 563 N.Y.S.2d 968, 975 (Sup. Ct. 1990).
35. *E.g., United States v. Virginia,* 116 S. Ct. 2264, 2277–78 (1996), quoting *Mississippi University for Women v. Hogan,* 458 U.S. 718, 725 (1985); *Craig v. Boren,* 429 U.S. 190, 197 (1976). *See Romer, supra* note 35, at 1628.
36. *United States v. Virginia, supra* note 36, at 2280, quoting *Mississippi University for Women, supra* note 36, at 725. The Court also said that "state actors may not rely on 'overbroad generalizations' to make 'judgments about people that are likely to . . . perpetrate historical patterns of discrimination.'" *Id.* at 2275.

37. 438 U.S. 265, 318 n.52 (1978) (Powell, J., concurring).
38. *E.g., Wittmer v. Peters,* 87 F.3d 916, 918 (7th Cir.), *reh'g and suggested for reh'g en banc denied* (1996); *McLaughlin by McLaughlin v. Boston School Committee,* 938 F. Supp. 1001, 1008 (D. Mass. 1996). These are cases that endorse racial affirmative action.
39. *E.g., Association of General Contractors of America v. City of Columbus,* 936 F. Supp. 1363, 1371 (S.D. Ohio 1996).
40. *E.g., Louisville Gas and Electric Co. v. Coleman,* 277 U.S. 32, 41 (Holmes, J.) (dissenting).

REFERENCES

Aristotle. 1941. Nichomachean ethics. In R. McKeon (ed.), *The basic works of Aristotle* (p. 1131a). New York: Random House.

Basler, R. P. (ed.). 1953–1955. *The collected works of Abraham Lincoln.* New Brunswick, NJ: Rutgers University Press.

Berlin, I. 1969. Two concepts of liberty. In *Four essays on liberty* (pp. 117–72). Oxford: Oxford University Press.

———. 1998. My intellectual path. *New York Review of Books,* May 14, pp. 53–60.

Cherniak, C. 1985. *Minimal rationality.* Cambridge: MIT Press.

Cohen, C. 1986. The case for the use of animals in biomedical research. *New England Journal of Medicine* 317: 867–70.

———. 1995. *Naked racial preference: The case against affirmative action.* Lanham, MD: Madison Books.

Correia, E. O. 1994. Moral reasoning and the due process clause. *Southern California Interdisciplinary Law Review* 3: 562–91.

D'Amato, A., and Chopra, S. K. 1991. Whales: Their emerging right to life. *American Journal of International Law* 85: 21–62.

Dworkin, R. 1977. *Taking rights seriously.* Cambridge, MA: Harvard University Press.

Eldredge, N., and Tattersall, I. 1982. *The myths of human evolution.* New York: Columbia University Press.

Feinberg, J. 1966. Duties, rights, and claims. *American Philosophy Quarterly* 3: 37.

Gewirth, A. 1978. *Reason and morality.* Chicago: University of Chicago Press.

Goodman, M., Porter, C., Czelusniak, C., Page, S., Schneider, H., Shoshani, J., Gunnell, G., Groves, C. 1998. Toward a phylogenetic classification of primates based on DNA evidence complemented by fossil evidence. *Molecular Phylogenetics and Evolution* 9: 585–98.

Hamilton, E. 1957. *The echo of Greece.* New York: W. W. Norton.

Herman, B. 1993. *The practice of moral judgment.* Cambridge, MA: Harvard University Press.

Holmes, O. W., Jr. 1897. The path of the law. *Harvard Law Review* 10: 457–78.

Johanson, D. 1998. Reading the minds of fossils. *Scientific American* 102: 10–11.

Kader, D. 1980. The law of tortious prenatal death since *Roe v. Wade. Missouri Law Review* 45: 639.

Katz, M. A. 1992. Ox slaughter and goring oxen: Homicide, animal sacrifice, and judicial process. *Yale Journal of Law & the Humanities* 4: 249–78.

Krings, M., Stone, A., Schmitz, R. W., Krainitzki, H., Stoneking, M., and Pääbo, S. 1997. Neanderthal DNA sequences and the origin of modern humans. *Cell* 90: 19–30.

Kull, A. 1992. *The color-blind constitution.* Cambridge, MA: Harvard University Press.

Langer, E. 1997. *The power of mindful learning.* Reading, MA: Addison-Wessley.

Larson, E. J., and Witham, L. 1999. Scientists and religion in America. *Scientific American* 281(3): 88–93.

Leakey, R. 1994. *The origin of mankind.* New York: Basic Books.

Lewin, R. 1993. *The origin of modern humans.* New York: Scientific American Library.

Lieberman, D. 1998. Sphenoid shortening and the evolution of modern cranial shape. *Nature* 393: 58.

McPherson, J. M. 1990. *Abraham Lincoln and the second American revolution.* Oxford: Oxford University Press.

———. 1994. Liberating Lincoln. *New York Review of Books,* April 21, pp. 7–10.

Miller, W. L. 1996. *Arguing about slavery—The great battle in the United States Congress.* New York: Alfred A. Knopf.

Minow, M. 1985. Beyond state intervention in the family: For Baby Jane Doe. *University of Michigan Journal Legal Reference* 18: 933–1014.

Mithen, S. 1996. *The prehistory of the mind.* London: Thames and Hudson.

Mommsen, T., Krueger, P., and Watson, A. (eds.). 1985. *The digest of Justinian.* Philadelphia: University of Pennsylvania Press.

Nino, C. S. 1993. *The ethics of human rights.* Oxford: Oxford University Press.

Oxford Dictionary of Quotations. 1979. (3rd ed.). Oxford: Oxford University Press.

Nussbaum, M. C. 1995. *Poetic justice.* Boston: Beacon Press.

Patterson, O. 1991. *Freedom: Freedom in the making of Western culture.* New York: Basic Books.

Paul, S. M. 1970. *Studies in the Book of the Covenant in light of cuneiform and biblical law.* London: E. J. Brill.

Rachels, J. 1990. *Created from animals.* Oxford: Oxford University Press.

Regan, T. 1983. *The case for animal rights.* Berkeley: University of California Press.

Russell, J. 1996. *Agency—Its role in mental development.* Hove, UK: Erlbaum.

Simons, K. W. 1989. Overinclusion and underinclusion: A new model. *UCLA Law Review* 36: 447–89.

Swisher, C. C., III, Rink, W. J., Anton, S. C., Schwarcz, H. P., Curtis, G. H., Suprijo, A., Widiasmoro, J. 1996. Latest *Homo erectus* of Java: Potential contemporaneity with *Homo sapiens* in Southeast Asia. *Science* 274: 1870–74.

Tattersall, I. 1995. *The last Neanderthal.* New York: MacMillan.

———. 1998. *Becoming human: Evolution and human uniqueness.* New York: Harcourt Brace.

Tribe, L. W. 1988 *American constitutional law.* Mineola, NY: Foundation Press.

van Vyer, J. D. 1991. Constitutional options for post-apartheid South Africa. *Emory Law Journal* 40: 771–835.

Watson, A. 1993. *Legal transplants—An approach to comparative law.* Athens: University of Georgia Press.

Wellman, C. 1995. *Real rights.* Oxford: Oxford University Press.

Wise, S. M. 1995. How nonhuman animals were trapped in a nonexistent universe. *Animal Law* 1: 15–43.

———.1996. The legal thinghood of nonhuman animals. *Boston College Environmental Affairs Law Review* 23(3): 471–546.

———. 1998a. Hardly a revolution—The eligibility of nonhuman animals for dignity-rights in a liberal democracy. *Vermont Law Review* 22: 793–915.

———. 1998b. Recovery of common law damages for emotional distress, loss of society, and loss of companionship for the wrongful death of a companion animal. *Animal Law* 4: 33.

———. 2000. *Rattling the cage—Toward legal rights for animals.* Cambridge, MA: Perseus.

World Charter for Nature. 1982. GA Res. 37/7 Annex UNGAOR, 37th Sess. Suppl. No. 51, UNDO A/37151 (Oct. 28) (preamble). In H. W. Wood, Jr., The United Nations World Charter for Nature: The developing nations' initiative to establish protections for the environment. *Ecology* 12: 977–92.

Wright, W. A. 1993. Treating animals as ends. *Journal of Value Inquiry* 27: 353–66.

PAUL WALDAU

16
INCLUSIVIST ETHICS

Will there ever be a time when the zoo establishment recommends that captivity is, for ethical reasons, wrong for some animals? This simple question frames fundamental issues that must be distinguished from technical problems, such as the inability to get some kinds of animals to reproduce in captive circumstances. Rather, our opening question raises *ethical* or *moral* concerns (these terms are used interchangeably in this chapter), and goes to the heart of cherished claims about what makes humans special. Phrasing our opening question more specifically, are we, especially those in the zoo establishment, capable of learning and then recommending that some animals simply do not belong in captivity because they, like humans, are so complex mentally and socially that they suffer when in captivity even if kept healthy and well fed?

ONE ATTEMPT TO ANSWER THIS QUESTION

The Great Ape Project answers this question in two ways. First, the project offers detailed arguments on why ethical considerations lead to the conclusion that certain animals should not be held captive (Cavalieri and Singer 1993; see also the "Declaration on Great Apes" at www.greatapeproject.org). These arguments rely heavily on evidence of the day to day realities of gorillas, chimpanzees, orangutans, and bonobos, supporting the conclusion that such complex individuals should never be treated as property or mere things. Thus the project argues that members of these species should be given fundamental moral protections that prevent humans from taking their lives, holding them captive, or inflicting bodily harm on them (as in biomedical experiments). There could be exceptions, of course, based on self-defense or public safety considerations. But such problems are individual-level considerations and can be handled much as we handle similar problems caused by certain humans. The arguments of the Great Ape Project are addressed to the propriety of holding members of these species in general—that is, without regard to individualized circumstances and problems.

Second, the project also answers this question by criticizing contemporary Western society's absolute dismissal of all other animals from basic moral protections. Such a radical exclusion of all other animals is implicit in the daily actions of many humans, and explicit in many philosophies and religions purporting to justify those daily actions.

The project's critique, then, is directed at humans living as if only humans really matter as individuals. Such practices and the underlying exclusivist value systems have been addressed in many ways, one of which is the challenge to what has been called "speciesism" (Ryder, 1970).

ETHICS AND THE QUESTION OF COMMUNITY

The title of this book forthrightly raises ethical dimensions of human interactions with gorillas, orangutans, bonobos, and chimpanzees. The title's phrasing also hints at some of the underlying problems, reflecting a tension in which people of goodwill must engage openly. The more accurate term "other great apes" might have allowed the issues to be framed more scientifically, because humans are, biologically speaking, great apes. Indeed, there is "no natural category that includes chimpanzees, gorillas and orangutans but excludes humans" (Dawkins 1993, 82).

It is particularly important to continue to associate the word *ethics* with nonhuman great apes, because how we speak about nonhuman animals is critical. Through the work of many twentieth-century philosophers, we now recognize that the words used in discussing issues go a long way toward determining how we and our cultural successors see those issues. The term *ethics* is from the Greek "ethos," meaning a custom or character. The term *morality* is similarly related to the Latin "mores," which also means customs. If we confined ourselves to what is customary in our culture, however, we almost certainly would not speak of our cousin great apes in terms drawn from traditional ethical discourse.

Yet the study of ethics goes beyond what has been customary. Even if "the laws of custom are very apt to be mistaken for the order of nature" (Paley 1788, 32), we are now aware that some customs were undeniably immoral. Among the many revolutions that have challenged customary practices, perhaps the best known are those that have led to repudiations of racist and sexist customs and views. A concern for "ethics," then, clearly pushes one to go beyond what is merely customary.

The title also invokes the important image of a "coexistence" because Western developed society is only now embarking on conversations about the important human ability to care about some animals beyond the human species. Using the prized word "ethics" with regard to this frontier is particularly appropriate because the word has come to have the general sense of individuals caring about other individuals, communities, populations, species, or ecosystems and, especially, taking responsibility for how their intentional actions affect those others.

But questions remain: Which others? Why just those living creatures and not others? There has been no agreement across history or cultures on answers to these questions. But even if no consensus exists, it is generally acknowledged that human individuals in all cultures normally take the concerns of some others into consideration when assessing what it is they "ought" to do.

The dominant religious and philosophical traditions of Western culture have long staked out this fascinating ability as uniquely human. Advocates within these traditions have not, however, been open-minded about whether any living beings outside the human sphere lived in social systems in which individuals cared about the interests of others. In fact, most Western ethicists have been totally disinclined to look at any other living beings to learn whether the traditional exclusivist claims about humans were accurate.

Looking at other animals carefully and with a rigorous commitment to describing their realities accurately, however, suggests that traditional ethicists' absolute dismissal of them must be seriously questioned. Consider, for example, that researchers such as de Waal (1996) have shown morality-like agreements and high levels of complex cognitive abilities among nonhuman primates and some other animals.

Because of their traditional refusal to engage the biological world carefully and systematically, the authority of philosophers and theologians about the nature and origins of ethics has been challenged, most notoriously by sociobiologists' recent attempts to "biologize" ethics (Wilson 1975). Sociobiologists' explanations have, in turn, been criticized as "essentially reductive" and unable to tolerate admissions and discussions of complexity (Midgley 1995, xxii).

The intensity and parochialism of such ethical debates, in particular the confusing vocabulary and controversial metaphysical assumptions, should not be allowed to obscure the fact that each of us is able to extend our concerns to at least some nonhuman lives and take responsibility for our intentional acts toward them.

Thus the real question is not whether we extend concern, but how far we might extend our concern.

SHOULD WE STOP AT THE SPECIES BOUNDARY? THE QUESTION OF SPECIES LOYALTY

One might assume that the question about how far we might extend our concern has an obvious answer—namely, that we humans naturally favor our own species (the term *species* is used in this context to mean both favoring the species as a whole as well as favoring each and every member of the species) over all other animals. In Western culture it is now assumed not only that we owe loyalty to the human species, as well as to each member thereof, but also that we naturally have such a loyalty to any and all humans.

In important respects, especially with regard to our relationship with other animals, loyalty to our species alone is a fiction, and it is found in only some human societies. To be sure, in a descriptive sense there are some specific ethical or moral sentiments that are widely found, perhaps even universally, in human cultures. For example, kinship loyalty and reciprocity in a local social group have been said to be perhaps the only "claimants to the title of universally accepted moral principles" (Singer 1995, 150).

But any linking of these "local" loyalties, so necessary for survival, to the existence of a broader, species-wide loyalty should be carefully scrutinized. Occurrence of loyalty to those who are truly "around us" (our kin and our local community) cannot automatically be equated with general species-wide loyalty, because humans who are halfway around the world are not "around us" in any meaningful sense.

So the claim that we naturally give, as well as ethically owe, loyalty to our species as a whole is not an empirically or scientifically derived claim. Rather, it is a cultural artifact that became established as a religiously derived postulate. It is currently a very influential claim because it is inculcated into children and taught in the Western world as if it were the order of nature.

But such a conclusion is not supported by day to day observations or history. In daily life, where one might expect "natural" tendencies to be evident, many people do not demonstrate such loyalties. Further, in many times and places, human cultures have not been typified by the view that each of us owes loyalty to any human above and beyond what we owe to nonhuman animals. Such speciesist claims come in both religious and nonreligious forms, such as "humans have been given dominion over the earth" or "humanity is the measure of all things." Rachels

(1991), Pluhar (1995), Midgley (1984), and Williams (1985) provide a range of views on speciesism.

In general, then, preferring some members of our own species cannot be equated with the very different phenomenon of a general species-wide loyalty. One most assuredly cannot convert preference for nearby humans to a claim that one must protect the interests of all members of one's own species to the exclusion of the interests of all beings outside the species.

If the individuals to which we as moral agents might pay attention are not, then, defined by mere membership in our own species, how inclusivist should we be?

THE GREAT APE PROJECT'S CHALLENGE

The Great Ape Project argues that the mentality that "all humans and only humans should matter to moral agents" is troubling. In fact, it should concern every moral agent, because even if the prevailing exclusivist claims are best understood as mere custom or cultural artifact, the effects of such claims are all too real. One needs to look only as far as factory farming, biomedical experimentation, and destruction of habitat under the euphemism "development" to see how absolute the dismissal has been.

Justification of species bias comes in many different, culturally relative forms. These are invariably bound up with other, more fundamental beliefs about which lives have importance, what kinds of relationships with other beings are possible, and how far a moral agent needs to extend her concern. The tying of species bias to these more fundamental beliefs often gives that bias the appearance of a moral claim. Even more radically, advancement of exclusivist, "humans only" concerns is sometimes characterized as a moral necessity.

The phenomenon of speciesism is also closely related to how we speak about other animals. As has happened with other exclusivist beliefs such as racism and sexism, speciesist views are brought to individuals in the language they learn as children, and then reinforced through constant repetition. The result is that the division referred to in phrases such as "humans and animals" is treated as if it is real and "natural." In fact, the division is merely an exclusivist, immoral custom masquerading as the order of nature.

The peculiar shortsightedness leading to the claim that only humans really matter and thus to the view that such exclusivist species bias is essentially and eminently moral now characterizes much ethical thinking in industrialized, "advanced" cultures. The Great Ape Project challenges this mentality, confronting directly the notion that mere membership in the human species establishes a definitive guideline regarding which biological beings should and should not be protected as to life, liberty, and freedom from torture. The arguments made by the project are not, however, an attempt to say definitively what the exact composition of the new "community" should be. Rather, the arguments are a challenge to the exclusivism and absolute dismissals found in speciesist attitudes, because the attitude "only humans really matter as individuals" requires not only rather odd assumptions but, indeed, flawed and unjustifiable ones.

A SHORT HISTORICAL AND BROAD CULTURAL TOUR

The idea of a community among all humans is old, having been first advanced more than 2,500 years ago. In India, it was advocated by the Buddhists and Jains. At about the same time, there was a glimpse of religious universalism found in the ancient Hebrew tradition (see, for example, Amos 9: 5–7). But the idea that a community was possible among only humans is, in the Western cultural sphere, primarily Greek in origin. First popularized by the early Stoics, it was but one option in a vibrant debate regarding the abilities of animals beyond the human species line (Sorabji 1993). Aristotle and the Stoics after him took the position that nonhuman animals could not have higher level mental processes, and therefore were not owed ethical protections humans were entitled to as mentally complex, intelligent beings. Many other philosophers, including Aristotle's immediate successors and thinkers in the Platonist, Cynic, and Pythagorean traditions, contended otherwise (chapter 8, this volume).

Recall that Greek life was centered on the city (*polis*), and it was on this basis that Aristotle pronounced his famous dictum "man is by nature a *political* animal"—that is, a city animal (Aristotle 1984, 1988, emphasis added). The Stoics emphasized this approach when advancing the notion of *oikeosis,* or "belonging" to a worldwide human city. They used this universalism among humans alone to explain why only humans could and should receive justice. The early Christians tried hard to fit into this Greek, all-too-urban world, and they borrowed extensively the vocabulary and concepts of the Greek culture that they eventually came to rule (Meeks 1993).

In the ongoing debate over the abilities and importance of nonhuman animals, the early Christian theologians sided with the Stoics in repudiating the importance of other animals. They also relied heavily on certain anthropocentric features of the Hebrew scriptures even as they played down the Hebrews' counterbalancing emphasis on humans' connectedness with the rest of life (Barbour 1996; Passmore 1974; Ruether 1993). Thus through a synthesis of human-centered Greek notions and only one side of their Hebrew inheritance, early

Christian theologians promoted the idea of a community of only human animals. Thereby, the related Aristotelian and Stoic denial of any high-level functions for any nonhuman animal became imprinted on Western intellectual and ethical thinking.

The debate about other animals' abilities, and therefore their moral standing, came to an end, for all practical purposes, with Augustine's work. He merely synthesized some previous views regarding which biological individuals mattered and effectively shut down the debate about some nonhuman animals having intelligence or other complex cognitive abilities (Sorabji 1993). From Augustine on in Western intellectual history, the view that humans were radically different from all other animals was entrenched in the potent combination of cultural datum and religious belief functioning as "common sense," making it hard to see the dismissal of all nonhuman animals as mere custom being mistaken for the order of nature.

The absolute dismissal of other animals was worked out philosophically in the modern era by René Descartes and Immanuel Kant. That their views have become so important on the issue of which animals matter suggests a great deal about the quality of the entrenched views, because neither of these important philosophers ever systematically researched other animals. Rather, each instead simply passed along those general dismissals of other animals that they imbibed as children growing up in an overwhelmingly anthropocentric culture.

Although Descartes purported to doubt everything, he never doubted his inherited views of other animals. Indeed, Descartes exhibits some of his shoddiest reasoning about nonhuman animals. Similarly, Kant purported to rely exclusively on reason, even as he refused to use reason to assess his own radical ignorance about the lives and mental complexities of many animals, human and otherwise. Kant's argument (1785/1993) is purely intellectual, not empirical, especially his claim that nonhuman animals are *unable* to be self-aware: "Animals are not self-conscious and are there merely as a means to an end. That end is man" (Kant 1963, 239).

In the end, Augustine's often profound religious insights and Descartes's and Kant's radical philosophical breakthroughs are, on the issue of other animals, ignorance-based and, thus, incredibly naive. What is particularly significant, however, is that we are all heirs to these respected thinkers' views, now sanctioned by our most cherished cultural institutions.

A SEEMINGLY RADICAL PROPOSITION

Thus our culture's views and practices that draw a radical line between all humans and all nonhuman animals are heirs to a problematic legacy, particularly as those views have been shaped by the religious and nonscientific values of mainline

Christian culture as it embodied prejudicial dismissals stemming from classical Greek thinkers. Because even our secular institutions are heirs to this exclusivism, speciesism remains fully embodied in cultural values that guide the way science is now done. What makes this challenging is that scientists have broadly claimed to be explicitly nonreligious. Indeed, it is commonly asserted that "science" overthrew religious views of the world.

Scientific method is indeed very different from the methods of religious sensibilities. But consider what some scientists do with "science" even as they maintain the pretense that the scientific community's approach is solely objective. Certain "sociological" features of modern biological sciences, and indeed of modern institutions that hold scientific findings and method as cherished values, are actually in service of religious but unscientific ideas that purport to be contemporary Western "common sense." Note, for example, that any type of nonhuman animal (except one's own companion animal) but no nonconsenting human animal is deemed fairly subjected to experimentation (for exceptions now deemed immoral see Grodin and Glantz 1994; Lammers and Verhey 1987; Lederer 1997).

Science has a value system that should permit scientists to question openly what goes on in any scientific circle, but scientists rarely do so because of economic and social factors, religious values, and various cultural biases. Rather, the contemporary science research community, in practice, docilely relies on a value system that favors some, but not other, animals. Thus even as we hear claims that science is objective, many segments of the science establishment of the late twentieth-century promote scientifically unsupportable values and discrimination.

Consider, for example, the use of chimpanzees in AIDS experiments. Because these same experiments are openly acknowledged as completely improper when forced on human research participants, researchers who recommend chimpanzees for these experiments face a dilemma that involves "a logical trap": "In order to defend the usefulness of research [researchers] must emphasize the similarities between the animals and the humans, but in order to defend it ethically, they must emphasize the differences" (Rachels 1991, 220). This leads to an irreconcilable tension between scientific justifications and moral justifications:

> If the cognitive abilities of humans and animals are so drastically different as to morally justify experimentation, then those differences will reflect and promote other biological differences which undercut inductions from animals to humans. On the other hand, if underlying biological mechanisms are sufficiently similar to justify scientific inferences from animals to humans, then the higher-order traits of the test subjects are sufficiently similar to human traits to make research morally problematic. (LaFollette and Shanks 1996, 56)

Some contemporary scientific experimentation is, then, a surreptitious affirmation of a fundamentally nonscientific world view, reminiscent of superstition-like prejudices perceived by many to be the sole province of religion. The Great Ape Project suggests that challenges to such unreflective acceptance of unscientific dualisms are required if we are to continue to assert that we are, by nature, ethical beings.

COMPARING THE POSITIONS OF THE GREAT APE PROJECT AND THE AMERICAN ZOO AND AQUARIUM ASSOCIATION

The historical view that other animals (and sometimes even humans) can be exhibited for human pleasure or education has led to many abuses. The American Zoo and Aquarium Association (AZA) has been instrumental in challenging some of these abuses, such that it is common for the AZA's own hierarchy to acknowledge that "zoos are in process." Indeed, modern zoos are changing as today's more responsible advocates of captivity implicitly, and even at times explicitly, admit that today's zoos are heirs to a complex legacy. Addressing the needed changes in an environment of extreme anthropocentrism has been difficult, and the AZA is to be commended for its attempts to address the shortcomings of the exhibits that earlier zoo managers created (chapter 18, this volume).

One aspect of that complex heritage is the belief that humans have a right to exhibit any animal other than humans. The AZA's commitment to "public education and scientific study" (the avowed goal of the AZA's Ape Taxon Advisory Group) reflects the common justification for the practice. Much is learned when viewing animals in captive circumstances, it must be acknowledged (chapters 5 and 18, this volume). Not all of it is good education, however. For example, exhibiting animals has at times clearly taught very negative messages. Consider the example of Charles Gore, the English bishop who succeeded Darwin's famous opponent Wilberforce. After viewing zoo chimpanzees, Gore commented that the sight made him

> return an agnostic. I cannot comprehend how God can fit those curious beasts into his moral order. . . . When I contemplate you [the chimpanzee], you turn me into a complete atheist, because I cannot possibly believe that there is a Divine Being that could create anything so monstrous. (quoted in Sagan and Druyan 1992, 272)

Gore's reaction was in part prejudiced by the captive circumstances, because the individuals he saw were being exhibited for Londoners' edification. He could not have missed the meta-message of the situation—namely, that his education was more important than their freedom. His extraordinary reaction might be explained by Midgley's general assessment, "We distance ourselves from the beast without for fear of the beast within" (quoted in Clark 1977, 120). It might also be explained by the collision between, on the one hand, the deep-seated conviction in Western culture that humans are set apart from the rest of nature, and, on the other hand, the obvious truth that many of the caged animals are extraordinary individuals and some are even very much like us. Gore's question about "how God can fit those curious beasts into his moral order" ignores the obvious fact that through the power relationship of captivity, we have subordinated them and thereby fit them into our moral order.

But even if captivity is the practice and mentality that we have inherited, it is just one option. This is implicitly acknowledged when some zoos on occasion educate the public that the best place for certain animals is not in our zoos but in their natural setting amid their social groups living a free life. This important message is, however, contradicted by the very circumstances in which it is given, because what child, let alone adult, is not mentally quick enough to receive the "educational" meta-message that our society sanctions as moral the captivity of all other animals for our benefit? Today the propriety of captivity for any nonhuman animal remains as adamantly asserted by the best zoos as it ever was by bad zoos, circuses, and roadside attractions.

WHAT IF CAPTIVITY IS NECESSARY?

To be sure, some form of life away from the wild may at times be necessary in the contemporary world because of humans' onslaught on the natural homes of many animals, including nonhuman great apes. The extraordinary problems being faced now or likely to be faced in the future are by no means easy to resolve. But when assessing the advisability of any action taken in a crisis, we need to recognize that the steps taken to resolve the crisis, as well as the guiding principles inherent in those steps, may well be invalid when the situation is normalized. That captivity may be necessary in some crises in no way validates captivity in general.

However, if we conclude that in some crisis circumstances captivity may be the best solution, there remains a complex set of questions regarding what kind of captivity is acceptable. The crisis-driven decision to approve captivity cannot be used as support for the right to exhibit the animals. Presumably the best interests of the individuals involved should dictate what should be done, and that might involve nonpublic sanctuaries.

By no means, then, does the necessity of some form of captivity for some animals in a time of crisis offer a justification for captivity in normal circumstances. Above all, such necessity does not justify any refusal of educators, as the AZA proudly claims it is, to raise with the public the question of whether captivity of some nonhuman animals, such as the nonhuman great apes, should, in the absence of a crisis, be a moral issue.

Consider the moral sensibilities implicit, and sometimes even explicit, in the AZA's deep concern for the serious problem of poor animal care in bad zoos, circuses, and roadside entertainment attractions. Implicit in such concern is a recognition that holding other animals is a sensitive matter that entails ethical obligations. Going beyond AZA's condemnation of bad captivity, the Great Ape Project has now argued that those same ethical sensibilities must be turned to the very propriety of holding nonhuman great apes when there is no crisis.

As a way of making a simple point about the mentality underlying the AZA's philosophy in general, the Great Ape Project challenges all of us to use our moral abilities when looking at the long-standing practice of treating only human individuals as having the special kind of lives and interests that make it immoral to hold them captive. If someone argued that captive populations of Tasmanians could have been held in the nineteenth century before that race of humans became extinct, or that members of some ethnic or disfavored race or sex could be held "for public education and scientific study," we would challenge these views as unethical. Why? Because we reason that no matter how significant the resulting advances in "public education and scientific study," they would not justify imprisoning these complex individuals. To refuse to use our moral reasoning abilities in the same way for other, demonstrably complex animals is a bias of the culture we have inherited. One might say, then, that the AZA position passes along, from at least one angle, a certain species bias.

ARE OUR VIEWS OF OTHER ANIMALS IMPOVERISHED? ON THE NATURE OF THE ETHICS OF COEXISTENCE

Consider how new the frontier is when we ask questions such as, "In what respects, if any, should human moral agents pay attention to the freedoms and lives of nonhuman great apes?" In contemporary Western culture, these issues are a genuine frontier. This is evident, for example, in the claim by the Catholic Church that "animals, like plants and inanimate things, are by nature destined for the common good of past, present and future humanity" (*Catechism of the Catholic Church* 1994, para. 2415). The same mentality, of course, has driven biomedical experimentation on nonhuman great apes for the benefit of humans and the tradition of exhibiting nonhuman great apes for humans' entertainment and education.

Very much in the manner of the Catholic Church, those who guide and control these secular activities invoke explicitly ethical reasons and language as justifications for such practices, such as benefits to the human species as a whole. It is noteworthy that the same reasoning could also operate as a justification of experiments on or captivity of humans, but the consensus of those who control experimentation and captivity is that the same practices involving human animals should be prohibited because they are obviously immoral. At a minimum, then, all of us share the belief that at least some animals, namely human animals, should not be subjected to experiments or captivity or be treated as property. Are these the only animals that should receive such protections? If so, why these animals alone?

It is wise to remember that extending the moral circle in the manner being discussed is not a frontier in many other cultures characterized by long-standing repudiations of such instrumental uses of nonhuman animals. Because the "frontier" we now encounter has been reached and crossed in many other cultures, certain ironies can be noted. Anthropologists in the nineteenth century, and indeed the imperialist European cultures before and after, demeaned the views of cultures outside the European sphere. One of the grounds was the respect that other cultures often exhibited for nonhuman animals. In the following passage, Northcote Thomas (1908, Vol. I., 483) provided a fascinating example of cultural arrogance and European ethnocentrism commingled seamlessly with an extraordinary species bias:

> In the lower stages of culture, whether they be found in races which are, as a whole, below the European level, or in the uncultured portion of civilized communities, the distinction between men and animals is not adequately, if at all, recognized. . . . The savage . . . attributes to the animal a vastly more complex set of thoughts and feelings, and a much greater range of knowledge and power, than it actually possesses. . . . It is therefore small wonder that his attitude towards the animal creation is one of reverence rather than superiority.

Midgley's (1984) comment about the inadequacy of Thomas's urbanized, European point of view highlights just how ironic such a claim is, especially in light of the so-called cognitive revolution in psychology (Gardner 1985; Griffin 1992).

> Ethological investigation . . . has shown that Western urban thought was (not surprisingly) often even more ill-informed than local superstition on many such questions [regarding capacities of nonhuman animals], and that it had consistently attributed to animals a vastly less complex set of thoughts and feelings, and a much smaller range of knowledge and power, than they actually possessed. (Midgley 1984, 123)

Thus respectful study of other humans' views suggests that "uncivilized" humans saw many features of the connection between humans and other animals that our city-based ethics have simply ignored. Indeed, for many indigenous peoples, it is our industrialized world's cultures, consumers, and captains of industry that fit the old charges of "barbarian" and "savage." The views of many indigenous peoples are now known to be more environmentally sensitive than are the views of the rapacious industrialized countries (Kinsley 1995).

Given that alternative views are available, does continued reliance on traditional Western, anthropocentric notions and practices excluding nonhuman great apes from the moral circle betray the same kind of "fundamentalism" that we roundly condemn in many other circles? Are the pervasive views of the propriety of captivity impervious to change? Are they a kind of unassailable revelation from our past? If we answer any of these questions affirmatively, we are arguably imprisoned in our own past. Being open to the possibility of including some other animals in our moral circle arguably elevates us, for such openness is a profound affirmation of our moral abilities. Those who resist frank consideration of these issues in effect repudiate this remarkable ability we have to be concerned for others beyond our own world.

A QUESTION OF LANGUAGE

Returning to the title of this volume, consider how contrasting "great apes" and "humans" removes humans from a very natural place. Such artificial, arguably unscientific divisions abound in our ordinary language. By effectively removing humans from our place in natural categories, such phrases none too subtly promote a separatist, dualist mentality. This is conducive to the speciesist predisposition. The persistence of these words in the scientific community, and in particular in the zoo community, reveals how the much-touted "scientific revolution" remains startlingly incomplete.

The terminology "humans and animals" has exactly the same shortcoming but is backed by a powerful tradition of discourse that validates the use of "animals" to mean only "all animals other than human animals." One need think only of the derogatory features of the observation "he acted like an animal" to see the negative implications of this dualism. The English philosopher Stephen Clark once commented, "One's ethical framework is determined by what entities one is prepared to notice or take seriously" (Clark 1977, 7). It is now broadly recognized that how we talk about a general subject determines in great measure how we and our cultural heirs will be prepared to see that subject. Our linguistic habits about other animals provide an excellent example of this phenomenon, because we are not well-prepared to see nonhuman animals as individuals who have lives that are distinctive and of value.

One potential foundation for "noticing or taking other animals seriously" is, of course, the realities of their individual lives. One of the great benefits of scientific method is that it leads to a commitment to accurate description of the realities of other animals. What makes the issue of accurate description relevant to any discussion of other animals is that the lives of other animals are independent of humans' descriptions of them. Indeed, reference to the realities of the lives of those individuals we describe provides one means of getting beyond our own prejudices.

Note how the tasks of ethics and science, broadly construed, work together. It has been said that "the best part of morality lies in eliminating prejudice, bias, fantasy and wishful thinking" (Clark 1977, 189). This negative task is not incompatible with, and indeed one could say it shares a certain spirit with, the verifiability and falsifiability criteria on which much scientific experimentation is based. By speaking more carefully, and by calling on the cherished value of not promoting any particular biased agenda under the guise of science, we can explore why it is that some institutions accept and even promote the captivity of nonhuman great apes while abhorring a similar practice with regard to human great apes.

NONHUMAN GREAT APES' PECULIAR ROLE

The realities of nonhuman great apes, when viewed without the distorting lens of human exclusivism, are such that no one can deny that the individuals and social groupings involved are far more complex than ever imagined by those who crafted Western ethical sensibilities. This suggests that those who rely either explicitly or implicitly on Western, traditional, anthropocentric ethical formulations and dualist language should pause before applying them to nonhuman animals and in particular to nonhuman great apes.

One cannot ignore, however, that attention has been paid to great apes, and indeed primates in general, because they are our closest evolutionary cousins. This is reflected in the common phenomenon of a scientist explaining that his or her work is important because it sheds light on human realities, suggesting how fully the specter of anthropocentrism continues to loom over certain biological sciences.

Because science regarding nonhuman great ape individuals and societies is well-developed, ethical arguments can be fully framed with regard to them. This puts nonhuman great apes in the unique position of being the animals best suited to demonstrating that complexities exist outside the human species. Ironically, in spite of our sometimes selfish concern to study our close evolutionary cousins for the principal purpose of illuminating our own history, the nonhuman great apes have become potential ambassadors for many other nonhuman animals.

Focusing on just the nonhuman great apes, however, has been criticized by some for being too conservative—that is, for simply moving the dividing line further out to the other great apes and thereby ignoring many other animals that should matter to any compassionate, moral being. This criticism is important, but it ignores the fact that the Great Ape Project Declaration on Great Apes (the text of which can be viewed at www.greatapeproject.org) clearly states, "This challenge seeks equal consideration for the interests of all animals, human and nonhuman." The project is not just about great apes, human and otherwise. It is also about human exclusivism as that affects all nonhuman animals.

INCLUSIVIST ETHICS IN THE NEW MILLENNIUM

A full, realistic affirmation of nonhuman great apes as valuable in and of themselves carries the prospect of opening up hearts and minds to an extension of ethical concerns outside the human species. Such an affirmation forces us to deal with the impoverished values regarding nonhuman animals that we have inherited from the Western intellectual and ethical traditions. In particular, it prompts questions about the nature of our moral abilities. For example, if morality is the extension of sympathies to others, and we can extend such sympathies to the species level, what is it that prevents extensions beyond the species line? Ethicists, if they are to maintain the exclusiveness of their earlier emphases on humans alone, must provide compelling arguments for why our loyalty should stop at our own species rather than somewhere else on the continuum from self to immediate family to extended family to local community group to ethnic group to nation to race to species to genus to taxonomic family, and eventually to any living being and its ecosystem context.

Advocates of antiracist and antisexist positions, and more generally those who promote universalism within the human species, have argued that those who fail to care about all members of the human species have committed the fallacy of misplaced community. What these same advocates must justify is why the "community" stops at the species level. I would argue that they, too, by relying so exclusively on the concept of species, commit the same error.

The relevant community may not be species at all but the life that surrounds each of us as a moral agent in the midst of her or his own bioregion. Those animals that are nearby, whether human or not, are in a very real sense members of our community. Affirming them as such need not diminish our sense of equality among humans around the world. In fact, such a forceful affirmation of life would likely produce greater awareness that other humans are often harmed by our lifestyles and consumption patterns as fully as are nonhuman animals.

Most relevant is that any expansion of concern must be guided by specific concern for the realities of individuals. Each of us, as a moral agent, is suited to this kind of inquiry about the living things immediately around us. It is crucial that in this new millennium, appraisals of the lives of nonhuman animals be reality-based. Such accuracy will allow perspectives on any neighboring animal to be informed, and thus permit a fair inquiry into whether we should, when considering the impact of our actions on that animal, use our prized ability to extend concern to other living beings.

THE GREAT APE PROJECT TODAY

The Great Ape Project takes the position that there is now sufficient evidence of several kinds to challenge the developed world's influential cultural datum that it is only and all humans who are entitled to be in the inner circle, where we offer fundamental protections of life, liberty, and bodily integrity. Those who assert that human interests are such that human interests rightfully eclipse nonhuman great apes' opportunity to enjoy these fundamental privileges must be forced to deal with the vast amount of information suggesting that this claim is a prejudice that is no more justifiable than racism or sexism.

Because any talk of the ethics of coexistence ought to include serious consideration of real-world solutions, this means challenging captivity in its many specific guises. Captivity determined along speciesist lines continues to manifest itself in so many ways, some far more troubling than others. But the Great Ape Project's approach challenges even the benign end of the range of practices. This challenges vested interests and traditional ways of doing things.

Although the principles underlying the project's challenge can be simply stated, the situations they address often are diverse, complex, and not easily amenable to simple solutions. The simplicity allows some vicious prejudices to be exposed, but it does not solve all of the real-world problems, some of which are truly intractable. But the project's direct challenge remains, beckoning us to elevate ourselves by questioning our unqualified acceptance of the prevalent mentality that supports the captivity of nonhuman animals simply because they are not members of the human species. What the Great Ape Project seeks is to become obsolete as an organization, because the focus has moved away from nonhuman great apes because of their inclusion, along with human great apes, as full members of the moral circle recognized by us as moral agents. This revolution will be a profound affirmation of humans' remarkable ability to care about animals, human and otherwise.

REFERENCES

Aristotle. 1984. *The complete works of Aristotle: The revised Oxford translation* (J. Barnes, ed.). 2 vols. Princeton, NJ: Princeton University Press.

Barbour, I. G. 1996. *Three paths from nature to religious belief, Farmington Papers, Number SC8 (Science and Christianity 8).* Oxford: Farmington Institute for Christian Studies.

Catechism of the Catholic Church. 1994. London: Geoffrey Chapman.

Cavalieri, P., and Singer, P. (eds.). 1993. *The Great Ape Project: Equality beyond humanity.* New York: St. Martin's Press.

Clark, S. R. L. 1977. *The moral status of animals.* Oxford: Clarendon Press.

Dawkins, R. 1993. Gaps in the mind. In P. Cavalieri and P. Singer (eds.), *The Great Ape Project: Equality beyond humanity* (pp. 80–87). New York: St. Martin's Press.

de Waal, F. 1996. *Good natured: The origins of right and wrong in humans and other animals.* Cambridge, MA: Harvard University Press.

Gardner, H. 1985. *The mind's new science: A history of the cognitive revolution.* New York: Basic Books.

Griffin, D. R. 1992. *Animal minds.* Chicago: University of Chicago Press.

Grodin, M. A., and Glantz, L. H. (eds.). 1994. *Children as research subjects: Science, ethics, and law.* New York: Oxford University Press.

Kant, I. 1963. Duties to animals and spirits. In Louis Infield (trans.), *Lectures on ethics* (pp. 239–41). New York: Harper Torchbooks/Harper and Row.

———. 1993. *Groundwork of the metaphysic of morals* (H. J. Patton, trans.). New York: Routledge. (Original work published 1785)

Kinsley, D. 1995. *Ecology and religion: Ecological spirituality in cross-cultural perspective.* Englewood Cliffs, NJ: Prentice-Hall.

Lafollette, H., and Shanks, N. 1996. The origin of speciesism. *Philosophy* 71: 41–61.

Lammers, S. E., and Verhey, A. (eds.). 1987. *On moral medicine: Theological perspectives in medical ethics.* Grand Rapids, MI: William B. Erdmans.

Lederer, S. E. 1997. *Subjected to science: Human experimentation in America before the second world war.* Baltimore: Johns Hopkins University Press.

Meeks, W. A. 1993. *The origins of Christian morality: The first two centuries.* New Haven, CT: Yale University Press.

Midgley, M. 1984. *Animals and why they matter.* Athens: University of Georgia Press.

———. 1995. *Beast and man: The roots of human nature* (rev. ed.). New York: Routledge.

Paley, W. 1788. *Principles of moral and political philosophy* (7th ed.). Philadelphia: Dobson.

Passmore, J. 1974. *Man's responsibility for nature.* New York: Scribner.

Pluhar, E. 1995. *Beyond prejudice: The moral significance of human and nonhuman animals.* Durham, NC: Duke University Press.

Rachels, J. 1991. *Created from animals: The moral implications of Darwinism.* New York: Oxford University Press.

Ruether, R. R. 1993. Men, women and beasts: Relations to animals in Western culture. In C. Pinches and J. McDaniel (eds.), *Good news for animals? Christian approaches to animal well-being* (pp. 12–23). Maryknoll, NY: Orbis.

Ryder, R. D. 1970. *Speciesism.* Privately printed leaflet, Oxford, UK.

Sagan, C., and Druyan, A. 1992. *Shadows of forgotten ancestors: A search for who we are.* New York: Random House.

Singer P. 1995. *How are we to live? Ethics in an age of self-interest.* Amherst, NY: Prometheus Books.

Sorabji, R. 1993. *Animal minds and human morals: The origins of the Western debate.* Ithaca, NY: Cornell University Press.

Thomas, N. W. 1908. Animals. In J. Hastings (ed.), *Encyclopedia of religion and ethics, vol. I* (pp. 483–535). New York: Charles Scribner's Sons.

Wilson, E. O. 1975. *Sociobiology: The new synthesis.* Cambridge, MA: Belknap Press of Harvard University Press.

Williams, B. 1985. *Ethics and the limits of philosophy.* London: Fontana.

MARY ANNE WARREN

17
THE MORAL STATUS OF GREAT APES

The great apes are, by all credible accounts, our closest living biological relatives. Our observable physical similarities led Thomas Henry Huxley to argue, in 1863, that the African great apes belong in the Hominidae, the same family as *Homo sapiens* (cited in Savage-Rumbaugh and Lewin 1994). Today we know that their genomes are highly similar to ours, from 98.5 percent identical in chimpanzees to 97 percent in orangutans. As a consequence, many taxonomists now agree with Huxley, and some argue for placing the bonobos, chimpanzees, and gorillas in our own genus (Savage-Rumbaugh and Lewin 1994). Given this biological proximity, it is entirely appropriate to refer to the great apes as fellow hominids and to *H. sapiens* as one of the five surviving great ape species.

The similarities between ourselves and the other living hominids are not merely physiological. The field work of Jane Goodall, Dian Fossey, Birute Galdikas, and other ethologists has revealed the complexity of apes' social and emotional lives. The research of the Gardners, David Premack, Sue Savage-Rumbaugh, Roger Fouts, and others has demonstrated the impressive, though still controversial, extent to which chimpanzees, bonobos, and gorillas can learn to understand and make meaningful use of human-created symbol systems (chapters 10, 13, this volume). Meanwhile, most of the remaining populations of free-living apes in Africa and Indonesia are declining because of habitat destruction, commercial hunting for the meat and pet trades, and private poaching (chapters 1 to 4, this volume). Chimpanzees are still being used in medical research projects that will cause them suffering and perhaps death. It is past time for a serious reconsideration of the assumption that other great apes are so different from us as to have no moral rights, or none that warrant legal recognition (chapter 15, this volume).

The drive to establish moral and legal rights for the other hominids has been led by the Great Ape Project (chapter 16, this volume), which has successfully campaigned to include basic rights for all great apes in New Zealand's new animal welfare bill: rights to life, freedom from cruel and degrading treatment, and freedom from participation in scientific research that will harm them (Nowak 1999). Nothing in this proposal suggests that the other great apes are entitled to all of the moral and legal rights that ought to be accorded to humans—for example, those listed in the United Nations Universal Declaration of Human Rights. However, the project's Declaration on the Great Apes accords the same basic moral rights, life, liberty, and freedom from torture, to humans and other great apes (Cavalieri and Singer 1993). This proposal raises fundamental questions about the necessary and sufficient conditions for the possession of basic moral rights. Are the other great apes entitled to exactly the same basic moral rights as human beings? Is such equality in their best interests? Or does their survival, both as individuals and as viable species, require that they be subject to a higher degree of paternalistic intervention than would be morally acceptable in the case of similarly endangered human populations?

I argue that the mental, social, and emotional capacities of the great apes, their endangered status, and their special scientific, aesthetic, and spiritual value to humans all support a level of protection for their lives and well-being equal to that to which humans are entitled. There are, nevertheless, morally relevant differences between humans and the other great apes, including the real or apparent differences between their cognitive and communicative abilities and ours. Few would deny that these differences make some nonbasic human rights, such as the right to attend college or to vote in political elections, inapplicable to the other great apes. Nonhumans would derive no benefit from such rights, because even the most gifted nonhuman ape cannot learn to read or write well enough to benefit from a college education or to understand human political affairs well enough to cast a meaningful vote (Rachels 1993). I argue that, although the basic moral and legal rights to life and liberty of the other great apes ought to be as strong as ours, and as morally binding, the content of those rights will need to be formulated somewhat differently.

THE CONCEPT OF MORAL STATUS

To say that a being has moral status is to say that we as moral agents can have moral obligations toward it. It is important to distinguish between having obligations *toward* an entity and having obligations *regarding* it. I have a moral obligation not to steal another person's coat, but this is an obligation to the person and not to the coat, which has no moral status. In contrast, my obligation not to kidnap another person's child is an obligation not only to the parent but also to the child, who has the same basic moral status as any other sentient human being. I argue (Warren

1997) that humans who have never been sentient, such as fetuses before some age of gestation, or those who will never be sentient again, such as the permanently vegetative, are nonsentient and thus do not have the same moral right to life as those who are now capable of sentience or of eventually returning to sentience.

I defend a multicriterial theory of moral status, which permits a plurality of factors to influence a being's moral status (Warren 1997). A being's intrinsic properties are relevant to its moral status, but its relational properties are also relevant. Sentience and mental sophistication are morally relevant intrinsic properties: The more highly sentient and mentally sophisticated a being, the more difficult it is to justify harming it. Social and environmental relationships can also alter what we owe to members of a given species. Not all human beings are equally self-aware, but all sentient human beings have equal basic rights as members of human social communities. As Mary Midgley (1983) has argued, nonhuman members of our mixed social communities, such as cats and dogs, are reasonably accorded a stronger moral status than that of many equally sentient animals who are not members of our social communities. The reciprocal relationships of care and affection between human beings and the nonhuman members of their social communities give rise to a moral obligation to respect these animals.

We also have good reasons to recognize moral obligations to nonhuman comembers of the earth's biological communities. The more endangered a species is by human activities, and the more important it is to the ecosystems of which it is part, the stronger the moral obligation to protect both the members of that species and the species itself. Some environmental ethicists construe this as an obligation not to nonhuman organisms or species but to those present and future human beings whose lives would be adversely affected by the loss of natural biodiversity. I argue, however, that there are sound pragmatic and psychological reasons for recognizing moral obligations to plants, animals, and species and not merely to other human beings. Finally, I argue that it is morally appropriate to extend special protections to animals of species that, in addition to their ecosystemic value, have special aesthetic, scientific, or spiritual value for human beings—for example, the snow leopard or the whooping crane.

All of these criteria of moral status point to the same conclusion: Great apes are entitled to a moral status far more similar to that of human beings than has commonly been supposed. I will begin with the case for according moral status to animals on the basis of their sentience and other mental capacities.

RESPECTING SENTIENCE AND SUBJECTHOOD

Peter Singer and Tom Regan are two contemporary philosophers who have presented powerful arguments for according an enhanced moral status to some nonhuman animals. Singer's criterion for full moral status is sentience—in other words, the capacity to experience pain and pleasure. His utilitarian theory requires us to give equal weight in our moral deliberations to the comparable interests of all the sentient beings who will be affected by our actions (Singer 1979). In contrast, Regan defends a narrower criterion of moral equality. His theory accords equal basic moral rights to all and only those beings who are "subjects-of-lives." Subjects-of-lives are beings that are not only sentient but possessed of additional and more sophisticated mental capacities, such as emotion, memory of the past, anticipation of the future, and the ability to act in pursuit of conscious goals (Regan 1983). These capacities are morally relevant, because a being with these capacities has a degree of self-awareness and awareness of a life that can go better or worse for it.

These theories lead to strikingly different conclusions about which animals are entitled to full and equal moral status. Singer's sentience criterion includes all vertebrate animals whose nervous systems are sufficiently functional and developed to permit the occurrence of pain, pleasure, and other conscious experiences. It probably includes many invertebrate animals as well—for example, those with complex sensory organs and central nervous systems, such as cephalopods, crustaceans, and arthropods. Regan's subject-of-a-life criterion applies, he has argued, to all mammals over 1 year of age, and possibly to some other vertebrate animals (Regan 1983).

Both these theories imply that the other great apes are entitled to the same moral status as human beings. However, neither provides a reason for according them a moral status stronger than that of any other sentient subject-of-a-life. This is not a problem for those animal rights advocates who are committed to equal moral status for all subjects-of-lives, or all sentient organisms. However, if we wish to convince those who are generally skeptical about animal rights to make an exception for the other hominids, we will need a theory that recognizes a wider range of morally relevant factors.

In my view, neither sentience nor being a subject-of-a-life is, in itself, sufficient for the possession of basic moral rights comparable in either strength or content to basic human rights. No morality can command our acceptance unless it is pragmatically possible to live by its terms. But it would be impossible to enforce a moral code that required us to accord the same protections to all sentient beings, or even to all subjects-of-lives. Rats, for instance, are highly sentient and intelligent animals, who are undoubtedly subjects-of-lives, even before they reach a year of age. But as Peter Singer has remarked, legal rights for rats are unlikely in the near future (Nowak 1999). This is not primarily because human beings are irrationally biased against rats, although many probably are. There are compelling reasons for not protecting rats nearly as carefully as, say, giant pandas. Giant pandas are a desperately endangered species, and they are very slow to reproduce. They are usually harmless to humans, and they are charming and beautiful creatures with a strong emotional appeal to us. These are sound reasons for protecting the life and reproductive potential of each giant panda. In contrast, rats exist in vast numbers, reproduce with extraordinary rapidity, and persistently threaten human lives by spreading disease and consuming or contaminating human food supplies. These facts make it reasonable, in the absence of better alternatives, to tolerate the killing of rats.

The view that all sentient animals, or all those that are subjects-of-lives, are entitled to the same moral status as human beings also conflicts in some instances with the need to protect endangered plant and animal species and natural ecosystems. The deliberate or accidental introduction of nonindigenous animals often endangers indigenous plants and animals, sometimes with ecosystemically catastrophic consequences (e.g., Quammen 1996). Perhaps the most notorious case is that of the introduction of European rabbits into Australia, which has no indigenous placental mammals other than bats (dingos are placental but were introduced 4,000 years ago). In the absence of natural predation, the rabbits multiplied prodigiously, contributing significantly to the decline of native flora and fauna throughout much of the continent. The only successful strategy for controlling rabbit populations has been the introduction of mixomytosis and calcivirus, pathogens lethal to rabbits but not to other animals (Anderson 1996). Although most people would agree that it would be morally wrong to use germ warfare to limit human population growth, for the rabbits this is arguably a lesser evil, so long as equally effective nonlethal means remain either impractical or unavailable. Warren (1983, 1986, 1997), Hutchins and Wemmer (1986–1987), and Hutchins, Stevens, and Atkins (1982) have discussed in greater detail the practical inconsistencies between the strong animal rights views of Regan and Singer and the need to control nonindigenous animals.

Indigenous herbivores can also cause severe ecosystemic damage, when human intervention has largely eliminated their natural predators. Aldo Leopold (1987) wrote movingly of the damage to lands in many parts of North America caused by the deliberate elimination of wolves and other large predators and the subsequent growth of deer populations. Systematic culling has often been the only feasible way of restoring a balance between the deer and the vegetation on which they feed. Nonlethal means, such as capture and relocation, are usually too expensive and can be more cruel than lethal means. Capture is traumatic and sometimes fatal, and relocated animals typically have a very low rate of survival (Hutchins et al. 1982). If it is reasonable to accord stronger moral status to pandas and whooping cranes than to rats, rabbits, and deer, then neither Singer's sentience criterion nor Regan's subject-of-a-life criterion provides a complete account of the basis of moral equality. Both sentience and subjecthood are relevant to moral status, but they are not the only morally relevant considerations. Moreover, it is not only the presence or absence of these capacities that matters, but also the degree to which these capacities are developed in animals of a given species.

Sentience is relevant to moral status, because sentient beings can experience pain, or be deprived of pleasure, or of the continuation of lives that they consciously enjoy. If rocks and microbes lack significant moral status, it is in large part because they are invulnerable to pain, suffering, or the loss of future pleasures. All and only sentient beings have a well-being that matters subjectively to them (Bernstein 1998). Because rocks and microbes are nonsentient, nothing that is done to them is cruel. The ordinary concept of cruelty applies both to causing unnecessary pain or suffering and to the needless killing of sentient beings. This is because both pain and premature death are harms that matter to sentient beings; pain lowers the quality of their experienced life, and death terminates all experience, thus precluding all future pleasures and satisfactions.

Such harms should not be inflicted on any sentient being, except under what Albert Schweitzer called "the compulsion of necessity"—when there is no other way to serve important human or ecosystemic needs. We owe it to sentient beings not to harm them without good reasons, reasons based on needs that cannot otherwise be met. Plowing, planting, and harvesting a field to grow food is likely to cause pain or death to an enormous number of small sentient invertebrate animals, including spiders, mites, insects, and worms, but these harms are justified by the compelling need to produce food. To take this view is not human chauvinism but the recognition of practical necessity. Such activities are cruel only when they are unnecessary.

In my view, Singer's sentience criterion delineates not the class of moral equals but rather the class of beings who are vulnerable to cruelty. Mere sentience is not sufficient for equal moral status; mental sophistication, self-awareness, intelligence, and sociability are also relevant. Although they disagree about many things, Singer and Regan agree that self-aware beings lose more that is of value to them when they die prematurely than do non–self-aware beings (Regan 1983; Singer 1975). Both say that this is why, in the lifeboat scenario offered by ethicists, it is better to throw a dog overboard to keep the boat afloat than to sacrifice a human passenger. Human beings have more complex plans and expectations for the future than do dogs, who in turn probably have more concern for their future than do ants. Highly self-aware, social, and intelligent beings are also likely to suffer more than simpler beings from capture, confinement, isolation, and boredom. Such beings can be harmed not only by pain and death but also by the disruption of their lives and the loss of the opportunity to live in freedom or in company.

If mental sophistication is relevant to moral status, then the other hominids have an unusually strong claim to high status. Their intelligence, self-awareness, and sociability have been amply documented. As Fouts (chapter 10, this volume) points out, it can be misleading to compare the mental abilities of an adult chimpanzee to those of a young human child, because adult chimpanzees have many mental and physical abilities that human children lack. Nevertheless, such comparisons can help undermine the stubborn assumption that no nonhuman animal can really use language, or form concepts, or reason about their environment, or possess a sense of self. If the other great apes can do these things about as well as, say, a 4-year-old human child, then the burden of proof is on those who deny that they are entitled to basic moral rights.

This point is relevant not only to the other hominids but also to many other primates, and to whales, dolphins, elephants, dogs, cats, and other relatively large-brained mammals. The highly developed sentience, sociability, and mental sophistication of these beings imposes a moral obligation on those who keep them in captivity to adequately provide for their psychological well-being (see Kreger, Hutchins, and Fascione 1998 for an excellent discussion of species-specific needs for environmental enrichment in zoos and aquariums) as well as their physical health and survival. Furthermore, the harms inevitably entailed by the capture and confinement of highly sentient and mentally sophisticated beings make it extremely difficult to justify taking free-living individuals of such species into long-term captivity, except when there is no other way to save them from an even worse fate.

RESPECTING ECOSYSTEMIC, SCIENTIFIC, AND AESTHETIC VALUE

For many people, the mental, emotional, and social capacities of the other great apes are sufficient to accord them basic moral rights to life, liberty, and well-being, rights of the same strength as the corresponding human rights. However, if the argument is allowed to rest entirely on these capacities, then it remains vulnerable to the response that the intelligence of the other great apes is, after all, considerably less impressive than human intelligence, and this difference may well be morally significant. What makes this response ultimately irrelevant to the case for equal basic moral rights is the existence of independent reasons, including environmentalist reasons, for protecting the lives and liberty of the other great apes just as carefully as those of human beings.

Environmental philosophers such as J. Baird Callicott (1989) have argued that we owe special protection to organisms of species that are important to the ecosystems of which they are part and that are endangered by human activities. Callicott's theory of moral status is based on the land ethic of Aldo Leopold. The land ethic holds that plants, animals, and even soil and water are fellow members of the biotic community, and we thus have moral obligations toward them. Endangered plant and animal species must be protected, because natural biodiversity is vital to the "integrity, stability and beauty" of the earth's ecosystems (Leopold 1987), and because the survival and flourishing of present and future human beings depends on the long-term health of those ecosystems.

Although all species are potentially important to the health of their respective ecosystems, some are more vital than others. For instance, large predatory or omnivorous animals are especially important, because there are relatively few individuals and species in any given ecosystem and because they play vital ecosystemic roles, such as limiting prey populations. These predatory species are everywhere vulnerable to human-caused extinction, because they are hunted for food or trophies or as threats to livestock and because often their low reproductive rates do not permit them to withstand much human predation. For many people, large indigenous mammals have special aesthetic and spiritual importance. Aesthetic and spiritual factors make up much of the value that human beings place on wilderness and provide much of the motivation for the heroic efforts being made on behalf of California condors and Bengal tigers, for example.

Great apes are near the top of the scale on these relational criteria of moral status. Their importance to their ecosystems makes protection of their remaining free-living populations an international priority. True, there is no reason to believe that these ecosystems would collapse altogether without these species; but they, and we, would be permanently impoverished. Because the other great apes are both very like us and very different, in ways we have just begun to understand, they have unique scientific value. They are among the few species some of whose members are able to communicate with us by means of conventional symbol systems. Moreover, their nonverbal modes of communication are enough like ours that we can readily learn to "read" their emotional states (and vice versa). Under favorable circumstances, humans can form affectionate and reciprocal social relationships with hominids of other species. For all of these reasons, our descendants will not forgive us if we allow the other hominid species to disappear from the earth or be eliminated from their natural habitats, surviving only in captivity.

The combined force of these arguments for a substantially equal moral status for the other great apes is tremendous. Their exceptional mental sophistication, their ecosystemic importance, and their special scientific, aesthetic, and spiritual value as our biological next of kin all provide reasons to protect them with the same urgency is appropriate as when human beings are threatened with murder or enslavement. In short, it makes good moral sense to accord them basic moral rights of the same strength as our own. But rights of the same strength need not be identical in content; and it is not in the interests of the great apes to be accorded rights to life and liberty that are entirely identical to the corresponding human rights.

THE RIGHT TO LIBERTY

Consider first the right to liberty as it applies to captive apes. Captive humans of sound mind who have been charged with no crime have an almost unconditional right to be released. (There are some defensible exceptions—for example, the quarantine of persons with highly infectious diseases.) The imprisonment of innocent humans who are no threat to themselves or their community cannot be justified on the grounds that it is for their own good, because it rarely is, and because the human right to liberty entails the right to choose a perilous freedom over a safer life in captivity.

The case of the other great apes is relevantly different. In most cases, long-term captives would not benefit from being immediately released into their native habitat, because they would not know how to find food or how to defend themselves against humans, natural predators, and aggressive conspecifics. Competent humans in comparable straights nevertheless have the right to be released, because they are presumably capable of understanding the comparative risks and benefits of freedom and confinement. But it is doubtful that long-term captive chimpanzees, bonobos, gorillas, or orangutans can understand the risks of release well enough to make an informed decision about it.

This is not to denigrate the overall intelligence of the other great apes. They are, when "wild," smart enough to survive with minimal technology in habitats where most civilized humans would quickly starve or be eaten. In their own worlds, free-living apes are capable of making highly informed decisions about the circumstances they are likely to encounter. Long-term captives, on the other hand, are apt to lack the knowledge and skills necessary for survival in the wild, and we do not know how to explain this reality to them. As long as this is true, it will be most respectful of the moral rights of long-term captive great apes to liberate them only in ways consistent with their survival and well-being.

It is important, nevertheless, to assert a qualified right to liberty for all the great apes, even long-term captives. Human beings who keep animals of any species in captivity have a moral obligation to provide them with enough space, privacy, exercise, social interaction, and mental stimulation to maintain their mental and physical health. With time and care, some captive apes can be taught to survive in their natural habitats. Those who are engaged in this important task deserve our admiration and support. Other captives, including chimpanzees who as research subjects were infected with HIV or other dangerous pathogens, will never be suitable for release into the wild. There are, however, ongoing efforts to provide refuges where research "retirees" can live out their lives in greater liberty than they have previously known (chapter 7, this volume). Such refuges may contribute to the conservation of ape species by helping to conserve genetic diversity. But even if they do not, they are morally justified as a partial compensation to individuals who have been wrongly deprived of liberty and who cannot be returned to freedom after such captivity.

The basic right to liberty implies that free-living apes ought not to be taken into long-term captivity, except in extreme circumstances. The preservation of free-living ape populations requires the preservation of the ecosystems that support them and provides important motivation for it. This is one reason for protecting mountain gorillas in situ. If the gorillas were removed from their natural habitats to protect them from poachers and civil strife, the political and economic incentives for protecting those habitats may vanish. Capture and confinement would also be physically and psychologically traumatic, would probably lead to some deaths, and would degrade the quality of their lives.

But what if the gorillas would inevitably be killed and the entire population lost? In that case, I think their capture would be justified, both for the good of the individuals and for the preservation of their subspecies. If so, then their liberty rights are somewhat different from ours, because it would be wrong to capture and imprison comparably threatened human beings. We are morally obliged to offer shelter to refugees. But it is wrong to remove forcibly people from their homelands simply because others believe that they would be safer in another place. The human right to liberty includes the right to choose a dangerous freedom, or even a certain death, in preference to exile or imprisonment. Other apes' right to liberty must be more circumscribed, because they cannot currently understand their situation well enough to make sound decisions about their individual and species' futures.

THE RIGHT TO LIFE

Similar considerations suggest that if the other great apes are granted a right to life of equal strength with our own, it will be in their interest that this right be legally protected in somewhat different ways. Consider, for instance, the case of a captive elderly chimpanzee who is terminally ill, suffering great pain, and clearly no longer enjoying life. If she were a currently competent human being, then it would be both immoral and illegal to kill her without her informed and voluntary consent, even if others regarded her life as no longer worth living. If she were a formerly competent human being, then it would be wrong not to take account of the preferences that she may have expressed while still competent.

It is possible that some of the chimpanzees, bonobos, and gorillas who have been taught human-style languages would be able to understand their options in this unhappy situation and to express a clear preference for or against euthanasia. If so, then they, like human patients, should not be euthanized without their voluntary and informed consent. However, most captive great apes cannot make informed choices in this situation, if only because there is no way adequately to explain the situation to them. Incompetent humans may not legally be actively euthanized. Even in the Netherlands, which has the most permissive laws respecting active euthanasia, this action is legally tolerated only with the informed and voluntary consent of a mentally competent and terminally ill patient. Some medical ethicists have argued that laws ought to be altered to permit active euthanasia for some mentally incompetent human beings—for example, severely abnormal neonates (Kuhse and Singer 1987), but such change does not appear imminent.

In the meantime, the question is whether the strictures against active euthanasia that currently apply to incompetent or unconsenting human beings should be extended to the other great apes. I strongly doubt that this would be a net benefit to them. The extension of a life dominated by pain and suffering is a doubtful benefit to the individual whose life it is, unless it is consciously desired by that individual. None of the other hominids ought to be killed for other than humane reasons or reasons of legitimate self-defense. But hastening an inevitable death to prevent protracted suffering ought to be tolerated as a necessary act of compassion. A greater degree of paternalism is justified toward captive apes than the law currently permits toward human beings.

There is another reason for not immediately demanding a right to life for the other hominids that is identical to the human right to life. In most societies, killing nonhuman primates has not been regarded as homicide, or anything like it. Until a majority of people in a given society come to believe that killing a nonhuman ape, for other than humane reasons or self-defense, is as serious a moral crime as murdering a human being, the passage of laws mandating equally severe punishment is bound to be perceived as unjust. Laws can perform an invaluable educational function, but only if they do not depart so severely from majority opinion as to inspire general resentment and defiance.

The premature passage of laws treating the killing of nonhuman apes as murder might not only appear unjust to the unenlightened but would in some instances be an injustice. People should not be classified as murderers because they still believe what the law and the most prestigious moral authorities have long told them: The other great apes have no moral status, and may legitimately be hunted for food or profit. Legally classifying the wrongful killing of nonhuman apes as murder would be particularly problematic in jurisdictions where a murder conviction may carry the death penalty. The death penalty is morally problematic in any circumstances, but doubly so when inflicted for an activity that has long been regarded as legitimate. Even long imprisonment is a harsh penalty for this crime. Our increased knowledge of the other great apes makes it possible to argue strongly against the view that there is nothing very wrong with killing them and with a realistic hope of changing minds. The recent public outcry over the resumption of legal whale hunting by some Native American nations demonstrates that such changes in attitude are possible.

A QUALIFIED EQUALITY

To argue for these differences in the basic moral rights of humans and other great apes is not to assume that humans are inherently superior or more valuable. The greater degree of paternalism necessary at the present time to protect the interests of both captive and free apes may not be due to any cognitive or other deficiency in them. All that can be said confidently is that we cannot now communicate with most of them well enough to be assured that they understand their options regarding such vital matters as medical treatments, protective captivity, or euthanasia. This situation may be a result as much of limitations as of theirs. It matters little to the argument that I am making whether the obstacles to communication lie primarily in their cognitive limitations or in our failure to devise more adequate means of communication. Either way, the result is that, for the present, it would not be a benefit to most captive great apes to be treated as if they were competent adult human beings, entitled always to decide such matters for themselves.

I have argued that, whenever possible, free-living apes should be protected in their freedom. But they too occasionally stand to benefit from paternalistic interventions that would be intolerable if imposed on competent adult human beings—for example, vaccinating them against infectious diseases without their consent or coercively relocating them when survival in the original habitat is no longer possible.

These distinctions between the ways that the rights of human and other hominids are formulated and protected can be accommodated without surrendering the claim that the apes should have basic moral and legal rights of the same strength as ours. We may, for instance, adopt a qualified equal rights principle for the great apes.

THE GREAT APES' RIGHTS PRINCIPLE

Within the limits of their own abilities, bonobos, chimpanzees, gorillas, and orangutans have the same basic rights to life, liberty, and freedom from harm as human beings.

This principle is necessarily vague regarding the differences in the content of the other hominids' basic rights that can be justified by the limits of their abilities. These differences must be based on empirical evidence, and changes in our knowledge of the other great apes and our ability to communicate with them are likely to lead to changes in the ways in which their basic rights are formulated. A Great Ape Project rights principle is, however, useful as a reminder that the rights of other hominids are to be given equal weight, even when they are formulated and protected in ways that are not appropriate in the case of competent adult human beings. Human children and incompetent adults are protected by a similar implicit moral principle, which insists on rights of equal strength while permitting limitations when and as necessary for the individual's long-term good.

The adoption of such a principle would be an empty gesture if it were not eventually backed by appropriate national and international legislation. Moreover, the extension of moral and legal rights to the other great apes could pave the way for the inclusion of other large-brained mammals, such as elephants and cetaceans. Yet reality forces the admission that moral and legal rights for nonhuman beings may not gain general acceptance, at least not soon enough. We must ask, therefore, whether a less ambitious rhetorical strategy might be more successful. Why speak at all of the rights of the other great apes, when so many regard the ascription of rights to nonhuman animals as absurd?

MORAL OBLIGATIONS TO NONHUMAN BEINGS

Perhaps the most plausible argument against the ascription of moral rights to terrestrial nonhumans is that they are incapable of moral agency and that only moral agents can have moral rights. Moral agents are beings who can regulate their own behavior by moral concepts and principles and accept moral responsibility for their actions. Rights are properly accorded to all and only moral agents, because only moral agents can accord rights to others. In other words, I am obliged to respect your life and liberty because and only because I can reasonably hope and expect that you will do the same for me. This is the core of the contractualist argument for restricting moral rights to human beings, for instance in John Rawls's *A Theory of Justice* (1971).

This argument suffers from an obvious problem: It appears to exclude from full moral status all those sentient human beings who are not yet, or not any longer, moral agents, or those who will never achieve moral agency at all. Young children and mentally disabled persons are not responsible moral agents, but it does not follow that they are not entitled to a basically equal moral status.

Furthermore, some people who have worked closely with great apes believe that they are indeed capable of moral agency. The issue turns in part on what we are willing to accept as moral agency. Does it require the ability to formulate explicit moral principles, rank their relative importance, and apply them impartially to each new situation? If so, then it seems unlikely that the other great apes are moral agents. Or does it require only the ability to care about the beings to whom one is socially related and to act altruistically on their behalf? On that definition, animals of many species, including the great apes, are certainly capable of moral agency. Until we achieve a better understanding of our own moral nature, and a greater ability to converse with other hominids about such abstract matters as morality, it is probably wise to maintain a principled agnosticism regarding their capacity for moral agency.

Despite the weakness of the argument from moral agency, the term "animal rights" is a sticking point for many people, including many who favor strong anticruelty laws and strong protections for endangered species. Although I favor the ascription of basic moral rights to some animals, I do not believe that the effective protection of animals, plants, species, and ecosystems necessarily requires the use of this terminology. Unlike oppressed human beings, the other great apes will not be offended or outraged if we protect them effectively, while declining to accord them equal moral or legal rights.

It is less important to maintain that other animals have moral rights than to maintain that we have moral obligations toward them. Many of the arguments I have given for accepting moral obligations toward the other great apes can be rephrased as reasons for regarding them as having significant instrumental value to human beings. For example, we can argue, with Immanuel Kant (1963), that it is bad for us to be unkind to them, that it degrades our moral character and makes us more likely to be unkind to humans. Or we can follow the so-called shallow environmentalists who deny that nonhuman animals have either intrinsic value or moral rights but nevertheless argue that some species have sufficient aesthetic, scientific, and economic value to living and future human beings to make it worth expending the resources required to preserve them. But something important is lost in these translations. If we reject the idea of obligations toward animals of other species, valuing them only instrumentally, then it is all but inevitable that our human psychology will prevent us from protecting them effectively.

To value a being only instrumentally is to value it only contingently. If, for example, we value apes only for their tourist-enticing potential, then we will stop valuing them once the tourists stop coming. If we value them only for their scientific interest, then we will consider them expendable in the name of science, and disposable once the scientists lose interest. Only the conviction that we have moral obligations to them can persuade us to protect them despite transitory competing human interests, such as the need to find a vaccine or a cure for AIDS.

Some ethicists have argued that human beings are psychologically incapable of accepting obligations to nonhuman animals. But even a cursory survey of cross-cultural attitudes toward the nonhuman world reveals that this claim is false. People on all of the six populated continents who live by gathering and hunting are likely to recognize moral obligations to animals and to other parts of the natural world (Callicott 1994). I am convinced that these moral attitudes are partially responsible for the remarkable success of the aboriginal peoples of Australia in protecting the lands and the species from which they have earned their subsistence for perhaps 60,000 years. The existence throughout the world of ancient moral traditions that accord moral status to animals, plants, and other parts of the natural world proves that human psychology does not preclude the acceptance of such obligations. If we are psychologically capable of accepting moral obligations to other species and ecosystems, then we ought to do it, because purely instrumental valuing will not adequately motivate the preservation of the other great apes and other life forms.

REFERENCES

Anderson, I. 1996. Australia's rabbits face all-out viral attack. *New Scientist* 7: 27.

Bernstein, M. H. 1998. *On moral considerability: An essay on who matters morally.* New York: Oxford University Press.

Callicott, J. B. 1989. *In defense of the land ethic: Essays in environmental philosophy.* Albany: State University of New York Press.

———. 1994. *Earth's insights: A multicultural survey of ecological ethics from the Mediterranean Basin to the Australian outback.* Berkeley: University of California Press.

Cavalieri, P., and Singer, P. (eds.). 1993. *The Great Ape Project: Equality beyond humanity.* New York: St. Martin's Press.

Hutchins, M., Stevens, V., and Atkins, N. 1982. Introduced species and the issue of animal welfare. *International Journal for the Study of Animal Problems* 3: 4, 318–36.

Hutchins, M., and Wemmer, C. 1986–1987. Wildlife conservation and animal rights: Are they compatible? In M. W. Fox and L. D. Mickley (eds.), *Advances in animal welfare science* (pp. 111–37). Washington, D.C.: Humane Society of the United States.

Kant, I. 1963. *Lectures on ethics* (Louis Infield, trans.). New York: Harper and Row.

Kreger, M. D., Hutchins, M., and Fascione, N. 1998. Context, ethics, and environmental enrichment in zoos and aquariums. In D. Shepardson, J. Mellen, and M. Hutchins (eds.), *Second nature: Environmental enrichment for captive animals* (pp. 59–82). Washington, D.C: Smithsonian Institution Press.

Kuhse, H., and Singer, P. 1987. *Should the baby live? The problem of handicapped infants.* Oxford: Oxford University Press.

Leopold, A. 1987. *A Sand County almanac, and sketches here and there.* Oxford: Oxford University Press.

Midgley, M. 1983. *Animals and why they matter.* Athens: University of Georgia Press.

Nowak, R. 1999. Almost human. *New Scientist* (Feb. 13): 20–21.

Quammen, D. 1996. *The song of the dodo: Island biogeography in an age of extinction.* New York: Scribner.

Rachels, J. 1993. Why Darwinians should support equal treatment for other great apes. In P. Cavalieri and P. Singer (eds.), *The Great Ape Project: Equality beyond humanity* (pp. 152–57). New York: St. Martin's Press.

Rawls, J. 1971. *A theory of justice.* Cambridge, MA: Harvard University Press.

Regan, T. 1983. *The case for animal rights.* Berkeley: University of California Press.

Savage-Rumbaugh, S., and Lewin, R. 1994. *Kanzi: The ape at the brink of the human mind.* New York: John Wiley and Sons.

Singer, P. 1975. *Animal liberation: A new ethic for our treatment of animals.* New York: Avon Books.

———. 1979. *Practical ethics.* Cambridge: Cambridge University Press.

Warren, M. A. 1983. The rights of the non-human world. In R. Elliott and A. Gare (eds.), *Environmental philosophy: A collection of readings* (pp. 109–34). University Park: University of Pennsylvania Press.

———. 1986. Difficulties with the strong animal rights position. In G. L. Bowie, K. Higgins, and M. W. Michael (eds.), *Between the species* 2:4. *Reprinted in Animal Experimentation: The Moral Issues* (pp. 163–73). New York: Harcourt Brace Jovanovich.

———. 1997. *Moral status: Obligations to persons and other living things.* New York: Oxford University Press.

MICHAEL HUTCHINS, BRANDIE SMITH,
RANDY FULK, LORI PERKINS, GAY REINARTZ,
AND DAN WHARTON

18
RIGHTS OR WELFARE: A RESPONSE TO THE GREAT APE PROJECT

The Great Ape Project (Cavalieri and Singer 1993) was heralded as a major advancement in the evolving relationship between humans and their closest living relatives, the great apes. The major conclusions were that apes should be afforded the same moral and legal status as humans and that an "international body" ought to be created to look out for the animals' interests and act on their behalf. The editors' goal was to present a "single challenging picture"—that is, "Visions of the apes living free in their native forests" contrasting with "descriptions of the miserable lives that many great apes are forced to live under human tyranny, whether in zoos, laboratories, or other captive conditions" (Cavalieri and Singer 1993, 2).

This may appear to be a commendable goal. The history of great apes in captivity has not always been admirable (Nichols and Goodall 1999). Furthermore, adult great apes have relatively large brains, may have a concept of self, are capable of using symbolic communication, make and use tools, experience a wide range of emotions, and can suffer both physical and psychological pain (Goodall 1986; Maple 1980; Maple and Hoff 1982; chapters 10, 11, 13, this volume). As such, they are certainly worthy of our moral consideration. However, in our opinion, the Great Ape Project has at least three major weaknesses. First, it fails to take into account the current global context in which great ape conservation must occur (Brooks and Leigh Smith 1998; Hutchins 1994; Wharton 1995). Second, it fails to distinguish between high quality and substandard captive facilities (Maple, McManamon, and Stevens 1995). Third, it fails to recognize the growing contributions of professionally managed zoological parks to great ape conservation (Koontz 1997; Wiese and Hutchins 1997). As a result, we believe that framing our ethical obligations to these creatures within the narrow and anthropomorphic perspective of rights may ultimately prove to be ineffective and even counterproductive to both the conservationist's and animal advocate's goals (see E. Wilson 1984, 134).

At the beginning of the twenty-first century, the situation for wild great apes has never been so precarious. Populations of all species are declining rapidly in nature as a result of a combination of many factors, including extensive deforestation (habitat loss) and hunting for the illegal commercial wildlife trade (Lee, Thornback, and Bennett 1988; Mittermeier 1986; Robinson and Bennett 2000; chapters 1 to 4, 12, this volume). For species such as great apes, all of which exist in developing countries, these problems are often exacerbated by sociopolitical and economic instability (Brooks and Leigh Smith 1998). Moreover, the human populations of these regions are projected to double within the next few decades (Brooks and Leigh Smith 1998; Ehrlich and Ehrlich 1990), continually forcing wildlife into smaller, fragmented habitats. The long-term effects of isolation and consequent loss of genetic diversity on great ape populations are yet to be determined. Thus any debate concerning our treatment of and moral obligations to great apes must take into account the current biological, sociopolitical, and economic context of wildlife conservation.

Animal rights advocates have responded to this emerging crisis with an approach that is perhaps well-meaning but unfortunately too rigid and often based on alternatives that are extreme or unrealistic (Hutchins and Wemmer 1991; Hutchins, Wiese, and Willis 1995; Lindburg 2000). They argue that all sentient animals have an inalienable "right" to life and liberty, and they are opposed to the very idea of zoos and wildlife management (Cavallieri and Singer 1993; Regan 1995). Zoo conservationists are also concerned about animal welfare (Mallinson 1995; Maple et al. 1995; Shepherdson, Mellen, and Hutchins 1998). However, unlike animal rights advocates, their *primary* concern is with securing a future for endangered species and their habitats in a world increasingly dominated by humans (Hutchins and Wemmer, 1986; Lindburg 2000).

There are profound differences between the philosophy of animal rights and animal welfare (Regan and Francione 1992) and also between the goals of animal rights advocates and conservationists (Hutchins and Wemmer 1986, 1991; Lindburg 2000; Norton 1987). Whereas animal rights advocates recognize the "moral inviolability of the individual," animal welfare advocates are committed to the pursuit of "gentle usage." One implication of the latter view is that it is morally acceptable to use animals for various justifiable reasons but that suffering and loss of life should be minimized. The focus of the conservationist's concern is the survival of populations, species and ecosystems, rather than any individual (Hutchins and Wemmer 1986; Lindburg 2000; Norton, 1987). There is certainly a spectrum of beliefs within the animal rights movement. However, one implication of the extreme rights view, as expounded by Regan (1983), is that when a conflict of interest exists, humans should allow a rare species to go extinct rather than violate the rights of even one individual of a common species. As a consequence, Hutchins, Dresser, and Wemmer (1995, 264) concluded that "the conservation ethic is generally compatible with the animal welfare ethic but largely incompatible with the animal rights ethic"; a similar view is expressed by Norton (1987).

Animal rights advocates, such as those who support the Great Ape Project, are opposed to zoos, regardless of their quality or of their contributions to wildlife conservation. Modern zoological parks have made the physical and psychological well-being of great apes a high priority. They also contribute to great ape conservation in a variety of ways, including public education, scientific research, nature travel programs, political action, and the direct support of national parks and reserves (discussed later in the chapter). Thus for those who believe in the mission of modern zoos, the critical question is not *whether* great apes should be maintained in captivity but where and how they will be maintained and for what purpose.

In this chapter we (1) review the current status of great apes in North American zoos, (2) explain how modern zoos manage great ape populations, (3) discuss the evolving role of zoos in great ape conservation, (4) summarize our current knowledge regarding great ape care and husbandry, (5) speculate about the future of great ape programs in zoos, and (6) respond to the Great Ape Project. We challenge the idea that great apes must live "miserable lives" in modern zoos and show how far the captive management of great apes has advanced, particularly in American Zoo and Aquarium Association (AZA)-accredited institutions.

STATUS OF GREAT APES IN ZOOS

In this section, we assess the number of great apes currently held in accredited North American zoos based on regional studbook databases (Fulk 1998; Perkins 1998; Reinartz 1998; Wharton 1997) and the number of great apes managed by other regional cooperative breeding programs worldwide based on International Species Information System records and international studbooks (Table 18-1). Population projections were calculated using DEMOG 4.2 (Bingaman and Ballou 1986), a demographic modeling program.

Bonobo

All bonobos, *Pan paniscus,* in North America are managed under the AZA Bonobo Species Survival Plan (SSP)™. There are currently 45 bonobos in six AZA-accredited institutions. There are 12 more at the Morelia Zoo in Mexico and at Georgia State University, which are approved nonmember participants in the SSP.

Table 18-1
Population of Apes in Regional Cooperative Breeding Programs Worldwide and in AZA Institutions

	Global Population	AZA Population (No. of Institutions)
Bonobo	141	45 (6)
Chimpanzee	1,752	253 (39)
Gorilla	660	355 (49)
Orangutan (Bornean)	385	74 (22)
Orangutan (Sumatran)	306	91 (33)
Orangutan (Hybrid/unknown)	189	46 (23)

If the population continues to grow as in the past, this number may increase to more than 80 animals within the next 20 years (Figure 18-1). Current AZA facilities are estimated to have space for as many as 120 animals. The global bonobo captive population is the smallest of all great ape species, with fewer than 150 animals (Leus and van Puijenbroeck 1997).

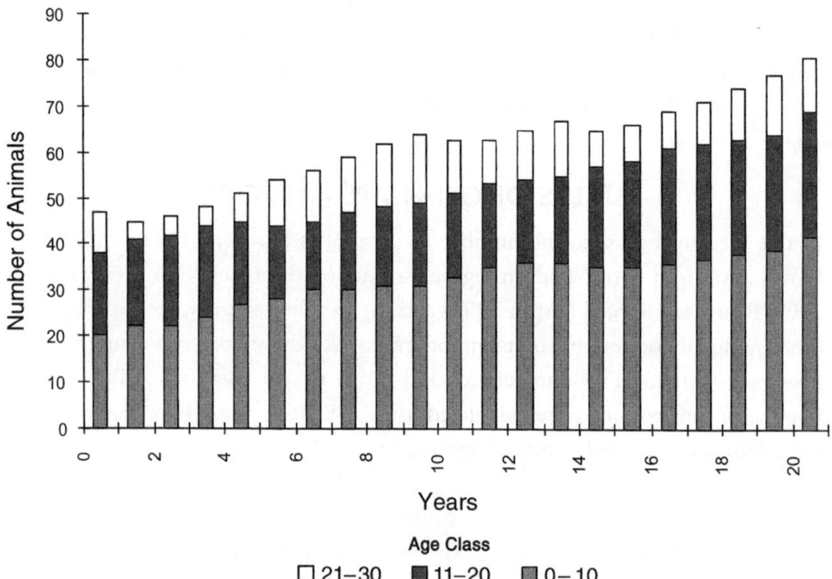

Figure 18-1. Bonobo 20-Year Population Projections

Chimpanzee

There are more than 1,700 chimpanzees, *Pan troglodytes,* in zoos throughout the world (ISIS 1998). The AZA Chimpanzee SSP currently manages 253 of these in 39 institutions. This is very close to the future estimated maximum SSP "carrying capacity" (i.e., the number of animals that can be held in current exhibition space) of 267 (AZA Great Ape TAG 1998, based on responses of 111 institutions), and reproduction is being carefully controlled. The population is not expected to expand much beyond this number in the next 20 years.

Gorilla

The current world zoo population of western lowland gorillas, *Gorilla gorilla gorilla,* is 660 (ISIS 1998). There are no mountain gorillas, *G. berengi berengi,* in captivity and only one eastern lowland gorilla, *G. berengi graueri.* The AZA Gorilla SSP manages 355 individuals in 49 institutions, and the population is growing at a rate in excess of 2 percent per year. If this growth rate continues, in 20 years there could be almost 500 individuals (see Figure 18-2). Although space does not currently seem to be a limitation for this species because many zoos want gorillas and are building exhibits to display them, future breeding may need to be limited.

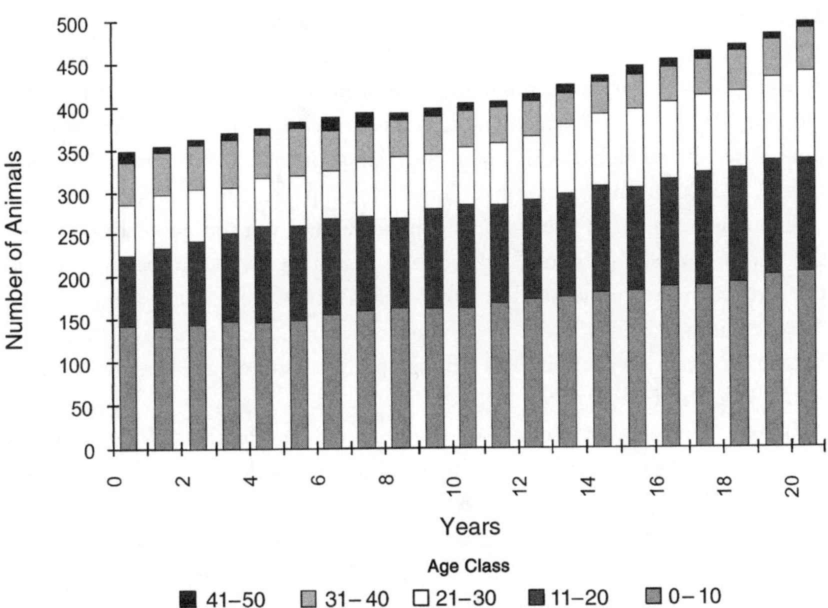

Figure 18-2. Gorilla 20-Year Population Projections

Orangutan

The world zoo population of Sumatran, *Pongo abelii*, and Bornean, *Pongo pygmaeus*, orangutans stands at 306 and 385, respectively (Perkins, personal communication). In addition, there are 189 subspecific hybrids and animals of unknown speices. The AZA Orangutan SSP encompasses both subspecies and the hybrid population. There are currently 234 individuals managed by the SSP program: 91 Sumatran in 31 institutions, 74 Bornean in 22 institutions, and 46 subspecific hybrids in 23 institutions.

Population projections based on recent performance of the SSP-managed population have not been made because the current demographic goal of the program is to reverse the historically negative population growth rates in both subspecies (Perkins 1997). However, space limitations will prevent any substantial growth in the population. There has been a moratorium on reproducing subspecific hybrids since 1985 (Perkins and Maple 1990). Many of these animals are still relatively young and will require space in holding institutions for many more years; more space will become available as the number of hybrids eventually drops through attrition. It is important to note that hybrid orangutans are managed with the same level of care as their "pure-bred" conspecifics (Perkins and Maple 1990).

POPULATION MANAGEMENT AND COLLECTION PLANNING

Accredited zoos have established specific procedures for managing populations of selected species, including great apes. Established in 1981, SSPs are cooperative breeding and conservation programs organized through the AZA (Wiese and Hutchins 1994). SSP management committees, representing all holding institutions, determine which animals are to breed based on genetic relatedness and other factors, such as behavioral compatibility and group structure. Animals are paired to maintain as much of the original genetic variation in the population as possible. Some animals, particularly those that are genetically overrepresented in the population, are recommended not to breed. Participating zoos cooperate with these recommendations. This type of cooperative management is essential for sustaining the genetic variability and demographic stability of small captive populations (Ballou and Foose 1996).

Genetic management also promotes animal welfare (Hutchins, Wiese, & Willis 1996). In unmanaged populations, inbreeding can produce individuals that are more susceptible to disease and have a higher incidence of birth defects (Ballou and Foose 1996; Smith 1986). Populations must also be managed to help maintain the population within its available space. This is critical because animal holding and exhibition space is limited (Soule et al. 1986). When captive populations are allowed to overshoot their available space, it is more likely that some animals could find their way into substandard facilities.

In the past, zoo collections often were assembled based on both availability and personal preference of curators and directors. This made it difficult to maintain captive animals in populations large enough to be sustainable over several generations, without allowing them to grow beyond their available space. In 1991, specialized committees called Taxon Advisory Groups (TAGs) were developed by the AZA to address this challenge (Hutchins and Wiese 1991). A TAG's primary function is to develop a regional strategic collection plan for a group of related taxa with similar biological and ethical considerations.

Strategic collection plans outline a set of objectives for the composition of animal collections in zoos, including the type and number of species and individuals to maintain (Hutchins, Willis, and Wiese 1995). Regional TAGs evaluate the present conditions surrounding their particular taxonomic group and prioritize species for zoo-based captive management and conservation programs (Wiese and Hutchins 1994). The *AZA Great Ape TAG's Strategic Collection Plan* (AZA Great Ape TAG 1995) sets forth general goals for the species in captivity and outlines ways in which they can be achieved. These plans are designed to maximize zoos' positive impact on wildlife conservation, while helping to maintain self-sustaining populations within available space (Hutchins, Wiese, et al. 1996; Hutchins, Willis, et al. 1995).

The population management goal of the AZA Great Ape TAG (and its associated SSPs) is to maintain healthy, self-sustaining great apes in captivity for a minimum of 100 years. Each SSP formulates an annual breeding master plan to ensure that population goals are met. Using GENES Version 11.82 (Lacy 1998), a software package for genetic analysis of studbook data, it was determined that SSP populations of chimpanzees, western lowland gorillas, and orangutans are all large enough and have enough founder representation to maintain more than 90 percent gene diversity for at least 100 years. Inbreeding levels in these populations are also very low. In chimpanzees, almost 99 percent gene diversity is still retained in the population today, primarily because most of the population is wild-caught and the program is still relatively young (Fulk 1997). The primary problem with the population is that it is at risk of becoming too large for the number of spaces available. To prevent this, a breeding moratorium was imposed in 1995; however, it was lifted in 1998 because new offspring were needed to ensure demographic stability. Only eight offspring per year are recommended for the entire population, and institutions that allow offspring also must commit to keeping them for a minimum of eight years.

The AZA Gorilla SSP population has retained in excess of 98 percent gene diversity, and although more than 100 founders have contributed to the population, there are still more potential founders available (Thompson, personal communication). Both SSP subpopulations of orangutans are genetically healthy, with gene diversity for both populations currently in excess of 97 percent. To retain as much of this genetic variation as possible, and because space is limited, only the most genetically valuable animals are paired, and efforts are being made to have all founders breed before they reach reproductive senescence (Perkins 1997).

The bonobo population in AZA facilities is not large enough to maintain 90 percent gene diversity for 100 years without intensive cooperative management. Thus a global master plan is being developed with the European equivalent of the SSP, the European Endangered Species Programme (EEP), to encourage cooperative interregional management. The combined subpopulations would maximize the genetic potential of the larger population while decreasing the number of animals needed in each region (Reinartz 1994, also personal communication).

EVOLVING ROLE OF ZOOS IN GREAT APE CONSERVATION

It is clear to us that proponents of the Great Ape Project have failed to understand the current situation for great apes in the wild; they also do not appear to appreciate the evolving role of accredited zoos in great ape conservation. The situation for wild great apes is likely to get worse before it gets better. All species are considered vulnerable, threatened, or endangered, and despite efforts to protect the animals and their habitats, their populations continue to decline because of widespread habitat destruction and hunting (Brooks and Leigh Smith 1998; Lee et al. 1988; Mittermeier 1986; Robinson and Bennett 2000; chapters 1 to 4, this volume). Recent fires have threatened orangutan populations in Indonesia, whereas ethnic conflicts, the growth of commercial logging, and the illegal commercial bushmeat trade have threatened chimpanzees, gorillas, and bonobos in Western and Central Africa (Eves and Ruggiero 2000; Pearce 1995, chapters 1–4 and 10, this volume). These and other factors have combined to take a tremendous toll on great ape populations, as well as causing untold suffering in many individual animals.

One role that zoos and captive ape populations play is to educate the public about great apes and the many threats to their existence (chapter 5, this volume). The AZA's 185 accredited member institutions have more than 130 million visitors annually, and for many, this will be their only chance to observe and learn about great apes. In the new Congo exhibit at New York's Bronx Zoo/Wildlife Conservation Park, visitors are given the opportunity to contribute funds to support fieldwork that could directly benefit great apes and other endangered species in the wild (Conway 1999a).

A role modern zoos could play in great ape conservation in the future is to serve as the source for any reintroduction effort that might be needed—that is, as a final insurance policy against extinction (Foose 1986). Although reintroduction is currently not a goal for any great ape SSP, ape populations in zoos are carefully managed to preserve the option (Wharton 1992).

Other contributions to conservation of today's professionally managed zoos include public education, scientific research, the development of relevant technologies, professional training and technology transfer, nature travel, political action, the direct support of national parks and reserves, and fundraising to support these activities (AZA Great Ape TAG 1995; Hutchins 1999; Hutchins and Conway 1995).

All AZA conservation and science committees have been asked to formulate a plan of action for relevant cooperative research and conservation projects to be supported by the zoo community (Hutchins and Conway 1995), and this has stimulated work in a variety of disciplines and geographic areas. For example, the Gorilla SSP and Zoo Atlanta have supported the work of Tom Butynski, who with his colleagues from the American Museum of Natural History has studied the taxonomy of mountain gorillas and the effects of tourism on free-ranging gorilla populations (Butynski and Kalina 1998; Sarmiento, Butynski, and Kalina 1996). The Bonobo SSP, led by the Zoological Society of Milwaukee, is initiating a field census of bonobos in the Democratic Republic of Congo (Reinartz and Boese 1997). The Orangutan SSP has developed a program whereby all ongoing in situ conservation and research projects on orangutans are surveyed annually. The updated listing is distributed to all 53 SSP institutions, and each institution is strongly encouraged to provide financial, material, or other direct support to the project of their choice. Preliminary analysis suggests that at least 35 percent of SSP participants use this listing to make informed, direct contributions to in situ conservation projects (Perkins, personal communication). For example, the Denver and Pittsburgh zoos are currently supporting fieldwork on orangutans in Sabah, Malaysia, by Drs. Isabelle Lackman-Ancrenaz and Marc Ancrenaz and the Sabah Wildlife Department (Read and Tilson 1998). Many independent initiatives are also underway. For example, Zoo Atlanta has developed a cooperative relationship with The Dian Fossey Gorilla Fund International, a major supporter of gorilla field conservation initiatives. The fund's offices are now based at Zoo Atlanta (DFGFI 1996). The Columbus Zoo has initiated a program called Partners in Conservation, which provides extensive support for mountain gorilla conservation and for educational outreach efforts (Jendry 1997).

The distinction between small protected areas (some African national parks are completely fenced) and larger, naturalistic zoos is rapidly disappearing (Adams and McShane 1992; Willcock 1964). As a consequence, the technologies developed to manage animals in zoos are becoming increasingly relevant to the intensive management required for animals in nature (Conway 1989; Hutchins 1994; Hutchins and Fascione 1993). For example, veterinarians with zoo training have treated mountain gorillas for injuries and vaccinated them against human diseases, such as measles (Hutchins, Foose, and Seal 1991). New organizational mechanisms will be necessary to facilitate cooperation between zoos and relevant government agencies and nongovernmental organizations in range countries. To help address this, the AZA developed the Conservation Action Partnership (CAP) concept in 1991 (Hutchins and Wiese 1991; Wiese and Hutchins 1994). CAPs are specialized committees designed to facilitate and coordinate in situ conservation activities by North American zoos. The AZA has developed East African and Southeast Asian CAPs, both of which include great ape conservation as one of their priorities (Read and Tilson 1998). The CAPs work closely with the Great Ape TAG and relevant SSPs to develop and maintain the link between AZA member institutions and the conservation of great apes and their habitats in nature. These committees are also closely linked to IUCN—the World Conservation Union's Species Survival Commission's Primate Specialist Group.

Zoos and aquariums and their regional associations are also experts at facilitating coalitions in support of conservation (Hutchins 1999). One recent effort that is expected to have a significant impact on great ape conservation in tropical Africa is the creation of the Bushmeat Crisis Task Force (BCTF). The BCTF is a broad coalition of major U.S. conservation organizations, zoological parks, museums, and animal protection groups that has the goal of ending the illegal commercial bushmeat trade in tropical Africa (Eves and Olson 2000). The bushmeat trade is considered among the most significant threats to wildlife, including great apes, in Africa today (Robinson and Bennett 2000; chapters 1, 3, and 4, this volume). The initial meeting that led to the formation of the BCTF was organized by the AZA, more than half of the BCTF's supporting organizations are AZA members, the BCTF chair is currently AZA's director of Conservation and Science, and the BCTF office and staff are housed at the AZA executive office. Other partners include Conservation International; the Wildlife Conservation Society; World Wildlife Fund–United States; The Jane Goodall Institute; and Dian Fossey Gorilla Fund International.

CARE AND HUSBANDRY OF GREAT APES IN ZOOS

The Great Ape Project also failed to make a distinction between quality and substandard captive facilities. In our opinion, this is based on the mistaken premise that it is impossible to meet the needs of great apes in any captive situation. In the following section, we summarize the many advances that have been made in great ape care and husbandry in AZA zoos in recent years.

Exhibition–Enclosure Design

Enclosure design for great apes has changed considerably in the past two decades. Early zoos sought to exhibit a wide variety of species, and the space available to each was comparatively small. In an effort to prevent disease, animals were kept in easily cleaned enclosures (Hancocks 1971). Great apes were no exception. Until the early 1980s, zoo apes were typically kept outdoors in concrete grottos, barred or wire-mesh cages, or in glass-fronted indoor exhibits with sterile tile walls and concrete floors (Hancocks 1971). In such cramped and inappropriate environments, the animals often exhibited atypical behaviors, such as lethargy, regurgitation and reingestion of food, and coprophagy (Akers and Schildkraut 1985; Erwin and Deni 1979; Gould and Bres 1986).

In recent years, the trend has been toward larger "landscape immersion" exhibits, designed to put both the animals and visiting public in a simulated natural environment. Designers use live vegetation, dead trees, rocks, running water, and a soil or grass substrate to create habitats that are interesting, functional, aesthetically pleasing, and educational (Coe 1985, 1996; Gold 1997; Lee and Coe 1988; Nash et al. 1997). Attempts have even been made to provide captive great apes with "ecologically relevant" sounds (Ogden, Lindburg, and Maple 1994). Great apes accustomed to living in more traditional zoo cages have taken some time to adjust to these more naturalistic settings (Ogden, Finlay, and Maple 1990). However, combined with other forms of environmental enrichment (see later discussion), outdoor, naturalistic zoo habitats have been shown to increase the range of natural behaviors exhibited by captive great apes and decrease the incidence of abnormal behaviors (Clark, Juno, and Maple 1982; Coe and Maple 1984; Goerke, Forthman, and Maple 1987; Hoff et al. 1994; Hutchins, Hancocks, and Crockett 1984; Ogden 1992; Pfeiffer and Koebner 1978; Poole 1991; Redshaw and Mallinson 1991).

Other aspects of the physical environment can be enhanced to provide opportunities for the animals to display natural behavior. For example, highly arboreal species, such as orangutans, can be encouraged to exhibit species-specific locomotor behavior by providing them with climbing structures, ropes, cables, or artificial lianas (Gippoliti 2000; Maple 1979; Maple and Perkins 1996; Perkins 1992). One of the most innovative climbing structures built to date is at the National Zoo's "Think Tank," where orangutans can rest on towers and climb across cable walkways suspended two stories above the ground (Hawes 1995). Although not naturalistic in appearance, this exhibit replicates many critical aspects of orangutan habitat and is thereby "functionally naturalistic."

Naturalistic exhibits also appear to have many positive effects on zoo visitors. Studies by Bitgood, Patterson, and Benefield (1988), Finlay, James, and Maple (1988), Shettel-Neuber (1988), and Bronwyn and Ford (1991) showed that zoo visitors preferred naturalistic zoo exhibits and spent considerably more time viewing these types of displays than traditional ones. These exhibits offer many learning opportunities, including interpretive graphics, interactive displays, videotaped and computer presentations, brochures, and on-site educators who inform visitors about the basic biology and conservation of great apes. At the National Zoo's "Think Tank," zoo visitors can observe ongoing studies of ape intelligence (Hawes 1995). In contrast, older, more traditional exhibits typically offered little more than a sign displaying the species' geographic range and diet.

In northern, temperate climates, it is often necessary for zoo apes to spend the winter indoors, where the temperature and humidity can be controlled. When seasonal indoor day rooms are necessary, efforts are now being made to provide functional and comfortable environments (Nash et al. 1997). Although it is usually impossible to replicate a large, naturalistic habitat under such conditions (e.g., plants are unlikely to thrive), it is still possible to meet the animals' basic physical and psychological needs. For example, climbing structures are fashioned from treated logs or ropes. Arboreal platforms and natural materials are provided for nesting, and visual barriers allow privacy.

Off-exhibit holding areas for great apes have also evolved over time. Although not as aesthetically pleasing as display areas, they are functional for the animals and much more complex and comfortable places. In the past many of these places were dingy; such areas typically now contain windows to allow natural light, artificial climbing structures, and natural bedding (Nash et al. 1997). Some environmental enrichment devices, inappropriate for naturalistic public display (e.g., toys, grooming boards), can also be provided in these off-exhibit spaces (Kreger, Hutchins, and Fascione 1998).

Environmental Enrichment

Shepherdson (1998, 1) defined environmental enrichment as "an animal husbandry principle that seeks to enhance the quality of captive animal care by identifying and providing the environmental stimuli necessary for optimal psychological and physiological well-being." In practice, enrichment covers a wide variety of devices and practices aimed at "keeping captive animals occupied, increasing the range and diversity of behavioral opportunities, and providing more stimulating and responsive environments" (Shepherdson 1998, 1).

Although much remains to be done in the planning and implementation of environmental enrichment programs in zoos (Mellen, Shepherdson, and Hutchins 1998), the professional zoo community has embraced the concept enthusiastically and actively (Tudge 1991). Great apes and other nonhuman primates are intelligent, social animals that require stimulating and responsive environments to maintain their psychological and physical health and to promote activity and reproduction (Poole 1992, 1998). As a consequence, many of the most innovative and imaginative enrichment techniques have been developed for these species.

Because wild apes spend much of their active time foraging, many enrichment techniques have focused on behaviors associated with food acquisition and feeding behavior (Maple and Perkins 1996). For example, wild chimpanzees use tools to extract termites from termite mounds (Goodall 1986), and several zoos have designed and constructed artificial termite mounds in chimp exhibits that effectively elicit tool use and feeding activity (Gilloux, Gurnell, and Shepherdson 1992; Nash 1982).

The simple provision of fresh browse has led to reductions in atypical behaviors, such as coprophagy and regurgitation and reingestion of food, and reduced aggression (Akers and Schildkraut 1985; Bloomsmith, Alford, and Maple 1988; Maple 1980; Ruempler 1990). In one study, the provision of browse increased time spent feeding in captive gorillas from 11 to 27 percent of the day (Gould and Bres 1986). For curious animals like great apes, whole trees can be uprooted and provided as a form of "megabrowse" (Maple and Perkins 1996). Maki and Bloomsmith (1986) found that uprooted oak trees were used extensively by captive chimpanzees for up to five months. More frequent feedings and scattering and hiding small food items throughout the enclosure also helps to encourage natural foraging activity and relieve boredom (Baker 1997; Chamove et al. 1982; Maple and Perkins 1996). The addition of straw and forage decreased the incidence of regurgitation and reingestion of food in indoor-housed chimpanzees by more than two thirds; levels of aggressive behavior were also reduced and locomotion and play increased (Baker 1997).

The availability of manipulable objects also appears to be important for captive great apes. S. Wilson (1982) studied the relationship between environmental variables and activity levels in captive gorillas and orangutans. The two factors most positively correlated with activity were the presence of moveable and stationary or temporary objects and the number of conspecifics sharing the enclosure. The presence of objects was more effective in eliciting activity than was the size of the enclosure, and movable objects were more effective than immovable objects for orangutans. In contrast, gorillas preferred stationary objects. Perkins (1992), who studied the activity patterns of 29 orangutans housed in nine American zoos, obtained similar results. Tripp (1985) investigated the effects of manipulable and edible objects on orangutan activity at the Topeka Zoo. The highest levels of manipulation and locomotion occurred when both were present. Apes habituate to novel objects over time, so that changing objects regularly also seems to be important (Paquette and Prescott 1988).

Because all ape species are social, the most critical aspect of their environmental enrichment is the presence of companions. Little was known about the social life of free-ranging great apes until the 1960s, when scientists began to study them in nature. Most zoos previously maintained gorillas inappropriately as pairs or singly. However, gorillas and all other great ape species are now routinely managed in larger groups, and this has had a positive effect on reproduction and on the quality of the animals' lives (see later discussion).

Veterinary Care

Captive great apes are at risk from a wide variety of health problems, including tuberculosis, gastrointestinal parasites, and bacterial infections (Munson and Montali 1990). Being so closely related to humans, they are also susceptible to many of our common childhood and adult diseases, such as colds, flu, and measles. Zoo veterinarians and pathologists have systematically studied the diseases of great apes over the past two decades (e.g., Benirschke and Adams 1980; Cousins 1983; Munson and Montali 1990; Snyder and Amand 1980); the data have yielded a wealth of information, resulting in a steady increase in longevity and reproductive success.

Around the turn of the century, zoo managers considered themselves lucky if gorillas survived more than one or two years, and the prospects of successful captive management seemed impossible (Hornaday 1922); now captive gorillas routinely live well into their 40s. As a consequence, many of the health challenges facing great apes in modern zoos are those associated with old age, such as rheumatoid arthritis, cardiovascular disease, and cancer (Janssen and Bush 1990). There is still much to be learned, however, and specific health and reproductive concerns may differ between species. For example, orangutans have a tendency to become obese and develop associated health problems (e.g., cardiac fibrosis), whereas gorillas have recurrent problems with male infertility (Munson and Montali 1990). Juvenile mortality rates continue to decline, and even the upper levels of captive mortality compare favorably with the lowest levels recorded in nature (e.g., chimpanzees: Courtnay 1988; Goodall 1983; gorillas: Schaller 1963; Weber and Vedder 1983; Wharton 1992).

The current focus of zoo veterinarians is on preventive medicine (Hinshaw, Amand, and Tinkerman 1996; L. Phillips 1997). Many modern medical techniques now used to diagnose and treat human illnesses are also used in the diagnosis and treatment of apes in zoos. These include computerized tomography, ultrasound, magnetic resonance imaging, viral screening, chemical anesthesia, vaccination, and antibiotics (Janssen and Bush 1990). In specialized cases, such as dental surgery, zoos call on recognized experts in human medicine to conduct or assist in the procedures (Robinson, Fagan, and Roffinella 1979). In fact, the Bonobo SSP strongly recommends that all holding institutions identify specialists in various aspects of human medicine to assist in any invasive veterinary procedures (Mills et al. 1997).

Feeding and Nutrition

Lack of adequate nutrition has often been a problem in the captive management of exotic species. Animal diets were once based on guesses of what was eaten in the wild and on what was available in the local grocery store. These diets were acceptable because they were sufficient for the animals to survive (Oftedal and Allen 1996a), but they often did not meet all nutritional requirements. Dietary deficiencies caused an increased susceptibility to disease, increased infant mortality, and decreased fertility and growth rates (Oftedal and Allen 1996a). For example Fulk and Garland (1992) cited a failure to meet the nutritional needs of chimpanzees as a primary factor in the high mortality in the early years of zoo management. Captive gorillas have been known to develop diseases similar to those found in industrialized human societies, because both species have been displaced from their natural diet and lifestyle (Popovich and Dierenfeld 1997). Health problems such as ulcerative colitis and cardiovascular disease often occur when foods containing appropriate amounts of dietary fiber, folic acid, and antioxidants, such as Vitamins C and E, are not fed (MacNamara et al. 1987; Popovich and Dierenfeld 1997).

As nutritional data were accumulated from field studies, and new problems relating to nutrient deficiencies were discovered, zoos began to improve their diets, formulating them more on the basis of good science rather than on anecdotal information (Ullrey 1996). As knowledge increased and diets became more intricate, some zoos hired or consulted with professional nutritionists to develop artificial diets that meet basic nutritional and behavioral needs, while taking cost and availability of food items into account (Ullrey 1996).

It is still often difficult to define a normal nutrient level for all species of great apes, whose diets in the wild are highly varied and diverse and whose nutritional needs change over time and vary according to sex. Juveniles need more protein than adults, and because of the demands of pregnancy, females often have nutritional needs that differ from those of males (Oftedal and Allen 1996b). Many data have accumulated on the nutrient composition of the diets of free-ranging great apes (e.g., gorillas: Calvert 1985). Although it is virtually impossible to replicate completely their natural diets in zoos, it is possible to provide a diet that will meet the animals' basic nutritional needs. Because it is difficult to provide adequate nutrients while staying within caloric limits, supplements are provided to make up for any deficiencies (Oftedal and Allen 1996b). A varied diet is important as a source of environmental enrichment.

Reproductive Biology and Contraception

A deeper understanding of great ape reproductive physiology has resulted from captive studies, and this has led to significant management improvements. For example, studies of the hormonal cycles of females have led to better methods of pregnancy detection and allowed veterinarians to detect any potential abnormalities that might lead to reproductive failure or be symptomatic of serious health problems (e.g., orangutans: Masters and Markham 1991).

The development of artificial reproductive techniques, such as cryopreservation of semen, artificial insemination, and in vitro fertilization and embryo transfer has also met with some success. For example, in 1981, the first gorilla was born as a result of artificial insemination (Douglas 1981), whereas the first "test tube" (in vitro fertilization and embryo transfer) gorilla was produced in 1995 (Pope et al. 1997). Such technologies should not be viewed as a panacea for the endangered species problem (Hutchins, Wiese, et al. 1996; Thompson 1993). However, they can help to overcome reproductive failure in individual animals, reduce the need to transport animals between zoos, and serve conservation by ensuring that the genetic diversity of managed taxa will be maintained into the future (Wildt 1989). Advanced genetic technologies, such as DNA fingerprinting, have been used to determine paternity, which is useful for facilitating genetic management in multimale groups (Ely and Ferrell 1990).

As many of the problems of breeding and maintaining great apes in captivity have been gradually solved, zoos now face the problem of controlling reproduction, so that captive populations do not outgrow their available holding space and individual animals do not become genetically overrepresented. To help control unwanted pregnancies in great apes, zoo veterinarians use many of the same birth control techniques now available for humans, such as Norplant implants (Bettinger et al. 1997) and oral contraceptives (Bettinger 1994).

Unfortunately, some forms of birth control have potentially detrimental side effects, which may vary by species. For example, oral contraceptives are known to cause an artificial cycle of genital swelling in female chimpanzees, which in turn may make them less attractive to males or have a negative affect on female proceptivity (Nadler et al. 1993). Oral contraceptives might not be the best alternative for this species. Similar studies of Norplant implants suggest that they provide effective contraception, although not inhibiting ovarian hormone secretion or altering the normal cycle of genital swelling (Bettinger et al. 1997). In contrast, female gorillas do not exhibit pronounced genital swelling, and oral contraceptives have proven effective in suppressing their ovarian cycles (Goodrowe, Wildt, and Monfort 1992).

The Bonobo SSP population is still relatively small, making intensive genetic and demographic management particularly important. To balance genetic considerations with social needs, males must be housed in social groups without reproducing. The Bonobo SSP has also experimented with reversible vas deferens ligation (tying the tubes that transport sperm to the urethra) to prevent unwanted conceptions (Reinartz, personal communication). The AZA Contraceptive Advisory Group monitors advances in animal and human contraception, makes annual recommendations to members, and lists on the AZA Website (*www.aza.org*) techniques that are most appropriate for various taxa, including great apes (Asa et al. 1996).

Behavioral Considerations

The management of captive wild animals includes providing them with an appropriate and stimulating social environment (Erwin 1986; van Hooff 1986). Fortunately, the basic social structure of all wild ape species is now better understood (Harcourt 1987; Kortlandt 1960; Tilson 1986; Watts 1990), and all SSP programs now carefully balance genetic needs with the animals' requirements for a stable and appropriate social environment.

The basic rule is to look to nature for clues (Hutchins et al. 1984), but behavior is flexible and exceptions are possible. For example, adult orangutans are relatively asocial in the wild, although they sometimes aggregate at rich sources of food. However, they can be considerably more sociable in captivity (van Hooff 1986), although Markham (1990) has suggested that "when given the choice," orangutans are "only slightly more sociable in captivity." Edwards and Snowdon (1980) found that adults at the Henry Vilas Zoo lived together harmoniously. The animals exhibited affiliative behaviors, such as social grooming, food sharing, and paternal care, which have never been reported in the wild. Poole (1987) obtained similar results with a multimale, multifemale group of 12 orangutans on an island at the Singapore Zoo. Older males had fewer social interactions than younger males and tended to spend more time alone. However, only one incident of serious aggression was observed, when a new male was attacked savagely by a resident male.

Great ape reproduction has improved considerably in zoos over the past two decades, primarily as a result of a greater understanding of behavioral development. For example, in the 1970s, gorilla births were still relatively rare, probably because of social deprivation in infancy and its later effects on adult sexual and maternal behavior (Maple 1980). Many apes brought into the United States before ratification of the Convention on International Trade in Endangered Species of Wild Fauna and Flora (CITES) in 1972 were imported as infants and therefore had little opportunity to become properly socialized. In addition, concerns for infant survival prompted some zoos to separate infants from their mothers for hand-rearing by human caretakers (Maple 1980). This created a self-perpetuating cycle of animals that exhibited inadequate sexual and maternal behavior. Beck and Power (1988) found that captive-born, hand-reared gorillas, particularly females, were consistently less successful reproducing than captive-born, mother-reared females. There was also some indication that gorillas of both sexes enjoy higher reproductive success when they have social access to conspecifics throughout life. This was especially true of females, where social access during the first year of life was correlated with higher reproductive success. Hannah and Brotman (1990) found that captive female chimpanzees with no opportunity to observe mothers with infants or to interact with infants show inappropriate maternal behavior as adults, particularly with their first born. This often led to the infant being removed for hand-rearing. King and Mellen (1994) found that the copulatory behavior of adult chimpanzees is affected by early rearing experiences. Nearly 90 percent of animals reared with at least one adult conspecific copulated successfully as adults, whereas individuals hand-reared in the total absence of conspecifics were the least likely to reproduce.

Beck and Power (1988) concluded that gorilla mothers should always rear their newborns unless there is a significant threat to the mother's health. They also concluded that if gorillas were to be hand-reared, they should be reared with other ape infants, preferably members of their own species. Both recommendations have been incorporated into day to day management by AZA institutions. In cases where the mother is not producing milk, techniques have been developed to provide supplemental feeding so that the infant can remain with the mother and hand-rearing can be avoided (Fontaine 1979; Thorpe 1988).

The commitment of AZA member zoos to maintain apes in larger social groups has led to an increased number of animals that exhibit normal behavior, and this in turn has resulted in improved reproductive success. Many training techniques have been developed to enhance maternal care (e.g., Hannah and Brotman 1990; Joines 1977; Keiter and Pichette 1977; Mager 1981; Markham 1990). In addition, when hand-rearing is unavoidable, young animals are integrated into a social situation as soon as possible. Several techniques have been developed to facilitate this process (e.g., Keiter, Reichard, and Simmons 1983; Meder 1985; Nadler and Green 1975). Since 1997, most infant gorillas in AZA institutions have been mother-reared (Wharton 1997). Those infants that are hand-reared are typically integrated into family groups within two years. The AZA Gorilla SSP is also pursuing direct adoption of infants by older or infertile adult female gorillas (Jendry 1996).

One of the most challenging aspects of great ape behavior, as it relates to captive management, is how to reduce serious aggression. For example, wild chimpanzees typically exhibit hierarchical promiscuity, where male access to receptive females is based on dominance (Tilson 1986). Although competition exists, intragroup relationships between males are usually friendly, and any aggressive encounters that occur are typically of short duration and seldom result in any serious injury. Males are not known to transfer between groups, but females may transfer either temporarily or permanently, especially when they are sexually receptive. In stark contrast, strangers are often met with hostility, and intergroup encounters can be violent, resulting in severe injury or even death (Alford et al. 1995; Brent, Kessel, and Barrera 1997; de Waal 1986; Goodall 1986). This xenophobic and aggressive response to strangers has also been documented in other great ape species (gorillas: Brown and Wagster 1986; Hoff et al. 1996; orangutans: Poole 1987), with the exception of the bonobo (Mills et al. 1997).

The potential for serious injury from aggression is often greater in captivity, because it is more difficult for subordinate animals to escape from dominants (Brent, Kessel, et al. 1997). In addition, the transfer of animals from one group to another to facilitate genetic and other forms of animal management can result in social instability, thus contributing to the problem (Alford et al. 1995; Brent, Kessel, et al. 1997). Conversely, changing group structure is also a source of social stimulation that can facilitate reproduction and enrich the lives of captive animals. The AZA Gorilla SSP actively promoted the movement of reproductively capable but unsuccessful adults to new settings where they have subsequently bred (Beck and Power 1988). Similarly, the presence of same-sex competitors can raise androgen levels in the blood and have a stimulating effect on male libido. In the absence of competition, male libido can decline and reproduction may be curtailed (Tilson 1986). Similar phenomena may also occur in females. For example, a group-living, nonreproductive female orangutan gave birth to her second infant after an interbirth interval of 12 years and 2 months (Zucker, Robinette, and Deitchman 1987). It is interesting to note that the event was correlated with the presence of another cycling female.

Several advances have been made in reducing serious aggression-related injuries among captive chimpanzees, particularly during the introduction of new individuals to established social groups. For example, to introduce new females into established groups, zoo managers mimic the pattern of female migration in the wild by introducing them when they are in estrus. The likelihood of infanticide (Goodall 1977) is reduced by avoiding the transfer of pregnant females or females with small infants into new groups (Alford et al. 1995). Gradual introductions, consisting initially of only visual and limited tactile contact before full contact, are generally considered better. However, direct introductions can work as well

(Brent, Kessel, et al. 1997). Introductions of adult male chimpanzees are the least likely to be successful, but stable multimale groups can be formed, with great benefits to the animals and to the educational and research potential of the exhibits (Brent, Kessel, et al. 1997). Each situation is different, however, and studies of the individual personality characteristics of great apes may assist zoo managers in making informed decisions (Brown and Wagster 1986; Gold and Maple 1994).

The development of training techniques based on positive reinforcement has resulted in a revolution in behavioral management in zoos (Laule and Desmond 1998; Mellen and Ellis 1996), and great apes have benefited. For management purposes, zoo apes must be transferred from one enclosure to another, examined, treated, and occasionally shipped to other facilities. In the past, this could be traumatic for both the animals and caretakers, but appropriate training has made these procedures easier and more humane. Training has been used to facilitate semen collection (Brown and Loskutoff 1998; VandeVoort et al. 1993) and the daily injection of insulin in diabetic individuals (Sodaro, personal communication). It has also been used to reduce the incidence of aggression during feeding (Bloomsmith et al. 1994) and induce animals to enter shipping crates calmly before transportation (Laule et al. 1992).

Behavioral stress is one of the most significant factors affecting great ape health, reproduction, and psychological well-being (Snyder 1975). Some stress is normal in wild animals, and the associated physiological responses are adaptive (Sapolsky 1990). However, in captivity, even low levels of stress over long periods can cause problems. For example stress is known to suppress the immune system, thus making an animal more susceptible to disease (Snyder 1975). Stress has also been implicated in infertility in both males and females (Janssen and Bush 1990).

Efforts to provide captive apes with more complex and interesting environments has gone a long way toward ameliorating stress. Some researchers have argued that the presence of zoo visitors is detrimentally stressful for primates (Chamove, Hosey, and Schaetzel 1988). However, some apes appear to be fascinated by visitors and will even attempt to interact with them. Further study is needed, because this has important implications for zoo exhibit design. Some advances have already been made. Traditional zoo exhibits contained few places where animals could remove themselves from public view. However, newer, more naturalistic exhibits usually incorporate private areas into their design (Coe and Maple 1984). Although these exhibits may also include areas where great apes and visitors can get remarkably close to one another (e.g., separated only by glass panels at ground level), the animals are free to move away from these areas should they choose to do so. Cook and Hosey (1995) found that the primary reason captive chimpanzees interact with people is the possibility that they will obtain food. This may make it even more important that zoos enforce their no-feeding policies and design exhibits that make unauthorized feeding difficult. However, the authors also state that "chimpanzees do not necessarily see humans as aversive in the way that many other primates do, and the opportunities to interact may even be enriching" (Cook and Hosey 1995, 439).

DISCUSSION: CURRENT TRENDS AND PREDICTIONS

Professionally managed zoos have made tremendous strides in the care and management of great apes over the past few decades (Fulk and Garland 1992; Mills et al. 1997; Ogden and Wharton 1997; Sodaro 1997). Steady improvement has resulted from intensive research and the willingness to experiment with innovative enclosure designs and animal management protocols (Maple and Finlay 1986). Field studies of great apes have provided a wealth of information that has influenced nearly every aspect of zoo ape management (Harcourt 1987; Kortlandt 1960; Maple and Hoff 1982; Watts 1990). Important lessons have been learned from laboratory studies as well (Crockett 1998; Erwin and Deni 1979).

The intelligent, highly arboreal, and physically powerful orangutan still provides the greatest challenge for zoo designers, and the "model" orangutan exhibit is yet to be built (Gippoliti 2000). We would challenge zoo architects to design an exhibit that is naturalistic, species-appropriate, and functional. The risk of injury or death is thought to be slightly higher in naturalistic zoo habitats compared with more traditional ones, but this may be outweighed by the substantial benefits to the animals' psychological and physical health (Baer 1998).

The AZA has an accreditation program designed to evaluate the quality of zoological facilities (AZA 2000c). AZA accreditation standards do not currently address taxon-specific husbandry requirements. At the request of the U.S. Department of Agriculture's (USDA) Animal and Plant Health Inspection Service (APHIS), the government agency responsible for enforcing the U.S. Animal Welfare Act, the AZA has formulated a set of minimum husbandry standards for all captive wild mammals, including great apes (AZA 1997). This provides USDA inspectors objective standards for evaluating captive facilities, especially those that are not AZA-accredited. We recommend that these standards be updated and adopted by the AZA for implementation at its own member institutions. Taxon-specific management and care guidelines are currently being developed by the AZA for elephants (AZA 2000a), and the great apes deserve similar consideration.

Research intended to address unresolved animal management or husbandry issues should also continue to be supported for all species. Continued collaboration between zoos and academic primatologists can greatly assist in this process (Wiese and Hutchins 1997). The AZA great ape SSPs have all developed husbandry manuals that summarize the current state of knowledge regarding great ape captive management (Fulk and Garland 1992; Mills et al. 1997; Ogden and Wharton 1997; Sodaro 1997). AZA members are strongly encouraged to adopt these protocols. SSP decisions on great ape placement are appropriately based on the quality of the facilities and staff expertise and on the potential of the institution to contribute to the program's overall mission.

Some accredited zoos do not currently have the resources to build state of the art ape facilities. In the interim, however, every effort should be made to provide the animals with stimulating and responsive environments, and environmental enrichment programs are particularly important in this context (Kreger et al. 1998). USDA regulations require zoos and other captive holding facilities to prepare a plan for enhancing the psychological well-being of nonhuman primates (Crockett 1998). The AZA has also added an institutional environmental enrichment plan as a requirement of accreditation, and we support this decision.

Great ape population management will continue to be an important issue. Uncontrolled population growth has the potential to be a problem with long-lived species such as great apes, and sending an unwanted animal to a substandard facility can result in serious consequences (discussed later). This is why all breeding is highly regulated in AZA facilities. As great ape carrying capacities are reached in zoos, their populations must be managed even more intensively. Cooperative management is particularly important for controlling reproduction while retaining gene diversity and maintaining an appropriate age and sex composition. The safety and efficacy of improved methods of contraception are being tested (Asa et al. 1996).

The problem of "surplus" males is partly one of perception and partly real. Ideally, genetic management of a population calls for an even sex ratio, with both sexes contributing equally to the next generation (Foose 1983). However, polygynous and promiscuous species like great apes present special challenges in balancing the genetic advantages of an even-sex ratio with limitations imposed by the social structure. Competition and aggression among males can be a problem, and as in nature, it becomes necessary for some males (i.e., younger, subordinate individuals) to spend some part of their lives outside mixed-sex family groups. To address this issue, the AZA Gorilla SSP has been asking participating institutions to experiment with the formation of "bachelor" groups, and several such groups have now been established in the United States (e.g., Porton and White 1996) and Europe (e.g., Downman 1998; Harcourt 1988). Alford et al.'s (1995) study indicates that bachelor groups are practical for both short- and long-term housing of chimpanzees. However, the success of all-male groups may ultimately depend on preparing high-quality, spacious, outdoor, off-exhibit facilities in which space and environmentally enriched habitats allow male groups to form and break apart naturally, thus minimizing the potential for severe intermale aggression while maximizing the opportunity for social interaction. These same males will remain socialized for the day when they may be needed in breeding groups.

Ethical issues surrounding the disposition of zoo-bred great apes will need to be discussed and resolved, particularly as populations of these animals reach their carrying capacities. The AZA Code of Professional Ethics requires members to "make every effort to assure" that "animals do not find their way into the hands of those not qualified to care for them" (AZA 2000b). Thus institutions or individual members who send great apes to substandard facilities could be subject to an ethics charge, which in turn could result in the loss of accreditation. In any case, the guidelines for participation in an SSP program dictate that no animals may be transferred out of the managed population without previous approval of the management committee. Zoos in developing countries where accepted care standards may be substantially lower than those in accredited AZA facilities, sometimes request great apes from SSPs. The AZA Great Ape TAG is encouraged to provide guidance to AZA members on international animal transfers.

AZA institutions and their international colleagues should continue their practice of loaning rather than selling great apes to one another. Prices of great apes otherwise are likely to be exorbitant, thus stimulating an illegal wildlife trade in Africa and Asia. Importation of wild-caught great apes into the United States dropped off sharply after the ratification of CITES nearly 30 years ago. In the early 1960s, before CITES, the International Union of Directors of Zoological Gardens (IUDZG) effected the first ban on gorilla importation from Africa, beginning with the vulnerable mountain gorilla. The ban is still in effect, with reinforcement from other conservation groups. Given proper cooperative management under the various SSPs, there should be no reason to bring additional wild apes into captivity. Exceptions might include orphaned animals that are brought into range-country sanctuaries for their own protection. New species (i.e., mountain gorillas) might also be brought into zoos someday if it were deemed in the species' best interest to do so (i.e., if the wild population became so small or fragmented that it could not survive without human intervention). Of course, any importation would require permits from the U.S. Fish and Wildlife Service, as well as the country of origin. Furthermore, this would involve providing a strong justification that the importation would benefit conservation and not have a detrimental effect on extant, viable wild populations (Koontz 1995).

Should zoos provide great apes for use by biomedical laboratories, circuses, other forms of entertainment, or for the pet trade? In our opinion, the answer is no. Although the best circuses, laboratories, and entertainers may care for their animals properly, their missions differ from those of modern zoos. The goals of accredited zoos include high-quality animal care, conservation, education, and science (Maple et al. 1995). It would therefore be inappropriate for them to breed and provide animals to institutions or individuals with incompatible or unrelated goals. The AZA Orangutan SSP has a policy against providing zoo-bred apes to any individual or organization "whose sole mission is entertainment, who routinely engage in aversive training methods, who display orangutans in costume, or otherwise degrade this species" (Perkins 1997, app. IV). The AZA Great Ape TAG supports efforts to extend this to all apes (AZA Great Ape TAG 1995).

As great ape zoo populations mature, the question arises of what to do with older, postreproductive individuals. Animal rights proponents argue that zoos have a responsibility to care for captive-bred animals from "the cradle to the grave." In the case of great apes, we agree. Despite arguments to the contrary (Graham 1996; Lacy 1995) and the fact that it is legal, euthanasia of healthy great apes is not generally accepted in the professional zoo community as an option for controlling populations. As a result, Lindburg (1991) and Lindburg and Lindburg (1995) have argued that AZA institutions pool their resources to construct and operate "retirement homes" for older animals. This is precisely what the U.S. government is considering for chimpanzees that have been used in biomedical and other research (Brent, Butler, and Haberstroh 1997; chapter 7, this volume). There are an estimated 600 surplus chimpanzees in biomedical laboratories in the United States. Although "retirement" facilities may sound like the best solution, the estimated cost to construct two such facilities for only 240 chimpanzees is more than $8,600,000, and nearly $1,000,000 per year would be needed to cover operating expenses (Brent and Koebner 1997).

We support the goals of retirement facilities for apes used in biomedical research and believe that we have a moral obligation to care for animals that have contributed so much to the advancement of human medicine. However, we do not believe that efforts to promote the welfare of captive apes should compete directly with efforts to protect and conserve great apes in nature. We recommend that if the U.S. government allocates funding for one or more "retirement" facilities for research chimpanzees, that it also establishes a fund to support field conservation efforts for chimpanzees or other great apes. Such a fund would be similar to those recently established to support the conservation of endangered rhinos, tigers, and elephants (M. Phillips 1998). Indeed, a Great Ape Conservation Act was recently passed by Congress and signed by President Bill Clinton in November 2000 (Pub. L. No. 106-411). Since then, Congress has appropriated $750,000 for the fund for fiscal year 2001.

The quality of the animal–caregiver relationship has been shown to have a major effect on the lives of captive wildlife, and is even correlated with reproductive success in some taxa (e.g., small cats: Mellen 1991). The issue of ape–caregiver relationships is addressed in SSP husbandry manuals (e.g., orangutans: Bond and Weber 1997). However, the training of great ape caregivers could be improved. Some accredited institutions have formal keeper-training programs, which is combined with on the job experience. Some SSP programs, notably the Red Panda SSP, have been experimenting with keeper training workshops as a method to more closely integrate cutting-edge research findings into day to day animal care (Roberts, personal communication). Great ape programs could profit from a similar approach.

The trend toward closer links between zoos and the conservation of great apes in nature must continue to grow (Conway 1999b; Hutchins 1999; Hutchins and Conway 1995; Mittermeier 1986). In particular, zoos need to expand their cooperative relationships with relevant government wildlife agencies and nongovernmental organizations and develop more "in-house" expertise in field conservation (Hutchins 1999).

The editors and some of the authors who contributed to the *Great Ape Project* publication are strongly opposed to any form of captivity. Indeed, proceeds from the book are to go toward "freeing all imprisoned chimpanzees, gorillas, and orangutans and returning them to an environment that accords with their physical, mental and social needs" (Cavalieri and Singer 1993, 3). Although we share the vision of great apes living in nature, we differ from animal rights activists in our view of how to ensure a future for these magnificent creatures in a human-dominated world. Even if everyone agreed on the Great Ape Project's approach, the immediate release of all captive great apes is not only impractical at present, but it may also prove to be inhumane. In fact, reintroduction techniques for these species are still in their infancy (Hannah and McGrew 1991), and the resultant mortality from such an experiment would likely be high. In addition, there are few places left where these animals could be released that are not already inhabited by free-ranging populations, and the potential risks associated with such efforts—for example, of the transmission of exotic diseases to immunologically naive individuals (Ballou 1993)—would have to be carefully assessed. It is equally important to put this unrealistic proposal into proper perspective. Currently, there are more gorillas killed each year in Western Africa as agricultural pests and for bushmeat than exist in the entire world's captive population of nearly 700 animals. In Cameroon alone, it is estimated that the yearly slaughter of western lowland gorillas exceeds the entire population of mountain gorillas in the Virunga Mountains (Rose 1996). We must acknowledge that great apes are in captivity to stay, at least for the near future, and accept our obligation to care for them properly. We also are obligated to put these precious beings to the best use for conservation. Only then can captive great apes be considered true ambassadors for their kind.

As stated earlier, one of the biggest weaknesses of the Great Ape Project is that it paints all captive facilities with the same brush. We believe, however, that recent advances in animal exhibit design and care programs have made it possible to maintain captive great apes humanely, and the situation is expected to continue to improve. The Great Ape Project also fails to recognize the significant contributions that modern professionally managed zoos are making to great ape conservation. Obviously, great apes should not be kept in substandard facilities or "roadside animal attractions," nor should they be kept in zoos that do not participate in cooperative, scientifically managed breeding programs. Any facility that displays great apes should also contribute to their conservation in some demonstrable way.

Rather than continuing the contentious and, in the current global context, largely meaningless debate pitting "captivity" versus "the wild" (Hutchins 1994), animal activists, conservationists, and zoo professionals should work together for the benefit of great apes. The problems facing our closest extant relatives are simply too insidious and pervasive to let perceived philosophical differences break down communication and cooperation. Ultimately, it has the potential to harm the animals for which we all share a concern (Hutchins and Wemmer 1986). However, any cooperation must be based on the inescapable and often stark realities of the current situation for free-ranging great apes. We agree with Burghardt (1997, 86), who in his review of the publication *The Great Ape Project,* said, "Good will, however passionate, will not save great apes. Excellent science, effective education (especially in ape homelands), stabilizing human population growth, and political reforms in countries with indigenous ape populations are the only hope for long-term success."

REFERENCES

Adams, J. S., and McShane, T. O. 1992. *The myth of wild Africa: Conservation without illusion.* New York: W. W. Norton.

Akers, J. S., and Schildkraut, D. S. 1985. Regurgitation/reingestion and coprophagy in captive gorillas. *Zoo Biology* 4: 99–109.

Alford, P. L., Bloomsmith, M. A., Keeling, M. E., and Beck, T. F. 1995. Wounding aggression during the formation and maintenance of captive multimale chimpanzee groups. *Zoo Biology* 14: 347–59.

Asa, C. S., Porton, I., Baker, A. M., and Plotka, E. D. 1996. Contraception as a management tool for controlling surplus animals. In D. G. Kleiman, M. E. Allen, K. V. Thompson, S. Lumpkin, and H. Harris (eds.), *Wild mammals in captivity: Principles and techniques* (pp. 451–67). Chicago: University of Chicago Press.

American Zoo and Aquarium Association (AZA). 1997. *Minimum husbandry guidelines for mammals.* Bethesda, MD: Author.

———. 2000a. AZA board of directors meeting, Gulf Breeze, FL. *AZA Communique:* 50–54.

———. 2000b. Code of professional ethics. In M. Moretti (ed.), *2000 AZA membership directory* (pp. 260–63). Silver Springs, MD: Author.

———. 2000c. *Guide to certification (and standardized guidelines).* Silver Springs, MD: Author.

AZA Great Ape Taxon Advisory Group (TAG). 1995. *Strategic collection plan.* Unpublished manuscript, National Zoological Park, Washington, D.C.

———. 1998. *Space survey.* Unpublished manuscript, National Zoological Park, Washington, D.C.

Baer, J. 1998. A veterinary perspective of potential risk factors in environmental enrichment. In D. Shepherdson, J. Mellen and M. Hutchins (eds.), *Second nature: Environmental enrichment for captive animals* (pp. 277–301). Washington, D.C.: Smithsonian Institution Press.

Baker, K. C. 1997. Straw and forage material ameliorate abnormal behaviors in adult chimpanzees. *Zoo Biology* 16: 225–36.

Ballou, J. D. 1993. Assessing the risks of infectious diseases in captive breeding and reintroduction programs. *Journal of Zoo and Wildlife Medicine* 24: 327–35.

Ballou, J. D., and Foose, T. J. 1996. Demographic and genetic management of captive populations. In D. G. Kleiman, M. E. Allen, K. V. Thompson, S. Lumpkin, and H. Harris (eds.), *Wild mammals in captivity: principles and techniques* (pp. 263–83). Chicago: University of Chicago Press.

Beck, B. B., and Power, M. L. 1988. Correlates of sexual and maternal competence in captive gorillas. *Zoo Biology* 7: 339–50.

Benirschke, K., and Adams, F. D. 1980. Gorilla diseases and causes of death. *Journal of Reproduction and Fertility (Suppl.)* 28: 139–48.

Bettinger, T. 1994. Effect of contraceptives on female chimpanzee genital swelling. *AZA Regional Conference Proceedings: American Zoo and Aquarium Association:* 9–17.

Bettinger, T., Cougar, D., Lee, D. R., Lasley, B. L., and Wallis, J. 1997. Ovarian hormone concentrations and genital swelling patterns in female chimpanzees with Norplant implants. *Zoo Biology* 16: 209–23.

Bingaman, L., and Ballou, J. 1986. *Demographic modeling program, Version 4.2.* Washington, D.C.: National Zoological Park.

Bitgood, S., Patterson, D., and Benefield, A. 1988. Exhibit design and visitor behavior: Empirical relationships. *Environment and Behavior* 20: 474–91.

Bloomsmith, M. A., Alford, P. L., and Maple, T. L. 1988. Successful feeding enrichment for captive chimpanzees. *American Journal of Primatology* 16: 155–64.

Bloomsmith, M. A., Laule, G., Thurston, R., and Alford, P. 1994. Using training to modify chimpanzee aggression during feeding. *Zoo Biology* 13: 557–66.

Bond, M., and Weber, B. 1997. Orangutan/caregiver relationships. In C. Sodaro (ed.), *Orangutan Species Survival Plan husbandry manual* (pp. 83–84). Chicago: Orangutan Species Survival Plan.

Brent, L., Butler, T. M., and Haberstroh, J. 1997. Surplus chimpanzee crisis: Planning for the long-term needs of research chimpanzees. *Lab Animal* 26: 36–39.

Brent, L., Kessel, A. L., and Barrera, H. 1997. Evaluation of introduction procedures in captive chimpanzees. *Zoo Biology* 16: 335–42.

Brent, L., and Koebner, L. 1997. *Chimp Haven executive summary.* San Antonio, TX: Chimp Haven. Unpublished plan.

Bronwyn, B., and Ford, J. 1991. Environmental enrichment in zoos: Melbourne Zoo's naturalistic approach. *Thylacinus* 16: 12–17.

Brooks, A. S., and Leigh Smith, J. N. 1998. Politics and problems of gorilla and chimp conservation. In R. Osterweis Selig and M. R. London (eds.), *Anthropology explored: The best of Smithsonian AnthroNotes* (pp. 24–37). Washington, D.C.: Smithsonian Institution Press.

Brown, C. S., and Loskutoff, N. 1998. A training program for noninvasive semen collection in captive western lowland gorillas (*Gorilla gorilla gorilla*). *Zoo Biology* 17: 143–51.

Brown, S. G., and Wagster, M. V. 1986. Socialization processes in a female lowland gorilla. *Zoo Biology* 5: 269–79.

Burghardt, G. B. 1997. *The Great Ape Project* (book review). *Animals and Society* 5: 83–86.

Butynski, T. M., and Kalina, J. 1998 Gorilla tourism: A critical review. In E. J. Milner-Gulland and R. Mace (eds.), *Conservation of biological resources* (pp. 280–300). Oxford: Blackwell Scientific.

Calvert, J. J. 1985. *Food selection by western gorillas in relation to food chemistry and selective logging in Cameroon, West Africa.* PhD dissertation, University of California, Los Angeles.

Cavalieri, P., and Singer, P. (eds.). 1993. *The Great Ape Project: Equality beyond humanity.* New York: St. Martin's Press.

Chamove, A. S., Anderson, J. R., Morgan-Jones, S. C., Jones, S. P. 1982. Deep woodchip litter: Hygiene, feeding and behavioral enhancement in eight primate species. *International Journal for the Study of Animal Problems* 3: 308–18.

Chamove, A. S., Hosey, G. R., and Schaetzel, P. 1988. Visitors excite primates in zoos. *Zoo Biology* 7: 359–69.

Clark, A. S., Juno, C. J., and Maple, T. L. 1982. Behavioral effects of a change in the physical environment: A pilot study of captive chimpanzees. *Zoo Biology* 1: 371–80.

Coe, J. C. 1985. Design and perception: Making the zoo experience real. *Zoo Biology* 4: 197–208.

———. 1996. What's the message? Education through exhibit design. In D. G. Kleiman, M. E. Allen, K. V. Thompson, S. Lumpkin, and H. Harris (eds.), *Wild mammals in captivity: Principles and techniques* (pp. 167–74). Chicago: University of Chicago Press.

Coe, J. C., and Maple, T. L. 1984. Approaching Eden: A behavioral basis for great ape exhibits. *AAZPA Annual Conference Proceedings:* 117–26.

Conway, W. G. 1989. The prospects of sustaining species and their evolution. In D. Western and M. Pearl (eds.), *Conservation for the twenty-first century* (pp. 199–363). New York: Oxford Press.

———. 1999a. Congo: A zoo experiment in participatory conservation. *AZA Annual Conference Proceedings:* 101–104.

———. 1999b. Linking zoo and field and keeping promises to dodos. In T. L. Roth, W. F. Swanson, and L. K. Blattman (eds.), *Proceedings of the seventh world congress on breeding endangered species: Linking zoo and field research to advance conservation* (pp. 5–11). Cincinnati, OH: Cincinnati Zoo and Botanical Garden.

Cook, S., and Hosey, G. R. 1995. Interaction sequences between chimpanzees and human visitors at the zoo. *Zoo Biology* 14: 431–40.

Courtnay, J. 1988. Infant mortality in mother-reared captive chimpanzees at Taronga Zoo, Sydney. *Zoo Biology* 7: 61–68.

Cousins, D. 1983. Mortality factors in captive gorillas (*Gorilla gorilla*). *International Zoo News* 30: 5–17.

Crockett, C. M. 1998. Psychological well-being of captive non-human primates: Lessons from laboratory studies. In D. Shepherdson, J. Mellen, and M. Hutchins (eds.), *Second nature: Environmental enrichment for captive animals* (pp. 129–52). Washington, D.C.: Smithsonian Institution Press.

de Waal, F. B. W. 1986. The brutal elimination of a rival among captive chimpanzee males. *Ethology and Sociobiology* 7: 237–51.

Dian Fossey Gorilla Fund International (DFGFI). 1996. Dian Fossey Gorilla Fund relocates to Zoo Atlanta from Denver, Colorado. *Endangered Species Update* 13: 15.

Douglas, E. M. 1981. First gorilla born using artificial insemination. *International Zoo News* 28: 9–15.

Downman, M. 1998. The formation of a bachelor group of gorillas at Loro Parque. *International Zoo News* 45: 208–11.

Edwards, S. D., and Snowdon, C. T. 1980. Social behavior of captive, group-living orangutans. *International Journal of Primatology* 1: 39–62.

Ehrlich, P. R., and Erhlich, A. H. 1990. *The population explosion.* New York: Simon and Schuster.

Ely, J., and Ferrell, R. E. 1990. DNA "fingerprints" and paternity ascertainment in chimpanzees (*Pan troglodytes*). *Zoo Biology* 9: 91–98.

Erwin, J. 1986. Environments for captive propagation of primates: Interaction of social and physical factors. In K. Benirschke (eds.), *Primates: The road to self-sustaining populations* (pp. 297–305). New York: Springer-Verlag.

Erwin, J., and Deni, R. 1979. Strangers in a strange land: Abnormal behaviors or abnormal environments? In J. Erwin, T. L. Maple, and G. Mitchell (eds.), *Captivity and behavior: Primates in breeding colonies, laboratories and zoos* (pp. 1–28). New York: Van Nostrand Reinhold.

Eves, H., and Olson, S. 2000. Crisis in Africa. *AZA Communique* (July): 15–16.

Eves, H., and Ruggiero, R. G. 2000. Socioeconomics and the sustainability of hunting in the forests of Northern Congo. In J. Erwin, T. L. Maple, and G. Mitchell (eds.), *Hunting for sustainability in tropical forests* (pp. 427–54). New York: Columbia University Press.

Finlay, T., James, L., and Maple, T. L. 1988. Zoo environments influence people's perceptions of animals. *Environment and Behavior* 20: 508–25.

Fontaine, R. 1979. Training an unrestrained orangutan mother to permit supplemental feeding of her infant. *International Zoo Yearbook* 19: 168–70.

Foose, T. J. 1983. The relevance of captive populations to the conservation of biotic diversity. In C. M. Shonewald-Cox, S. M. Chambers, B. MacBryde, and L. Thomas (eds.), *Genetics and conservation: A reference for managing wild animal and plant populations* (pp. 374–401). Menlo Park, CA: Benjamin Cummings.

———. 1986. Riders of the last ark: The role of captive breeding in conservation strategies. In L. Kaufman and K. Mallory (eds.), *The last extinction* (pp. 141–65). Cambridge: MIT Press.

Fulk, R. (ed.). 1997. *Chimpanzee* (Pan troglodytes) *species survival plan master plan.* Ashboro: North Carolina Zoological Society.

———. 1998. *Chimpanzee studbook.* Ashboro: North Carolina Zoological Park.

Fulk, R., and Garland, C. (eds.). 1992. *The care and management of chimpanzees* (Pan troglodytes) *in captive environments.* Ashboro: North Carolina Zoological Society.

Gilloux, I., Gurnell, J., and Shepherdson, D. 1992. An enrichment device for great apes. *Animal Welfare* 1: 279–89.

Gippoliti, S. 2000. Orangutans in zoos: Husbandry, welfare, and management in an atypical arboreal solitary mammal. *International Zoo News* 47: 356–69.

Goerke, B., Fleming, L., and Creel, M. 1987. Behavioral changes of a juvenile gorilla after transfer to a more naturalistic environment. *Zoo Biology* 6: 283–95.

Gold, K. 1997. The conservation role of primate exhibits in the zoo. In J. Wallis (ed.), *Primate conservation: The role of zoological parks* (pp. 43–61). Norman, OK: American Society of Primatologists.

Gold, K., and Maple, T. L. 1994. Personality assessment in the gorilla and its utility as a management tool. *Zoo Biology* 13: 509–22.

Goodall, J. 1977. Infant killing and cannibalism in free-living chimpanzees. *Folia primatologica* 28: 259–82.

———. 1983. Population dynamics during a 15-year period in one community of free-living chimpanzees in the Gombe National Park, Tanzania. *Zeitschrift für Tierpsychologie* 61: 1–60.

———. 1986. *The chimpanzees of Gombe: Patterns of behavior.* Cambridge, MA: Harvard University Press.

Goodrowe, K. L., Wildt, D. E., and Monfort, S. L. 1992. Effective suppression of ovarian cyclicity in the lowland gorilla with an oral contraceptive. *Zoo Biology* 11: 261–69.

Gould, E., and Bres, M. 1986. Regurgitation and reingestion of food in captive gorillas: Description and intervention. *Zoo Biology* 5: 241–50.

Graham, S. 1996. Issues of surplus animals. In D. G. Kleiman, M. E. Allen, K. V. Thompson, S. Lumpkin, and H. Harris (eds.), *Wild mammals in captivity: Principles and techniques* (pp. 290–96). Chicago: University of Chicago Press

Hancocks, D. 1971. *Animals and architecture.* New York: Praeger.

Hannah, A. C., and Brotman, B. 1990. Procedures for improving maternal behavior in captive chimpanzees. *Zoo Biology* 9: 233–40.

Hannah, A. C., and McGrew, W. C. 1991. Rehabilitation of captive chimpanzees. In H. O. Box (ed.), *Primate responses to environmental change* (pp. 167–86). London: Chapman and Hall.

Harcourt, A. H. 1987. Behavior of wild gorillas and their management in captivity. *International Zoo Yearbook* 26: 248–55.

———. 1988. Bachelor groups of gorillas in captivity: The situation in the wild. *Dodo* 25: 54–61.

Hawes, A. 1995. Mindful of thought. *Zoogoer* (National Zoological Park) 24: 22–23.

Hinshaw, K., Amand, W. B., and Tinkerman, C. L. 1996. Preventative medicine. In D. G. Kleiman, M. E. Allen, K. V. Thompson, S. Lumpkin, and H. Harris (eds.), *Wild mammals in captivity: Principles and techniques* (pp. 16–24). Chicago: University of Chicago Press.

Hoff, M. P., Forthman, D. L., and Maple, T. L. 1994. Dyadic interactions of infant lowland gorillas in an outdoor exhibit compared to an indoor exhibit. *Zoo Biology* 13: 245–56.

Hoff, M. P., Hoff, K. T., Horton, C., and Maple, T. L. 1996. Behavioral effects of changing group membership among captive lowland gorillas. *Zoo Biology* 15: 383–93.

Hornaday, W. 1922. *The minds and manners of wild animals.* New York: Charles Scribner and Sons.

Hutchins, M. 1994. What do "wild" and "captive" mean for large ungulates and carnivores now and into the twenty-first century? In A. N. Rowan (ed.), *Wildlife conservation, zoos and animal protection* (pp. 1–18). Boston: Tufts Center for Animals and Public Policy.

———. 1999. Why zoos and aquariums should increase their contributions to in situ conservation. *AZA Annual Conference Proceedings:* 126–39.

Hutchins, M., and Conway, W. G. 1995. The evolving role of modern zoos and aquariums in field conservation. *International Zoo Yearbook* 34: 117–30.

Hutchins, M., Dresser, B., and Wemmer, C. 1995. Ethical considerations in zoo and aquarium research. In B. Norton, M. Hutchins, E. F. Stevens, and T. L. Maple (eds.), *Ethics on the ark: Zoos, animal welfare and wildlife conservation* (pp. 253–76). Washington, D.C.: Smithsonian Institution Press.

Hutchins, M., and Fascione, N. 1993. What is it going to take to save wildlife? *Proceedings: American Association of Zoological Parks and Aquariums Regional Conference:* 5–15.

Hutchins, M., Foose, T., and Seal, U. S. 1991. The role of veterinary medicine in endangered species conservation. *Journal of Zoo and Wildlife Medicine* 22: 277–81.

Hutchins, M., Hancocks, D., and Crockett, C. 1984. Naturalistic solutions to the behavioral problems of captive animals. *Zoologishe Garten* 54: 28–42.

Hutchins, M., Paul, E., and Bowdoin, J. 1996. Contributions of zoo and aquarium research to wildlife conservation and science. In J. Bielitzki, J. Boyce, G. Burghardt, and D. Schaffer (eds.), *Well-being of animals in zoo and aquarium-sponsored research* (pp. 23–29). Greenbelt, MD: Scientists Center for Animal Welfare.

Hutchins, M., and Wemmer, C. 1986. Wildlife conservation and animal rights: Are they compatible? In M. Fox and L. Mickley (eds.), *Advances in animal welfare science 1986/87* (pp. 111–37). Boston: Martinus Nijhoff.

———. 1991. In defense of captive breeding. *Endangered Species Update* 8: 5–6.

Hutchins, M., and Wiese, R. 1991. Beyond genetic and demographic management: The future of the Species Survival Plan and other AZA conservation efforts. *Zoo Biology* 10: 285–92.

Hutchins, M., Wiese, R., and Willis, K. 1996. Why we need captive breeding. *AZA Regional Conference Proceedings:* 77–86.

Hutchins, M., Willis, K., and Wiese, R. 1995. Strategic collection planning; Theory and practice. *Zoo Biology* 14: 2–22.

International Species Information System (ISIS). 1998. Abstracts. Apple Valley, MN: Author.

Janssen, D., and Bush, M. 1990. Review of the medical literature of great apes in the 1980s. *Zoo Biology* 9: 123–34.

Jendry, C. 1996. Utilization of surrogates to integrate hand-reared infant gorillas into an age/sex diversified group of conspecifics. *Applied Animal Behaviour Science* 48: 173–86.

———. 1997. Partners in conservation: Establishing in situ partnerships to aid mountain gorillas and people in range countries. In J. Wallis (ed.), *Primate conservation: The role of zoological parks* (pp. 199–213). Norman, OK: American Society of Primatologists.

Joines, S. 1977. A training programme designed to induce maternal behaviour in a multiparous female lowland gorilla, *Gorilla g. gorilla,* at the San Diego Zoo. *International Zoo Yearbook* 17: 185–88.

Keiter, M., and Pichette, P. 1977. Surrogate infant prepares a lowland gorilla for motherhood. *International Zoo Yearbook* 17: 185–88.

Keiter, M., Reichard, T., and Simmons, J. 1983. Removal, early hand-rearing and successful reintroduction of an orangutan (*Pongo pymaeus pygmaeus*) to her mother. *Zoo Biology* 2: 55–59.

King, N. E., and Mellen, J. 1994. The effects of early experience on adult copulatory behavior in zoo-born chimpanzees (*Pan troglodytes*). *Zoo Biology* 13: 51–59.

Koontz, F. W. 1995. Wild animal acquisition ethics for zoo biologists. In B. Norton, M. Hutchins, E. F. Stevens, and T. L. Maple (eds.), *Ethics on the ark: Zoos, animal welfare and wildlife conservation* (pp. 127–45). Washington, D.C.: Smithsonian Institution Press.

———. 1997. Zoos and in situ primate conservation. In J. Wallis (ed.), *Primate conservation: The role of zoological parks* (pp. 63–81). Norman, OK: American Society of Primatologists.

Kortlandt, A. 1960. Can lessons from the wild improve the lot of captive chimpanzees? *International Zoo Yearbook* 2: 76–80.

Kreger, M., Hutchins, M., and Fascione, N. 1998. Context, ethics and environmental enrichment in zoos. In D. Shepherdson, J. Mellen, and M. Hutchins (eds.), *Second nature: Environmental enrichment for captive animals* (pp. 59–82). Washington, D.C.: Smithsonian Institution Press.

Lacy, R. 1995. Culling surplus animals for population management. In B. G. Norton, M. Hutchins, E. F. Stevens, and T. L. Maple (eds.), *Ethics on the ark: Zoos, animal welfare and wildlife conservation* (pp. 187–94). Washington, D.C.: Smithsonian Institution Press.

———. 1998. *Genes, Version 11.82*. Brookfield, IL: Chicago Zoological Society.

Laule, G., and Desmond, T. 1998. Positive reinforcement training as an enrichment strategy. In D. Shepherdson, J. Mellen, and M. Hutchins (eds.), *Second nature: Environmental enrichment for captive animals* (pp. 302–13). Washington, D.C.: Smithsonian Institution Press.

Laule, G, Keeling, M., Alford, P., Thurston, R., Bloomsmith, M., and Beck, T. 1992. Positive reinforcement techniques and chimpanzees: An innovative training program. *AZA Regional Conference Proceedings:* 713–22.

Lee, G. H., and Coe, J. C. 1988. Greenspace for primates. *International Zoo News* 35: 19–21.

Lee, P. C., Thornback, J., and Bennett, E. 1988. *Threatened primates of Africa: The IUCN red data book*. Gland, Switzerland: International Union for the Conservation of Nature and Natural Resources.

Leus, K., and Van Puijenbroeck, B. 1997. *Bonobo* (Pan paniscus) *International studbook*. Antwerp: Royal Zoological Society of Antwerp.

Lindburg, D. 1991. Zoos and the "surplus" problem. *Zoo Biology* 10: 1–2.

———. 2000. Zoos and the rights of animals. *Zoo Biology* 18: 433–48.

Lindburg, D., and Lindburg, L. 1995. Success breeds a quandry: To cull or not to cull? In B. G. Norton, M. Hutchins, E. F. Stevens, and T. L. Maple (eds.), *Ethics on the ark: Zoos, animal welfare and wildlife conservation* (pp. 195–208). Washington, D.C.: Smithsonian Institution Press.

MacNamara, T., Dolensek, E. P., Liu, S., and Dierenfeld, E. 1987. Cardiomyopathy associated with Vitamin E deficiency in two lowland gorillas. *Proceedings of the American Association of Zoo Veterinarians:* 89–91.

Mager, W. 1981. Stimulating maternal behaviour in the lowland gorilla (*Gorilla g. gorilla*) at Apledorn. *International Zoo Yearbook* 21: 138–43.

Maki, S., and Bloomsmith, M. A. 1986. Uprooted trees facilitate the psychological well-being of captive chimpanzees. *Zoo Biology* 8: 79–88.

Mallinson, J. C. 1995. Zoo breeding programmes: Balancing conservation and animal welfare. *Dodo* 31: 66–73.

Maple, T. L. 1979. Great apes in captivity: The good, the bad and the ugly. In J. Erwin, T. L. Maple, and G. Mitchell (eds.), *Captivity and behavior: Primates in breeding colonies, laboratories and zoos* (pp. 239–72). New York: Van Nostrand Reinhold.

———. 1980. *Orangutan behavior.* New York: Van Nostrand Reinhold.

Maple, T. L., and Finlay, T. W. 1986. Evaluating the environments of captive non-human primates. In K. Benirschke (eds.), *Primates: The road to self-sustaining populations* (pp. 479–88). New York: Springer-Verlag.

Maple, T. L., and Hoff, M. P. 1982. *Gorilla behavior.* New York: Van Nostrand Reinhold.

Maple, T. L., McManamon, R., and Stevens, E. F. 1995. Defining the good zoo: Animal care, maintenance and welfare. In B. G. Norton, M. Hutchins, E. F. Stevens, and T. L. Maple (eds.), *Ethics on the ark: Zoos, animal welfare and wildlife conservation* (pp. 155–63). Washington, D.C.: Smithsonian Institution Press.

Maple, T. L., and Perkins, L. 1996. Enclosure furnishings and structural environmental enrichment. In D. G. Kleiman, M. E. Allen, K. V. Thompson, S. Lumpkin, and H. Harris (eds.), *Wild mammals in captivity* (pp. 212–22). Chicago: University of Chicago Press.

Markham, R. J. 1990. Breeding orangutans at Perth Zoo: Twenty years of appropriate husbandry. *Zoo Biology* 9: 171–82.

Masters, A. M., and Markham, R. J. 1991. Assessing reproductive status in orangutans by using urinary estrone. *Zoo Biology* 10: 197–207.

Meder, A. 1985. Integration of hand-reared gorilla infants in a group. *Zoo Biology* 4: 1–12.

Mellen, J. 1991. Factors influencing reproductive success in small captive exotic felids (*Felis* spp.): A multiple regression analysis. *Zoo Biology* 10: 95–110.

Mellen, J., and Ellis, S. 1996. Animal learning and husbandry training. In D. G. Kleiman, M. E. Allen, K. V. Thompson, S. Lumpkin, and H. Harris (eds.), *Wild mammals in captivity* (pp. 88–99). Chicago: University of Chicago Press.

Mellen, J., Shepherdson, D., and Hutchins, M. 1998. The future of environmental enrichment. In D. Shepherdson, J. Mellen, and M. Hutchins (eds.), *Second nature: Environmental enrichment for captive animals* (pp. 329–36). Washington, D.C.: Smithsonian Institution Press.

Mills, J., Reinartz, G., De Bois, H., Van Elsacker, L., Van Puijenbroeck, B. (eds.). 1997. *The care and management of bonobos in captive environments.* Milwaukee, WI: Zoological Society of Milwaukee County.

Mittermeier, R. A. 1986. Who will pilot the ark? In K. Benirschke (ed.), *Primates: The road to self-sustaining populations* (pp. 985–87). New York: Springer-Verlag.

Munson, L., and Montali, R. J. 1990. Pathology and diseases of great apes at the National Zoological Park. *Zoo Biology* 9: 99–105.

Nadler, R. D., Dahl, J., Collins, D., and Gould, K. 1993. Effects of an oral contraceptive on sexual behavior of chimpanzees (*Pan troglodytes*). *Archives of Sexual Behavior* 22: 477–500.

Nadler, R. D., and Green, S. 1975. Separation and reunion of a gorilla (*Gorilla g. gorilla*) infant and mother. *International Zoo Yearbook* 15: 198–201.

Nash, N., Ogden, J., Meller, L., and Wall, V. 1997. Design. In J. Ogden and D. Wharton (eds.), *Management of gorillas in captivity* (pp. 217–65). Atlanta, GA: Gorilla Species Survival Plan and Atlanta/Fulton County Zoo.

Nash, V. J. 1982. Tool use by captive chimpanzees at an artificial termite mound. *Zoo Biology* 1: 211–22.

Nichols, M., and Goodall, J. 1999. *Brutal kinship.* New York: Aperature Foundation.

Norton, B. 1987. *Why preserve natural variety?* Princeton, NJ: Princeton University Press.

Oftedal, O. T., and Allen, M. E. 1996a. The feeding and nutrition of omnivores with emphasis on primates. In D. G. Kleiman, M. E. Allen, K. V. Thompson, S. Lumpkin, and H. Harris (eds.), *Wild mammals in captivity* (pp. 148–57). Chicago: University of Chicago Press.

———. 1996b. Nutrition and dietary evaluation in zoos. In D. G. Kleiman, M. E. Allen, K. V. Thompson, S. Lumpkin, and H. Harris (eds.), *Wild mammals in captivity* (pp. 109–116). Chicago: University of Chicago Press.

Ogden, J. J. 1992. *A comparative evaluation of natural habitats for captive lowland gorillas* (Gorilla g. gorilla). PhD dissertation, Georgia Institute of Technology, Atlanta.

Ogden, J. J., Finlay, T. W., and Maple, T. L. 1990. Gorilla adaptations to naturalistic environments. *Zoo Biology* 9: 107–21.

Ogden, J. J., Lindburg, D. G., and Maple, T. L. 1994. A preliminary study of the effects of ecologically relevant sounds on the behaviour of captive lowland gorillas. *Applied Animal Behaviour Science* 39: 163–76.

Ogden, J., and Wharton, D. (eds.). 1997. *Management of gorillas in captivity.* Atlanta, GA: Gorilla Species Survival Plan and Atlanta/Fulton County Zoo.

Paquette, D., and Prescott, J. 1988. Use of novel objects to enhance environments of captive chimpanzees. *Zoo Biology* 7: 15–23.

Pearce, J. 1995. *Slaughter of the apes: How the tropical timber industry is devouring Africa's great apes.* London: World Society for the Protection of Animals.

Perkins, L. A. 1992. Variables that influence the activity of captive orangutans. *Zoo Biology* 11: 117–86.

Perkins, L. A., and Maple, T. L. 1990. North American orangutan Species Survival Plan: Current status and progress in the 1980's. *Zoo Biology* 9: 135–39.

———. 1997. *Species Survival Plan masterplan for the orangutan (*Pongo pygmeaus *sp.)* (9th ed.). Atlanta, GA: Atlanta/Fulton County Zoo.

———. 1998. *Orangutan studbook.* Atlanta, GA: Atlanta/Fulton County Zoo.

Pfeiffer, A., and Koebner, L. 1978. The resocialization of single caged chimpanzees and the establishment of an island colony. *Journal of Medical Primatology* 7: 70–81.

Phillips, L. 1997. Preventive medical program for captive gorillas. In J. Ogden and D. Wharton (eds.), *Management of gorillas in captivity* (pp. 181–90). Atlanta, GA: Gorilla Species Survival Plan and Atlanta/Fulton County Zoo.

Phillips, M. 1998. African Elephant Conservation Act. *Endangered Species Bulletin* 23: 38–40.

Poole, T. B. 1987. Social behavior of a group of orangutans (*Pongo pygmaeus*) on an artificial island in Singapore Zoological Gardens. *Zoo Biology* 6: 315–30.

———. 1991. Criteria for the provision of captive environments. In H. O. Box (ed.), *Primate responses to environmental change* (pp. 357–74). London: Chapman and Hall.

———. 1992. The nature and evolution of behavioural needs in mammals. *Animal Welfare* 1: 300–311.

———. 1998. Meeting a mammal's psychological needs. In D. Shepherdson, J. Mellen, and M. Hutchins (eds.), *Second nature: Environmental enrichment for captive animals* (pp. 83–94). Washington, D.C.: Smithsonian Institution Press.

Pope, C. E., Dresser, B. L., Chin, N. W., Liu, J. H., Loskutoff, N. M., Behnke, E. J., Brown, C., McRae, M. A., Sinoway, C. E., Campbell, M. K., Cameron, K. N., Owens, O. M., Johnson, C. A., Evans, R. R., and Cedars, M. I. 1997. Birth of a western lowland gorilla (*Gorilla gorilla gorilla*) following in vitro fertilization and embryo transfer. *American Journal of Primatology* 41: 247–60.

Popovich, D., and Dierenfeld, E. S. 1997. Nutrition. In J. Ogden, and D. Wharton (eds.), *Management of gorillas in captivity* (pp. 139–46). Atlanta, GA: Gorilla Species Survival Plan and Atlanta/Fulton County Zoo.

Porton, I., and White, M. 1996. Managing an all-male group of gorillas at the Saint Louis Zoological Park. *AZA Regional Conference Proceedings:* 720–28.

Read, B., and Tilson, R. 1998. *Southeast Asian Fauna Interest Group: Results from 1997 survey of projects.* Orlando, FL: Disney's Animal Kingdom.

Redshaw, M. E., and Mallinson, J. C. 1991. Stimulation of natural patterns of behaviour: Studies with golden lion tamarins and gorillas. In H. O. Box (ed.), *Primate responses to environmental change* (pp. 217–38). London: Chapman and Hall.

Regan, T. 1983. *The case for animal rights.* Berkeley: University of California Press.

———. 1995. Are zoos morally defensible? In B. Norton, M. Hutchins, E. F. Stevens, and T. L. Maple (eds.), *Ethics on the ark: Zoos, animal welfare and wildlife conservation* (pp. 38–51). Washington, D.C.: Smithsonian Institution Press.

Regan, T., and Francione, G. 1992. The animal "welfare" versus "rights" debate. *Animal's Agenda* 12: 45.

Reinartz, G. E. (ed.). 1994. *Bonobo* (Pan paniscus) *Species Survival Plan master plan 1994–1996.* Milwaukee, WI: Zoological Society of Milwaukee County.

———. 1998. *Bonobo studbook.* Milwaukee, WI: Zoological Society of Milwaukee County.

Reinartz, G., and Boese, G. 1997. Bonobo conservation: The evolution of a zoological society program. In J. Wallis (ed.), *Primate conservation: The role of zoological parks* (pp. 215–25). Norman, OK: American Society of Primatologists.

Robinson, J. G., and Bennett, L. (eds.). 2000. *Hunting for sustainability in tropical forests.* New York: Columbia University Press.

Robinson, P. T., Fagan, D. A., and Roffinella, J. P. 1979. Surgical removal of impacted molar teeth in an orangutan. *Journal of the American Veterinary Association* 175: 1000–1001.

Rose, A. L. 1996. The African forest bushmeat crisis: Report to the ASP. *African Primates* 2: 32–34.

Ruempler, U. 1990. Verhaltensanderungen von flachlangorillas in Zoologischen Garten Koln nach futterumstellung. *Zeitschrift des Kolner Zoo* 2: 75–84.

Sapolsky, R. M. 1990. Stress in the wild. *Scientific American* 262: 116–23.

Sarmiento, E. E., Butynski, T. M., and Kalina, J. 1996. Gorillas of the Bwindi-Impenetrable Forest and Virunga Volcanoes: Taxonomic implications of morphological and ecological differences. *American Journal of Primatology* 40: 1–21.

Schaller, G. B. 1963. *The mountain gorilla.* Chicago: University of Chicago Press.

Shepherdson, D. 1998. Tracing the path of environmental enrichment in zoos. In D. Shepherdson, J. Mellen, and M. Hutchins (eds.), *Second nature: Environmental enrichment for captive animals* (pp. 1–12). Washington, D.C.: Smithsonian Institution Press.

Shepherdson, D., Mellen, J., and Hutchins, M. (eds.). 1998. *Second nature: Environmental enrichment for captive animals.* Washington, D.C.: Smithsonian Institution Press.

Shettel-Neuber, J. 1988. Second- and third-generation zoo exhibits: A comparison of visitor, staff and animal responses. *Environment and Behavior* 20: 452–73.

Smith, D. G. 1986. Incidence and consequences of inbreeding in three captive groups of rhesus macaques (*Macaca mulatta*). In K. Benirschke (ed.), *Primates: The road to self-sustaining populations* (pp. 857–74). New York: Springer-Verlag.

Snyder, R. L. 1975. Behavioral stress in captive animals. In *Research in zoos and aquariums* (pp. 41–76). Washington, D.C.: National Academy of Sciences.

Snyder, R. L., and Amand, W. 1980. Diseases of gorillas and orangutans in the Philadelphia Zoological Gardens between 1918–1979. In *Proceedings of the international symposium on the diseases of animals* (pp. 69–72). Berlin: Akademie Verlag.

Sodaro, C. 1997. *Orangutan Species Survival Plan husbandry manual.* Atlanta, GA: Orangutan Species Survival Plan.

Soule, M., Gilpin, M., Conway, W. G., and Foose, T. J. 1986. The millennium ark: How long a voyage, how many staterooms, how many passengers? *Zoo Biology* 5: 101–13.

Thompson, S. 1993. Zoo research and conservation: Beyond sperm and eggs toward the science of animal management. *Zoo Biology* 12: 155–59.

Thorpe, L. 1988. Supplemental feeding of a western lowland gorilla at Audubon Park Zoological Garden (abstract). *American Journal of Primatology* 14: 488.

Tilson, R. L. 1986. Primate mating systems and their consequences for captive management. In K. Benirschke (ed.), *Primates: The road to self-sustaining populations* (pp. 361–73). New York: Springer-Verlag.

Tripp, J. K. 1985. Increasing activity in captive orangutans: Provision of manipulable and edible objects. *Zoo Biology* 4: 225–34.

Tudge, C. 1991. The buzz word in zoos is "behavioral enrichment" or how to make a captive environment more like the wild. *New Scientist* 129: 26–30.

Ullrey, D. E. 1996. Skepticism and science: Responsibilities of the comparative nutritionist. *Zoo Biology* 15: 449–53.

van Hooff, J. A. R. A. M. 1986. Behavior requirements for self-sustaining primate populations—Some theoretical considerations and a closer look at social behavior. In K. Benirschke (ed.), *Primates: The road to self-sustaining populations* (pp. 307–19). New York: Springer-Verlag.

VandeVoort, C. A., Neville, L. E., Tollner, T. L., and Field, L. P. 1993. Noninvasive semen collection from an adult orangutan. *Zoo Biology* 12: 257–65.

Watts, D. P. 1990. Mountain gorilla life histories, reproductive competition, and sociosexual behavior and some implications for captive husbandry. *Zoo Biology* 9: 185–200.

Weber, A. W., and Vedder, A. 1983. Population dynamics of the Virunga gorillas. *Biological Conservation* 26: 341–66.

Wharton, D. 1992. *The biology of gorilla SSP recommendations.* Paper presented at the Gorilla Workshop, Milwaukee, WI.

———. 1995. Zoo breeding efforts: An ark of survival? *Forum for Applied Research and Public Policy* 10: 92–96.

———. 1997. *North American regional studbook for the western lowland gorilla* (Gorilla g. gorilla). New York: Wildlife Conservation Society.

Wiese, R., and Hutchins, M. 1994. *Species survival plans: Strategies for wildlife conservation.* Bethesda, MD: American Zoo and Aquarium Association.

———. 1997. The role of North American zoos in primate conservation. In J. Wallis (ed.), *Primate conservation: The role of zoological parks* (pp. 29–41). American Society of Primatologists.

Willcock, C. 1964. *The enormous zoo: A profile of the Uganda National Parks.* New York: Harcourt Brace and World.

Wildt, D. 1989. Reproductive research in conservation biology: Priorities and avenues for support. *Journal of Zoo and Wildlife Medicine* 20: 391–95.

Wilson, E. O. 1984. *Biophilia: The human bond with other species.* Cambridge, MA: Harvard University Press.

Wilson, S. F. 1982. Environmental influences on the activity of captive great apes. *Zoo Biology* 1: 201–209.

Zucker, E. L., Robinette, D. S., and Deitchman, M. 1987. Sexual resurgence and possible induction of reproductive synchrony in a captive group of orangutans. *Zoo Biology* 6: 31–39.

ARNOLD ARLUKE

19
PERSPECTIVES ON THE ETHICAL STATUS OF GREAT APES

We are concerned about conserving, protecting, and studying great apes. It is often assumed that these approaches entail different and perhaps conflicting assumptions and strategies about how great apes should be viewed and treated. If this is true, these differences would impede communication among interested parties and make the formation and enactment of great ape policy more difficult. However, it is not clear whether these differences are real or mere stereotypes.

To explore these presumed differences, all participants at the 1998 workshop on Great Apes at an Ethical Frontier were asked to respond to a number of scenarios and statements, potentially eliciting different ethical perspectives toward the use and management of great apes. Approximately two months before the workshop, participants were sent a questionnaire to examine these perspectives. Respondents were asked to react to hypothetical situations and statements dealing with the moral equivalency of great apes and humans, to the human management of and responsibility toward great apes, and to the similarity of humans and great apes.

The mailed, self-administered, anonymous questionnaire had a response rate of 96 percent (22 of 23 attendees). Seven respondents were from zoo–conservation organizations (in this chapter called "conservationists"), six were from animal protection organizations ("protectionists"), and nine were from other organizations, most specifying university or university research–teaching ("academics"). It should be noted that there are missing cases in many tables because respondents either mistakenly omitted their responses or deliberately chose not to answer certain questions. Given the small number of respondents and the exploratory nature of this project, statistical tests of significance were not used; rather, results were eyeballed for general trends in responses.

RESULTS

Overall, these trends suggest that conservationists, protectionists, and academics were often in general agreement on many questions regarding the ethical status and management of great apes. Nevertheless, there were some differences among these groups when it came to viewing humans and great apes as morally equivalent, taking responsibility for great apes, and seeing humans and great apes as similar.

Moral Equivalency of Humans and Great Apes

The first scenario involved killing a chimpanzee for organ transplant to humans:

> Your spouse has suffered several major heart attacks and is now in congestive heart failure. There are no traditional medical interventions that can prevent death in the near future. You have recently discovered that Columbia Presbyterian Hospital in New York City has an experimental heart transplant program that would accept your spouse as a recipient, free of cost to you. Hospital researchers explain that they would transplant a heart from a healthy chimpanzee to your spouse. They also explain that the chimpanzee and your spouse have similar blood types, making the transplant safe. Choose either option below to deal with this scenario, even if your choice is "the lesser of two evils."
> (A) Sacrifice the chimpanzee to save your spouse.
> (B) Do not perform the transplantation and let your spouse die.

Answer	Conservationists	Protectionists	Academics	All
A	6	0	1	7
B	1	5	6	12

Slightly more than a third of the respondents expressed a willingness to sacrifice a chimpanzee to save a spouse. Most of this support came from conservationists, and there was virtually no support for such use from protectionists or academics.

Although there was some support for the medical use of chimpanzees, there was little support for the use of great apes in entertainment.

> Do you strongly approve/disapprove of Las Vegas stage shows that use great apes?

Strongly approve				Neutral			Strongly disapprove		
1	2	3	4	5	6	7	8	9	10

Conservationists	Protectionists	Academics	All
10.0	9.5	8.6	9.3

All respondents, despite affiliation, opposed such use, with strongest opposition coming from conservationists. The latter finding is interesting, given that conservationists provided the bulk of support for the use of chimpanzees in the medical scenario. Apparently, for some respondents willingness to regard great apes as objects to be used for human interest depends on the situation, whereas for other respondents, a more universal principle appears to operate.

Respondents also were given scenarios involving the management of a human and a chimpanzee having deadly, infectious diseases.

> An HIV-infected man is determined to infect other humans because of his anger about his condition. Which option would you prefer? (select only one).
> (A) Arrest and incarcerate him on the strength of his anger.
> (B) Wait until he engages in unprotected sex, then arrest him.
> (C) Let society take its chances.
> (D) Impose a death penalty for intentional, malicious infection.

Answer	Conservationists	Protectionists	Academics	All
A	1	4	2	7
B	3	1	5	9
C	0	0	1	1
D	2	1	0	3

> You are responsible for a 40-year-old AIDS-infected chimpanzee who also has A+ and B+ hepatitis. His rehabilitation chances are very low. Which option would you choose? (select only one).
> (A) Release him in to a park in Africa.
> (B) Euthanize him.
> (C) Continue to house him at $5,000 a year.

Answer	Conservationists	Protectionists	Academics	All
A	0	0	0	0
B	4	1	1	6
C	3	5	8	16

Almost all respondents did not want either the human or the chimpanzee to be allowed to infect others, and there was almost no support for letting "society take its chances" with the human or for releasing the chimpanzee to a park. If "A" and "B" are combined in the human-HIV scenario, approximately 80 percent of the respondents would opt for arrest and incarceration of the human, compared to 73 percent of the respondents opting for continued housing of the chimpanzee. However, although only 15 percent of the respondents would impose a death penalty on the human, 27 percent of the respondents would euthanize the chimpanzee. There were no marked differences among the three affiliation groups when it came to the management of the human HIV case (with "A" and "B" collapsed); however, protectionists and academics were more supportive of continued confinement of the chimpanzee, whereas conservationists were divided between confinement and euthanizing the chimpanzee.

A final way of examining perceived moral equivalency of great apes and humans was to ask for respondents' agreement with aims of the Great Ape Project (Cavalieri and Singer 1993), an organization devoted to improving the humane treatment of other great apes.

One such goal is to prevent great apes from being categorized as property and to include them within the category of persons. Attendees were asked about the degree of their support for the inclusion of great apes within the category "human" or vice versa.

Would you support reclassifying genus Pan to genus Homo or vice versa?

	Strongly approve			Neutral			Strongly disapprove		
1	2	3	4	5	6	7	8	9	10

Conservationists	Protectionists	Academics	All
5.4	2.0	4.6	4.1

Overall, attendees were slightly supportive of this notion, but protectionists were most supportive, although conservationists and academics were neutral.

Respondents also were directly asked about the extent of their agreement with the three Great Ape Project "principles."

> How do you feel about giving rights to great apes?
> (A) Great apes may not be killed except in very strictly defined situations, such as self-defense.
> (B) Great apes may not held captive without due legal process.
> (C) Great apes may not be used for invasive experiments that cause physical harm having no benefit to the animals involved.

1	2	3	4	5	6	7	8	9	10
Strongly approve				Neutral			Strongly disapprove		

	Conservationists	Protectionists	Academics	All
A	3.0	1.3	2.7	2.4
B	6.7	2.8	4.1	4.6
C	4.0	1.5	2.8	3.1

Overall, the strongest support was for A (not killing great apes except in very strictly defined situations), and the weakest support was for B (not holding great apes captive without due legal process). Protectionists showed the strongest support for all three principles, and conservationists the least support.

In summary, these responses suggest that conservationists are at least willing and protectionists quite eager to grant great apes the same moral status as humans, with academics falling in between. More specifically, although all three groups opposed the use of great apes for human entertainment, conservationists, as opposed to protectionists and academics, strongly supported the use of chimpanzee organs for human medical needs. Also, conservationists were least supportive of the Great Ape Project principles, such that of the three groups studied, they most opposed reclassifying great apes with genus *Homo* and granting them the right not to be killed, held captive, or used for experiments.

Human Responsibility for Great Apes

A second set of scenarios and questions examined respondents' views toward the human management of great apes. The first scenario pitted the protection of individual apes against the protection of larger populations of apes.

> There are 144 chimpanzees at an air force base in New Mexico. The Air Force no longer wants them and plans to euthanize the chimpanzees. Some

animal groups want to raise $14 million for an endowment to maintain these chimpanzees for life. Would you support this effort or would you prefer that the $14 million were used to buy land in Africa to create a reserve around Gombe Park?

(A) Use the money for the endowment.

(B) Use the money to buy land.

Answer	Conservationists	Protectionists	Academics	All
A	2	5	3	10
B	5	1	5	11

Conservationists and academics supported efforts to protect chimpanzee populations, and protectionists supported efforts to help individual chimpanzees.

Another scenario pitted human concerns against those of great apes.

To obtain local support for parks to protect apes, it might be necessary to allow local people to use parks as a reserve for bushmeat (not including apes) or wood. Would you approve of a plan with appropriate safeguards that would allow local people to use parks in this manner?

	Strongly approve			Neutral			Strongly disapprove		
1	2	3	4	5	6	7	8	9	10

Conservationists	Protectionists	Academics	All
2.4	5.0	3.9	3.7

Conservationists were the most willing to allow humans to share parks used by great apes, although this was not true of protectionists and academics, who were more neutral.

Respondents were also asked about their agreement with sentiment going beyond the three Great Ape Project principles discussed earlier.

If your response to the last statement was 1 through 4 (i.e., leaning toward approval), could you please respond to the following suggestions.

(A) Humans must accept a positive responsibility to actually protect apes in the wild from human predation and environmental destruction.

(B) Humans should provide territories for great apes.

(C) Humans should provide education for indigenous people near great apes.

(D) Humans should repatriate captive chimpanzees when it is in the chimpanzees' interest.

	Strongly approve				Neutral			Strongly disapprove		
	1	2	3	4	5	6	7	8	9	10

	Conservationists	Protectionists	Academics	All
A	1.3	1.0	1.1	1.1
B	1.6	1.9	1.2	1.6
C	1.4	1.2	1.1	1.2
D	5.1	2.0	1.9	3.0

There was very strong support across all three groups for humans to accept responsibility for protecting great apes, especially for providing territories to great apes and for providing education to indigenous people. There was less support for repatriating captive chimpanzees, especially from conservationists.

Finally, respondents were given a list of many of the organizations responsible for or influential in the conservation and protection of great apes, and were asked to rate the strength of their trust in these groups to manage a gorilla sanctuary:

To what extent would you trust the following organizations to manage a gorilla sanctuary? Please check the appropriate box for each organization.

	Strongly trust				Neutral			Strongly distrust			DK
	1	2	3	4	5	6	7	8	9	10	

	Conservationists	Protectionists	Academics	All	
African Wildlife Federation	4.0	5.3	3.0	4.0	12
Agency for International Development (AID)	6.6	5.8	4.8	5.8	7
American Zoo and Aquarium Association (AZA)	2.7	4.2	4.3	3.3	4
Humane Society of the United States	5.3	3.0	2.3	3.8	5

continued

continued

	Strongly trust 1 2 3	4 Neutral 5 6	7 Strongly distrust 8 9	10	DK
	Conservationists	Protectionists	Academics	All	
International Fund for Animal Welfare (IFAW)	4.6	3.2	2.0	3.4	8
World Conservation Union	4.7	3.2	4.3	3.8	4
Rwandan Government	6.1	5.5	4.9	5.5	1
United Nations Environment Program (UNEP)	6.2	6.5	3.7	5.7	10
World Society for Protection of Animals	5.8	3.4	3.7	4.4	9
World Wildlife Fund	4.1	4.3	2.1	3.5	2

Note: DK (don't know) refers to respondents that were not familiar enough with the organization to make a judgment.

Overall, respondents were not markedly different in their trust of these groups. Mean scores were concentrated between moderate trust and neutral. Least trusted were AID and UNEP, whereas the AZA and the IFAW were the most trusted. Conservationists appeared to be least trusting of many of these organizations, and protectionists appeared to be most trusting.

When looking at the results of this second set of scenarios and questions, conservationists had somewhat different responses than did protectionists regarding human management of great apes. Conservationists were more inclined to help populations of great apes than individual animals, to aid humans as well as great apes, and to kill, hold captive, and experiment on great apes under specified (and limited) conditions. They also had the least trust in organizations involved with the conservation and protection of great apes.

Human Similarity to Great Apes

To the extent that members of the groups did not share the same assumptions about the moral status of great apes and how best to manage them, it was important to ask what underlying beliefs might account for these differences. A variety of morally relevant traits, if attributed to great apes, might be linked to respondents' willingness to expand the "moral circle" so that ethical value is accorded to more than just humans. To examine this, respondents were first asked to rate various nonhuman great apes (and gibbons, as "lesser apes," Byrne 1995, were used for comparison) according to their similarity to adult humans.

On the following scale, where would you place the following animals?

Alzheimer's patient		Preverbal human infant (9 months)			Human child (4 years)			Verbal adult human	
1	2	3	4	5	6	7	8	9	10

	Conservationists	Protectionists	Academics	All
Bonobo	5.8	7.5	6.4	6.6
Chimpanzee	5.7	7.5	5.8	6.4
Gibbon	4.4	5.5	4.6	4.9
Gorilla	5.8	6.7	5.8	6.1
Orangutan	6.0	7.0	6.0	6.3

Apart from gibbons, respondents rated all apes very similarly. Yet protectionists saw all apes as most like verbal adult humans than did academics or conservationists.

Respondents' perceptions of the similarity of great apes with adult humans might be linked to other views that respondents have toward great apes. The results in this regard were mixed. On one hand, respondents who were more likely to see great apes as similar to adult humans also tended to approve more strongly of giving great apes the right not to be held captive without due legal process and the right not to be used for invasive experiments that cause physical harm having no benefit to the animals involved. However, there was no apparent trend between seeing great apes as like adult humans and support for the right of great apes not be to be killed except in strictly defined situations. Also, respondents' perception of the similarity of great apes with adult humans appeared to be slightly related to respondents' support for reclassifying genus *Pan* to genus *Homo* or vice versa. However, there was no apparent relationship between the perception of similarity and respondents' views on medically sacrificing a chimpanzee or managing an HIV-infected human versus an AIDS-infected chimpanzee.

When asked to rate whether great apes possessed certain "human" faculties, respondents were in fairly strong agreement that apes (except gibbons) possessed these faculties (responses to "bonobo" and "chimpanzee adult" could not be dichotomously recoded because responses were too similar).

The following animals can make plans for the future, are self-conscious, and are able to experience emotions such as love, hate, jealousy, or disappointment.

Strongly agree				Neutral			Strongly disagree		
1	2	3	4	5	6	7	8	9	10

	Conservationists	Protectionists	Academics	All
Bonobo	2.4	2.8	1.9	2.3
Chimpanzee	2.6	2.7	2.0	2.4
Gibbon	4.6	6.0	3.1	4.4
Gorilla	2.6	3.3	2.4	2.7
Orangutan	2.4	3.3	2.4	2.7

Respondents who more strongly agreed that gibbons, gorillas, and orangutans could make plans for the future, are self-conscious, and are able to experience various emotions were also more approving of granting them the right not to be held captive and the right not to be used for invasive experiments. However, there was no apparent trend between this perception of gibbons, gorillas, and orangutans and approval of their right not to be killed. There also was no apparent connection between this perception and respondents' views on classifying genus *Pan*, medically sacrificing a chimpanzee, or managing an HIV-infected human/HIV-infected chimpanzee.

Finally, respondents also were asked whether certain faculties should serve as a basis for moral consideration to see if these views might be related to other views of great apes.

Not everyone agrees which faculties should form the basis for moral consideration. Indicate below which of the following facts are relevant to whether or not an individual should be accorded moral consideration.

Strongly agree				Neutral			Strongly disagree		
1	2	3	4	5	6	7	8	9	10

	Conservationists	Protectionists	Academics	All
Sentience	3.3	1.7	2.3	2.4
Feel emotions	3.9	1.5	2.3	2.6
Needs/interests	4.3	1.3	2.3	2.6
Intelligence	4.4	2.3	3.2	3.4
Defenselessness	5.6	6.2	3.0	4.7
Pain sensitivity	3.9	2.7	2.3	2.9
Self-awareness	3.2	2.3	2.7	2.8
Future planning	3.7	3.0	3.3	3.4

Findings were mixed. Although all respondents were in general agreement, conservationists agreed least that these faculties should serve in this manner, and protectionists (except for defenselessness and sensitivity to pain) agreed the most. (Of the eight items in this question, only "intelligence" could be recoded dichotomously because responses were so strongly skewed at one end of the scale for other faculties.) Respondents who more strongly agreed that intelligence should serve as a basis for moral consideration to great apes were also more approving of granting them the right not to be held captive and the right not to be used for invasive experiments. However, there did not appear to be any link between respondents' opinions about the moral significance of intelligence and their views on granting the right of not being killed to great apes, reclassifying genus *Pan,* medically sacrificing a chimpanzee, or managing an HIV-infected human/AIDS-infected chimpanzee.

The results from this third set of questions suggest two things. First, that conservationists were the least willing to see great apes as similar to adult humans and to possess human faculties. And second, viewing humans and great apes as similar might be linked to some, but not all, of the attitudes expressed by respondents, such that regardless of their background, respondents who were least willing to see great apes as similar to humans and as possessing human faculties were least willing to grant certain rights to great apes.

DISCUSSION

The results are mixed regarding the similarity of attitudes of conservationists, protectionists, and academics toward the ethical status of great apes. On the one hand, members of these groups hold generally similar views. Even for those questions that produced somewhat different responses, respondents' answers were often in the same ballpark. This finding is consistent with an earlier study (Rowan 1995) that examined the views of individuals from these groups on topics relating to captive and wildlife animal management and found them to be largely in agreement. Moreover, all respondents in this study were remarkably similar when comparing their demographic background and general attitudes. For instance, the vast majority of respondents owned pets, felt they were capable of empathizing with animals, and considered themselves to be politically liberal.

Despite these similarities, findings point to differences, perhaps subtle but nonetheless real, among conservationists, protectionists, and academics. Although many answers were in general agreement, even small differences may hint at formidable problems for consensus building or compromise in the construction of policy and programs in the future. Conservationists, compared to protectionists and academics, were somewhat more willing to maintain great apes in captivity, to sacrifice–euthanize great apes, to allow human contact with them, and to resist taxonomic reclassification that would place great apes and humans in the same genus. Conservationists, compared to protectionists and academics, were somewhat less willing to grant certain rights to great apes, to see great apes as similar to adult humans, and to award moral consideration to great apes based on certain faculties.

These differences might be a methodological artifact. The forced-choice approach, used in many of the questions, might exaggerate differences among the groups studied. For example, when respondents were *not* forced to pit human interests against those of great apes, there was a marked willingness to seek a middle ground by trying to help both humans and apes. When given the choice to share resources to help both Rwandan refugees and mountain gorillas, respondents overwhelmingly (90 percent) chose to do so rather than electing to give aid only to help humans or only to gorillas. Of course, it could be argued that the forced-choice approach was necessary to flush out genuine differences in perspective. Clearly, future research is needed on the perspectives of conservationists, protectionists, and academics to provide definitive answers about the disparity of their views.

REFERENCES

Byrne, R. 1995. *The thinking ape: Evolutionary origins of intelligence.* Oxford: Oxford University Press.

Cavalieri, P., and Singer, P. (eds.). 1993. *The Great Ape Project.* London: Fourth Estate.

Rowan, A. (ed.). 1995. *Wildlife conservation, zoos and animal protection: A strategic analysis.* Grafton, MA: Tufts Center for Animals and Public Policy.

INDEX

A

aesthetic factors, 315, 320, 326
Africa. *See also specific countries:* chimpanzee sanctuaries, 136–141, 143; human population pressures, 23–24; political instability and wars, 32, 235; total great apes killed for meat trade, 27
African Wildlife Federation, 373
Agency for International Development, 373
aggressive or violent behavior, 35, 37, 169–170, 236–238, 347
agricultural harvesting, 231, 236, 238
AIDS. *See* HIV and AIDS
American Zoo and Aquarium Association (AZA), 144; accreditation, 225, 331, 349; Bushmeat Crisis Task Force, 28, 338; Code of Professional Ethics, 351; Conservation Education Committee, 120, 121, 128; education by, 113, 128; on ethics of captivity, 303–304, 305; ethics questionnaire results, 373; number of accredited institutions, 336; source and disposition of animals, 351; Species Survival Plan. *See* Species Survival Plan; Taxon Advisory Groups, 335, 351, 352
Angola, 6, 12, 13
animal consciousness, 266–270, 289, 301, 315–319, 376
Animal Legal Defense Fund, 146
animal rights. *See* rights
animal rights advocates, 207–208, 316, 330
animal welfare, *vs.* animal rights, 330, 353
Animal Welfare Act, 156, 349
Anmann, Karl, 255
anthropocentrism, 174, 194–196, 205–206, 275, 297–301, 305–308
anthropological research, 169–170, 171–172, 181–182, 185–186
anthropomorphism, 253
Ape Alliance, 28
ape-human boundaries, 163–175, 178–181, 191–194, 203–204, 245–255, 370–371
ape refugees. *See* sanctuaries
apes rights. *See* rights
Ardipithecus, 182, 284
area of occupancy, *vs.* geographic range, 4–5, 10
Atlanta zoo case study, 124
attacks, 35, 37, 236–238. *See also* aggressive or violent behavior
Australopithecus, 180, 182–183, 284
autonomy, rights derived from, 280–283
awareness. *See* consciousness

AZA. *See* American Zoo and Aquarium Association

B

behavior: aggressive or violent, 35, 37, 169–170, 236–238, 347; atypical, in zoos, 114, 339, 341; cognitive ethology, 262–272; conservation-related, 117; of early man, 185, 284; emergent, 249–250; extrapolation from apes to humans, 165, 167–168, 171; foraging, 341; gestural dialects, 200–201; grooming, 201, 340; maternal, 346; moral, and empathy, 261; in natural environment, 169; play, 199; reproductive, 201, 345–347; stress, 36–37, 348
bipedalism, 184, 246
bonobo. See *Pan paniscus*
breeding programs: National Chimpanzee Breeding and Research Program, 150–153, 155, 156, 158; for research, 226; software, 335; in zoos, 119, 332, 334–336, 351–352
Bronx Zoo, Congo Forest Exhibit, 125–126, 336
Brookfield Zoo, "Quest" exhibit, 126–127
Burundi, 6, 140
bushmeat. *See* commercial hunting
Bushmeat Crisis Task Force (BCTF), 28, 338
"by-catch," great apes as, 87, 234

C

Cameroon: chimpanzee population estimate, 6; chimpanzee sanctuaries, 139; commercial hunting, 27, 76–79, 353; gorilla population estimate, 13; lack of political will, 80–81
captivity: as the best solution, 304–305; breeding programs for research, 226; breeding programs in zoos, 119, 332, 334–336, 351–352; care and husbandry, 338–349, 350; cognitive ethological approach, 271–272; and inclusivist ethics, 295–310; justification for, and the right to liberty, 321–322, 331; management in, 143, 145, 151–153; National Chimpanzee Breeding and Research Program, 150–153, 155, 156, 158; and transmission of disease, 35–36; zoo education, 113–128
Central African Republic (CAR): chimpanzee population estimate, 6; gorilla population estimate, 12, 13
Chicago zoo case study, 126–127
chimpanzee: central. See *Pan troglodytes troglodytes;* cladistic analysis, 180–181, 192; cognitive capability, 224–226; common. See *Pan troglodytes;* conceptual capacity, 215–226; eastern. See *Pan troglodytes schweinfurthii;* gorilla-like. See *Pan troglodytes koolokamba;* gracile. See *Pan paniscus;* Nigeria. See *Pan troglodytes vellerosus;* pet trade, 225; pygmy. See *Pan paniscus;* as research subjects. *See* research subjects; robust. See *Pan troglodytes;* sanctuaries, 135–141; sign language research, 196–201; similarity of DNA with humans, 12, 186, 245; Species Survival Plan, 225, 333; toolmaking capability, 171–172; western. See *Pan troglodytes verus;* wild, moral issues, 230–242; zoo population, 332, 333
Chimpanzee Health Improvement Maintenance and Protection Act, 141–142, 153–154, 156, 158
CITES (Convention on International Trade in Endangered Species of Wild Fauna and Flora), 21–22, 150, 345, 351
cladistic analysis, 180–181, 192, 284
Code of Professional Ethics, 28, 338
cognitive capability, 187, 246–247; and brain size, 250, 251; to choose a dangerous freedom, 322; conceptual capacity, 215–226; discrimination, 268–269; ethics questionnaire results, 374–377; evolutionary aspects, 263; experience of pain, 329; learning systems, 270; number concepts, 216–217; quantity judgement, 217–220; "subjects-of-lives" sentience, 316–317, 318, 326; toolmaking, 171–172, 185, 252, 329; use of symbols, 172, 219, 221, 224, 251, 320, 329
Columbus Zoo, 337

commercial hunting, 26–29; apes as "by-catch," 87, 234; Cameroon, 27, 76–79, 353; Congo Basin, 86–102; Congo-Brazzaville, 74–76, 82; Democratic Republic of Congo, 72–74; Gabon, 79; inclusion in zoo education, 120; income generated by, 91–93; and loggers, 79–80, 255; mitigation options, 94–101; multinational agreements to limit, 255; and political corruption, 69, 75, 76, 80–82; poor law enforcement, 60, 61, 68, 71, 78; quotas, 100–101; Uganda, 233
community development, 240
conceptual capacity, 215–226
Congo Basin, 86–102
Congo-Brazzaville: chimpanzee sanctuaries, 139–140; commercial hunting, 74–76, 82; lack of political will, 82
consciousness, animal, 266–270, 289, 301, 315–319, 376
conservation. *See also* nature reserves; sanctuaries: community-based, 99–100; ecosystem, 62, 68; ethical issues, 38–39; "flagship species," 3; forest management in Uganda, 236; global context, 330; hunting management in the Congo Basin, 93–101; individual rights *vs.* population survival, 230, 287–288, 330–331; Kibale National Park chimpanzee issues, 230–241; "precautionary principle," 4; recommendations, 39–42, 83–84; research in nature reserves, 240–241; "tragedy of the commons," 63; *vs.* rehabilitation, 64–65; zoos and education, 116–117, 118, 121–127, 329, 336–338
Conservation Action Partnership, 338
Conservation Education Committee, 120, 121, 128
Conservation International, 338
Convention on International Trade in Endangered Species of Wild Fauna and Flora (CITES), 21–22, 26, 150, 345, 351
Côte d'Ivoire, 6, 138
crime. See illegal practices
crop-raiding, 231, 236, 238
cross-fostering, 196–197, 198

D

Darwinian theory, 166, 191–208, 262–263; a modern model, and fuzzy logic, 201–208; natural selection, 194; perception of nature, 167, 195; theory of evolution, 191, 247
Dawkins, Richard, 202–203, 261
Declaration on Great Apes, 38, 186, 309, 314
deforestation. *See* logging
Democratic Republic of Congo (DRC): chimpanzee population estimates, 4, 6; commercial hunting, 72–74; distribution, 4; gorilla population estimate, 12, 13
dentition, 182, 183, 184
Descent of Man, The, 166, 263
de Waal, Frans, 169
Dian Fossey Gorilla Fund International, 337, 338
diet: feeding and nutrition in zoos, 343; human, role of great apes in. *See* subsistence hunting
disease. *See also* HIV and AIDS: apes as research subjects. *See* research subjects; effects on endangered species, 29–36; safety recommendations, 41; and sanctuary animals, 142–143, 152; transmission from apes to humans, 29, 34, 152, 238; transmission from humans to apes, 30–34, 236, 239, 240; transmission risk from reintroduction, 142–143, 156, 353; vaccinations for gorillas, 338
Disney's Animal Kingdom, 126
distribution. *See* geographic range
DNA analysis: *Gorilla gorilla,* 12; *Hominoidea,* 178; *Pan troglodytes vellerosus,* 5; percent similarity between chimpanzees and humans, 12, 186, 245; use in zoos, 344
DRC. *See* Democratic Republic of Congo

E

ecological aspects: approach to conservation, 62; of behavior, 171; biodiversity and forest management, 25; great apes as flagship species, 3; invasive species, 317; land ethic

ecological aspects (*cont.*)
and moral obligation, 319–320; naturalistic exhibit design, 339–340
ecotourism, 239–240
education: cognitive learning, 115–116; objectives, 122–123; the role of sanctuaries, 144; the role of zoos, 113–128, 303–304, 330, 340; use of virtual reality, 124
emergent behavior, 249–250
endangered, definition, 22
Endangered Species Act, 278
endangerment status, 21–23; chimpanzee, 16; effects of disease, 29–36; gorilla, 16; obligation toward, 314–315, 317, 326–327; orangutan, 4
England, sanctuary, 136
entertainment industry: age limitations, 226; apes in films and television, 168, 169, 225; ethics questionnaire results, 368–369; refugees from, 136, 142, 156; zoo-bred apes in, 351–352
environmental impact assessment, 39
environment enrichment, in captivity, 156–157, 205, 319, 339, 340–342, 349
equality right, 285–288, 324–325, 368–371
Equatorial Guinea, 6, 12, 13
ethical issues. *See also* moral issues; rights: anthropocentrism, 194–196, 205–206, 275, 297–301, 305–308; and conservation of free-living apes, 38–39; disposition of zoo-bred apes, 351; euthanasia, 151, 152, 323, 352; evolutionary biology, 186; inclusivist ethics, 295–310; individual rights *vs.* population survival, 230, 287–288, 330–331; and phylogenetic relatedness, 262; workshop questionnaire results, 367–378
ethology, 262–272
European Endangered Species Programme (EEP), 336
European Union: conservation recommendations for, 39; and deforestation, 25, 28
euthanasia, 151, 152, 323, 352
evolution: adaptation and emergent behavior, 249, 268; cladistic analysis, 180–181, 192, 284; of cognitive ability, 263; of common law, 276–277; convergent, 262, 263–264; Darwinian theory. *See* Darwinian theory; human, 166–167, 179–186; natural selection, 194; phylogeny, 178–187; of the primate brain, 250
exhibit-enclosure design, 339–340
extinction: and justification for captivity, 321–322; timeframe estimates, 113, 133

F

feeding and nutrition, 343
financial aspects: bushmeat tax, 97; conservation, 41, 68, 125; ecotourism, 239; hunting management options, 96; sanctuaries, 143, 144–145
foraging, 341
Fossey, Dian, 169. *See also* Dian Fossey Gorilla Fund International
fossil record, 179–186
Friends of the Animals, 143
Friends of the Earth, 28

G

Gabon: chimpanzee population estimate, 6; chimpanzee sanctuaries, 139; commercial hunting, 79; gorilla population estimate, 13
Gambia, 6, 136–137
generation, years in one, 22
genetic comparisons. *See* DNA analysis
gene transfer, 174
geographic range, 3–17; *Gorilla beringei,* 3, 13–14, 15–17; *Gorilla gorilla,* 12–14; *Pan paniscus,* 8, 10–12; *Pan troglodytes,* 3, 4, 5–10; *vs.* area of occupancy, 4–5, 12
Ghana, 6, 138
global market: bushmeat, 26–27; pets, 57–58; timber, 25
Goodall, Jane, 168, 171. *See also* Jane Goodall Institute
gorilla: Bwindi. See *Gorilla beringei* (ssp?); cladistic analysis, 180–181; Cross River. See *Gorilla gorilla diehli;* eastern. See *Gorilla beringei;* enriched environment in zoos,

341; Grauer's. See *Gorilla beringei graueri;* mountain. See *Gorilla beringei beringei;* population estimate, 332; projected population, 20-year, 333; reproduction in captivity, 345–346; sanctuaries, 135; social groups, 342, 346, 350–351; Species Survival Plan, 333, 335, 337, 346, 350; status in zoos, 332, 333; western. See *Gorilla gorilla;* western lowland. See *Gorilla gorilla gorilla;* zoo population, 332, 333
Gorilla beringei: geographic range, 3, 13–14, 15–17; habitat, 15; population estimates, 16, 21
Gorilla beringei (ssp?): endangerment status, 23; geographic range, 13, 15; population estimates, 16, 21
Gorilla beringei beringei: endangerment status, 23; geographic range, 13–14, 15; population estimates, 16, 21; tourism and disease, 31–32
Gorilla beringei graueri: area of occupancy *vs.* geographic range, 4–5; geographic range, 13–14, 15–17; one in captivity, 333; population estimates, 16, 21
Gorilla gorilla: geographic range, 3, 12–15; habitat, 12; hunting, 27; population estimates, 16, 20–21
Gorilla gorilla diehli: endangerment status, 23; geographic range, 12–14, 15; population estimates, 16, 21
Gorilla gorilla gorilla: geographic range, 12–14; population estimates, 16; slaughter in Cameroon, 353; status in zoos, 332, 333
Gorilla gorilla uellensis, 14
Great Ape Conservation Act, 352
Great Ape Project: on ape-human boundaries, 261; on apes as moral agents, 299–300; current positions, 310; on ethics of captivity, 295–296; ethics questionnaire results, 367–378; impact on public policy, 186–187; response to the, 329–354; rights principle, 325, 371; *vs.* positions of the AZA, 303–304
Great Ape Project, The, 261, 354
grooming, 201, 340
Guinea, 6, 137

H

H. R. 106-3514 (Chimpanzee Health Improvement Maintenance and Protection Act), 141–142, 153–154, 156, 158
habitat loss: deforestation. *See* logging; effects on population estimates, 4; motivation to avoid, 322; relocation, 324; reserves as biological islands, 241; Uganda, 235–236
habitats, natural, 3, 5, 10, 12, 15
habituation, 36
historical background: Darwinian theory, 166; dignity-rights, 279, 281–283; human community, 300; legal rule and nonhuman animals, 274–276, 323–324; of primate science, 163–167; zoos, 339, 343
HIV and AIDS: chimpanzees as research subjects, 151, 205, 302, 322, 369–370; transmission to humans, 29, 34
home range, 232, 271
Hominidae, 178–179, 181–182, 191, 313
Homo, in primate hierarchy, 163–167, 171–172, 180–187, 191–194, 246–248, 370–371
human-ape boundaries. *See* ape-human boundaries
human-ape interactions. *See also* disease; conservation issues, 230–241; guidelines, 142; relationship with caregiver, 353
Humane Society of the United States, 143, 373
human population pressures, 23–24, 330. *See also* disease; habitat loss; global, 133; Indonesia, 57, 61; stress behavior, 36–37; Uganda, 235
human rights, 279–282, 315
hunting. *See* commercial hunting; subsistence hunting
Huxley, Thomas, 166

I

illegal practices: hunting, 27, 29, 235, 330. *See also* political aspects, corruption; logging, 25; orangutan captivity, 57, 67
inclusivist ethics, 295–310

Indonesia, 57–69, 336
infection. *See* disease
informed consent, 323
intelligence. *See* cognitive capability; conceptual capacity; linguistic capability
International Code of Zoological Nomenclature, 179
International Fund for Animal Welfare, 374
International Monetary Fund, 83
International Primate Protection League, 143
International Primatological Society, 254
International Union for the Conservation of Nature and Natural Resources (IUCN), 4; endangerment status by, 16; Primate Specialist Group, 5, 22–23, 144; Reintroduction Specialist Group, 145; *World Conservation Strategy,* 61
International Union of Directors of Zoological Gardens (IUDZG), 351
IUCN Red Data Book, 16, 22
IUCN Red List of Threatened Animals, 22, 23

J

Jane Goodall Institute, 143, 144, 338

K

Kenya, 83–84, 140
koolokamba. See *Pan troglodytes koolokamba*

L

laws: Animal Welfare Act, 156, 349; Chimpanzee Health Improvement Maintenance and Protection Act, 141–142, 153–154, 156, 158; common, 276–277; Endangered Species Act, 278; Great Ape Conservation Act, 352; historical background, 274
legal issues: of liberty, 278–279; rights for nonhuman animals, 274–289, 286–287, 323–324, 325; for sanctuaries, 146
Liberia, 138
liberty: legal issues, 278–279; the right to, 321–322
life, the value of, 277–278
lifespan, 225, 342
linguistic capability, 204, 246; development in infants, 251–252; of early man, 185–186, 284; and informed consent, 323; as measurement of consciousness, 266; Native Americans, 193; primates, 163–164, 170–174, 186, 196–201, 204–205; sign language, 196–201, 204, 251
Linnaeus, 163–165, 178, 191
locomotion, bipedalism, 184, 246
logging, 24–26, 330; and commercial hunting, 79–80, 255; and hunting management, 97–98; Indonesia, 57, 58, 60–63; roads, and increased access, 133; Uganda, 235

M

Mammal Species of the World, 191
markets. *See* global market
maternal behavior, 36
media and the lure of rehabilitation, 66, 67
mental abilities. *See* cognitive capability; conceptual capacity
metaphysics and apes, 165, 166, 173–174, 203
Miocene Epoch, 178–181
mitigation options: demand-side, 94–95; supply-side, 96–101; three categories, 102
moral issues. *See also* ethical issues; rights; about wild chimpanzees, 230–242; animal impulses from primate ancestry, 167–168; bestiality, 163; cognitive ability and animal rights, 261–262; equality right, 285–288, 324–325, 368–371; individual *vs.* group benefits, 230, 287–288; moral agency as prerequisite to rights, 325–326; moral status of great apes, 313–327; of research, 206, 226; rights *vs.* obligation toward, 326–327, 371–372; role of empathy, 261
movies, apes in. *See* entertainment industry

N

National Chimpanzee Breeding and Research Program, 150–153, 155, 156, 158

National Institutes of Health. *See* National Chimpanzee Breeding and Research Program

National Research Council, report on chimpanzees in research, 150–153

National Zoological Park "Think Tank," 124–125, 339, 340

natural selection, 194

nature reserves, 40; animal management, 337–338; Congo, 75; Gabon, 79; research, 240–241

Nigeria, 6, 13, 138–139

non-governmental organizations (NGO): Cameroon activities, 28; collaboration with private sector, 99; ethics questionnaire results, 373–374; human population policy, 39; sanctuary support, 143

O

orangutan: asocial nature, 345; Bornean. See *Pongo pygmaeus;* conservation in Indonesia, 57–64, 68; enriched environment in zoos, 341, 349; hybrids, population estimates, 332, 334; in Indonesia, 57–69, 336; pet trade, 57–58; sanctuaries, 135, 136; Species Survival Plan, 334, 335, 337, 352; status in zoos, 334; Sumatran. See *Pongo abelii;* zoo population, 332

Orangutan Foundation International, 143

Orangutan Survival Programme, 68, 69

organ transplantation, 174, 205, 368

orphans. *See* rehabilitation; sanctuaries

P

Paleolithic Period, 185

paleontology, 179–186

Pan paniscus: endangerment status, 23; geographic range, 3, 8, 10–12; habitat, 10; population estimates, 20, 332; projected population, 20-year, 331–332; Species Survival Plan, 331, 337; status in zoos, 331–332, 336, 344–345; toolmaking ability, 252

Pan troglodytes: endangerment status, 23; geographic range, 3, 4, 5–10; habitat, 5; population estimates, 4, 6–7, 16, 17–20, 23, 332; taxonomy, 5–7; zoo population, 332, 333

Pan troglodytes koolokamba, 5, 7

Pan troglodytes schweinfurthii, 5; geographic range, 9–10; population estimates, 6–7, 16, 19

Pan troglodytes troglodytes, 5; geographic range, 9; population estimates, 6, 16, 18–19

Pan troglodytes vellerosus, 5; geographic range, 7, 9; population estimates, 6, 16, 18

Pan troglodytes verus: elevation to *Pan verus,* 5; geographic range, 7, 9; population estimates, 6, 16

Pan verus, species status pending, 5

Paranthropus, 183, 185

parasites. *See* disease

People's Republic of Congo (PRC): chimpanzee population estimate, 6; commercial hunting, 26–27, 28; gorilla population estimate, 13

perception of apes. *See* ape-human boundaries

pet trade, 34, 133; chimpanzee, 225, 234; orangutan, 57–58; zoo-bred apes in the, 351–352

philosophical and religious aspects: anthropocentrism of the Catholic Church, 305; of ape-human boundary, 164–166, 174, 192–194, 203–204, 247–248; community, 300, 309; early man, 185, 284; of ethics for nonhuman animals, 297; moral status, 313–327; of scientific investigation, 264–265

phylogeny, 178–187

play, 199

political aspects: conservation, 82–84; corruption, 69, 75, 76, 80–82; exploitation of instability by logging companies, 25; poor law enforcement, 60, 61, 68, 71, 78; Rwanda, 32; Uganda, 235; war, 32, 168, 235

Pongo abelii: conservation activities, 60; endangerment status, 4, 59; geographic range, 58; population estimates, 58–59, 332; zoo population, 332, 334

Pongo pygmaeus: endangerment status, 4, 59; geographic range, 58; phylogeny, 181; population estimates, 58, 332; zoo population, 332, 334
population density, Congo Basin estimates, 86, 90
population estimates, 16, 17–21. *See also specific great apes;* accuracy problems, 4–5; and endangerment status, 16; great apes in zoos, 331–334
population genetics, 335–336
population management, in zoos, 334–336, 350–351
PRC. *See* People's Republic of Congo
Primarily Primates sanctuary, 136
Primate Specialist Group, 5, 22–23, 144
proportionality rights, 286
protein, great apes as source. *See* subsistence hunting
psychological aspects: attitudes toward animals, 114, 116, 134, 144, 314–315; cognitive learning in zoos, 116; discrimination, 268–269; effects of environment, 246; loyalty to our species, 298–299; of rehabilitation, 66; of violence and peacefulness, 169–170
public perception of apes. *See* ape-human boundaries

R

ranching, bushmeat privatization, 100
range. *See* geographic range
recommendations: conservation, 39–42, 83–84; disease safety precautions, 41; education objectives, 122–123; sanctuaries, 144–146
Red Data Book, 16, 22
Red List of Threatened Animals, 22, 23
refugees. *See* sanctuaries
rehabilitation: chimpanzees, 225; enriched environment during, 156–157, 205, 319; of former pets, 225; methodology, 143, 145; orangutans, 57, 58, 64–69; in sanctuaries, 133–146, 152; *vs.* conservation, 64–65
reintroduction: of chimpanzees from sanctuaries, 142–143, 145; disease transmission risk, 142–143, 156, 353; ethics questionnaire results, 373; low survival rate, 156, 321; and the right to liberty, 321; zoos as source for, 336–337
Reintroduction Specialist Group, 145
religion. *See* philosophical and religious aspects
relocation, 324
reproduction: artificial, 344; behavior, 201, 345–347; contraception, 145–146, 344–345; of research chimpanzees, 151; in zoos, 344–345
rescue projects. *See* rehabilitation; sanctuaries
research on great apes: conceptual capacity, 220–224; cross-fostering, 196–197, 198; by host-country scientists, 240–241; in nature reserves, 240–241; philosophical aspects of scientific investigation, 264–265; seven U.S. centers, 150; sign language, 196–201, 204, 251; in zoos, 118, 330
research subjects: chimpanzee refugees, 126, 141–142, 150–151; and the Darwinian perspective, 205–208; enriched environment for, 206; housing and management standards, 155, 157; retirement. *See* retirement; rights. *See* rights; zoo-bred apes as, 351–352
retirement. *See also* euthanasia; sanctuaries: ethical considerations, 369–370; research subjects, 150–158, 322; from zoos, 352
rights, 305–307. *See also* ape-human boundaries; Great Ape Project; derived from practical autonomy, 280–283; dignity-, 187, 279, 281–283, 315; equality, 285–288, 324–325, 368–371; Great Ape Project principle, 38, 325, 371; human, 279–282, 315; human attitudes and abuse, 134, 205–208; individual *vs.* population, 230, 287–288, 330–331; laboratory care policy, 154, 155; legal, 274–289, 323–324, 325; liberty, 321–322; life, 322–324; moral status of great apes, 313–327; "poster mammals" for the movement, 186; proportionality, 286, 287; sanctuary inspec-

tions, 156; species *vs.* ecosystem, 317; U.S. legislation, 141–142, 153–154; *vs.* welfare, 330, 353
Rwanda, 6, 32, 374

S

San Antonio sanctuary, 136
sanctuaries, 133–146; ethics questionnaire results, 373–374; inspections, 156; limited space, 225; for retired research subjects, 153–158, 322
Senegal, 6, 137
sentience. *See* consciousness
Sierra Leone, 6, 138
sign language, 196–201, 204, 251
Sivapithecus, 181
snare wounds, 234–235
Société Industrielle et Forestière Congo Allemand (SIFORCO), 72, 73, 74
South Africa, 141
Species Conservation Priorities in the Tropical Forests of Southeast Asia, 62
speciesim, by humans, 194–196, 205, 206, 275, 297–299, 307
Species Survival Plan, 119; bonobo, 331, 337; chimpanzee, 225, 333; European equivalent, 336; gorilla, 333, 335, 337, 346, 350; number of participating institutions, 337; orangutan, 334, 335, 337, 352
speech. *See* linguistic capability
stress behavior, 36–37, 348
subsistence hunting, 27, 86; harvest estimates, 90–91; income from. *See* commercial hunting; meat consumption estimates, 89, 91–93; moral obligation and attitudes with, 327; sustainable harvest estimates, 88–91; technology development, 133; and transmission of disease, 34
Sudan, 6
sustainable practices: forest management, 25; hunting harvest quotas, 100–101; hunting in the Congo Basin, 88–91, 93–101; and moral attitude of aboriginal peoples, 327; *World Conservation Strategy,* 61–62
Systema naturae, 163, 164

T

Tanzania: chimpanzee attacks on humans, 236; chimpanzee population estimate, 6; chimpanzee sanctuaries, 140–141
Taxon Advisory Groups (TAG), 335, 351, 352
taxonomic issues, 171–172, 178–179
teeth, 182, 183, 184
toolmaking capability, 171–172, 185, 252, 329
tourism: ecotourism, 239–240; habituation and stress behavior, 36–37; and the lure of rehabilitation, 66; and transmission of disease, 29–34, 41, 239
trade. *See* global market

U

Uganda: chimpanzee attacks on humans, 236–238; chimpanzee population estimate, 6; chimpanzee sanctuaries, 140; commercial hunting, 233; gorilla population estimate, 13; habitat loss, 235–236; Kibale National Park conservation issues, 230–241
United Nations: Environment Programme, 61, 374; World Charter for Nature, 278; *World Conservation Strategy,* 61
urbanization. *See* habitat loss

V

veterinary care, 342–343
violence. *See* aggressive or violent behavior
vulnerable status, definition, 22

W

walking, 184, 246
Wildlife Conservation Society, 239, 338
World Bank: conservation recommendations for, 39; and deforestation, 25, 28, 80
World Conservation Strategy, 61
World Conservation Union, 338, 374
World Society for the Protection of Animals (WSPA), 27–28, 72, 255, 374

World Wide Fund for Nature, 18
World Wildlife Fund, 61, 338, 374
Wrangham, Richard, 169

Z

Zaire. *See* Democratic Republic of Congo
Zambia, 141
Zoo Atlanta, 124, 337
zoonotic disease, 29, 34, 152, 238
zoos: animal acquisition policy, 119; case studies, 123–127; collection planning, 334–336; and conservation-related behavior, 116–118, 336–337; and the Darwinian perspective, 205–208; disposition of zoo-bred apes, 351–352; enriched environment, 156–157, 205, 319, 339, 340–342, 349; ethical issues, 205, 351–352; exhibit-enclosure design, 339–340; feeding and nutrition, 343; population management, 334–336, 350–351; private areas, 348; relationship with caregiver, 353; "roadside," 225, 354; role in conservation education, 113–123, 303–304, 330, 336–338, 340; role in research, 117–118, 330; social interaction, 345–348; status of great apes in, 330–334; trends and predictions, 349–354; veterinary care, 342–343